PHYSICIANS' DESK REFERENCE®

Supplement B

www.PDR.net

Web | PDA | Print

IMPORTANT NOTICE

Supplements to *Physicians' Desk Reference* are published twice yearly to provide readers with significant revisions of existing product listings as well as comprehensive information on new drugs and other products not included in the current annual edition. Before prescribing or administering any product described in *Physicians' Desk Reference*, be sure to consult this supplement to determine whether revisions have occurred since the 2009 edition of *PDR* and 2009 Supplement A went to press.

NEW AND REVISED PRODUCT LISTINGS INDEX

Listed below are new *PDR* listings first appearing in 2009 *PDR Supplement B*, as well as listings that have been revised since the publication of the main 2009 edition. New listings are in **bold type** and include comprehensive descriptions of new pharmaceutical products, new dosage forms of previously described products, and existing products not described in the main 2009 edition. Revised listings are in light type and may include new research data or clinical findings of importance.

NEW PRODUCT LISTINGS

This section contains comprehensive descriptions of new pharmaceutical products introduced since publication of the 2009 *PDR*, new dosage forms of products already described, and existing pharmaceutical products not described in the 2009 *PDR*.

Centocor Ortho Biotech Inc.
800 RIDGEVIEW DRIVE
HORSHAM, PA 19044
USA

Direct General Inquiries to:
Ph: (610) 651-6000
Fax: (610) 651-6100
Medical Emergency Contact:
Ph: (800) 457-6399
For Medical Information/Adverse Experience Reporting Contact:
Medical Information
Ph: (800) 457-6399

SIMPONI™ ℞
(golimumab)
Injection, solution for subcutaneous use

HIGHLIGHTS OF PRESCRIBING INFORMATION
These highlights do not include all the information needed to use SIMPONI safely and effectively. See full prescribing information for SIMPONI.
SIMPONI (golimumab) Injection, solution for subcutaneous use
Initial U.S. Approval: 2009

WARNING: RISK OF SERIOUS INFECTIONS
See full prescribing information for complete boxed warning

- **Serious infections leading to hospitalization or death including tuberculosis (TB), bacterial sepsis, invasive fungal, and other opportunistic infections have occurred in patients receiving SIMPONI (5.1).**
- **SIMPONI should be discontinued if a patient develops a serious infection or sepsis (5.1).**
- **Perform test for latent TB; if positive, start treatment for TB prior to starting SIMPONI (5.1).**
- **Monitor all patients for active TB during treatment, even if initial latent TB test is negative (5.1)**

-------------INDICATIONS AND USAGE-------------

SIMPONI is a tumor necrosis factor (TNF) blocker indicated for the treatment of:

- **Moderately to severely active Rheumatoid Arthritis (RA) in adults, in combination with methotrexate (1.1)**
- **Active Psoriatic Arthritis (PsA) in adults, alone or in combination with methotrexate (1.2)**
- **Active Ankylosing Spondylitis in adults (AS) (1.3)**

----------DOSAGE AND ADMINISTRATION----------

Rheumatoid Arthritis, Psoriatic Arthritis, and Ankylosing Spondylitis (2.1)

- 50 mg administered by subcutaneous injection once a month.

-----------DOSAGE FORMS AND STRENGTHS-----------

- 50 mg/0.5 mL in a single dose prefilled SmartJect autoinjector (3)
- 50 mg/0.5 mL in a single dose prefilled syringe (3)

-----------------CONTRAINDICATIONS-----------------

- None (4)

-----------WARNINGS AND PRECAUTIONS-----------

- Serious Infections—Do not start SIMPONI during an active infection. If an infection develops, monitor carefully, and stop SIMPONI if infection becomes serious (5.1).
- Invasive fungal infections—For patients who develop a systemic illness on SIMPONI, consider empiric antifungal therapy for those who reside or travel to regions where mycoses are endemic (5.1).
- Hepatitis B reactivation—Monitor HBV carriers during and several months after therapy. If reactivation occurs, stop SIMPONI and begin anti-viral therapy (5.1).
- Malignancies—The incidence of lymphoma was seen more often than in the general U.S. population. Cases of other malignancies have been observed among patients receiving TNF-blockers (5.2).
- Heart failure—Worsening, or new onset, may occur. Stop SIMPONI if new or worsening symptoms occur (5.3).
- Demyelinating disease, exacerbation or new onset, may occur (5.4).

---------------ADVERSE REACTIONS---------------

Most common adverse reactions (incidence > 5%): upper respiratory tract infection, nasopharyngitis (6.1).

To report SUSPECTED ADVERSE REACTIONS, contact Centocor Ortho Biotech Inc. at 1-800-457-6399 or FDA at 1-800-FDA-1088 or www.fda.gov/medwatch

---------------DRUG INTERACTIONS---------------

- Abatacept—increased risk of serious infection (5.1, 5.5, 7.2)
- Anakinra—increased risk of serious infection (5.1, 5.6, 7.2).
- Live vaccines—should not be given with SIMPONI (5.8, 7.3).

See 17 for PATIENT COUNSELING INFORMATION and Medication Guide

Revised: 4/2009

FULL PRESCRIBING INFORMATION: CONTENTS*

* Sections or subsections omitted from the full prescribing information are not listed

FULL PRESCRIBING INFORMATION

WARNING: RISK OF SERIOUS INFECTIONS **Patients treated with SIMPONI™ are at increased risk for developing serious infections that may lead to hospitalization or death** *[see Warnings and Precautions (5.1)]*. **Most patients who developed these infections were taking concomitant immunosuppressants such as methotrexate or corticosteroids.**
SIMPONI should be discontinued if a patient develops a serious infection.
Reported infections include:

- **Active tuberculosis, including reactivation of latent tuberculosis. Patients with tuberculosis have frequently presented with disseminated or extrapulmonary disease. Patients should be tested for latent tuberculosis before SIMPONI use and during therapy. Treatment for latent infection should be initiated prior to SIMPONI use.**
- **Invasive fungal infections, including histoplasmosis, coccidioidomycosis, and pneumocystosis. Patients with histoplasmosis or other invasive fungal infections may present with disseminated, rather than localized, disease. Antigen and antibody testing for histoplasmosis may be negative in some patients with active infection. Empiric anti-fungal therapy should be considered in patients at risk for invasive fungal infections who develop severe systemic illness.**
- **Bacterial, viral, and other infections due to opportunistic pathogens.**

The risks and benefits of treatment with SIMPONI should be carefully considered prior to initiating therapy in patients with chronic or recurrent infection.

Continued on next page

Simponi—Cont.

Patients should be closely monitored for the development of signs and symptoms of infection during and after treatment with SIMPONI, including the possible development of tuberculosis in patients who tested negative for latent tuberculosis infection prior to initiating therapy *[see Warning and Precautions (5.1)].*

1.0 INDICATIONS AND USAGE

1.1 Rheumatoid Arthritis

SIMPONI, in combination with methotrexate, is indicated for the treatment of adult patients with moderately to severely active rheumatoid arthritis.

1.2 Psoriatic Arthritis

SIMPONI, alone or in combination with methotrexate, is indicated for the treatment of adult patients with active psoriatic arthritis.

1.3 Ankylosing Spondylitis

SIMPONI is indicated for the treatment of adult patients with active ankylosing spondylitis.

2.0 DOSAGE AND ADMINISTRATION

2.1 Rheumatoid Arthritis, Psoriatic Arthritis, Ankylosing Spondylitis

The SIMPONI dose regimen is 50 mg administered by subcutaneous (SC) injection once a month.

For patients with rheumatoid arthritis (RA), SIMPONI should be given in combination with methotrexate and for patients with psoriatic arthritis (PsA) or ankylosing spondylitis (AS), SIMPONI may be given with or without methotrexate or other non-biologic DMARDs. For patients with RA, PsA, or AS, corticosteroids, non-biologic DMARDs, and/or NSAIDs may be continued during treatment with SIMPONI.

2.2 Monitoring to Assess Safety

Prior to initiating SIMPONI and periodically during therapy, patients should be evaluated for active tuberculosis and tested for latent infection *[see Warnings and Precautions (5.1)]*.

2.3 General Considerations for Administration

SIMPONI is intended for use under the guidance and supervision of a physician. After proper training in subcutaneous injection technique, a patient may self inject with SIMPONI if a physician determines that it is appropriate. Patients should be instructed to follow the directions provided in the Medication Guide *[see Medication Guide (17.3)]*.

To ensure proper use, allow the prefilled syringe or autoinjector to sit at room temperature outside the carton for 30 minutes prior to subcutaneous injection. Do not warm SIMPONI in any other way.

Prior to administration, visually inspect the solution for particles and discoloration through the viewing window. SIMPONI should be clear to slightly opalescent and colorless to light yellow. The solution should not be used if discolored, or cloudy, or if foreign particles are present. Any leftover product remaining in the prefilled syringe or prefilled autoinjector should not be used. NOTE: The needle cover on the prefilled syringe as well as the prefilled syringe in the autoinjector contains dry natural rubber (a derivative of latex), which should not be handled by persons sensitive to latex.

Injection sites should be rotated and injections should never be given into areas where the skin is tender, bruised, red, or hard.

3.0 DOSAGE FORMS AND STRENGTHS

SmartJect™ Autoinjector

Each single dose SmartJect autoinjector contains a prefilled glass syringe (27 gauge ½ inch) providing 50 mg of SIMPONI per 0.5 mL of solution.

Prefilled Syringe

Each single dose prefilled glass syringe (27 gauge ½ inch) contains 50 mg of SIMPONI per 0.5 mL of solution.

4.0 CONTRAINDICATIONS

None.

5.0 WARNINGS AND PRECAUTIONS

5.1 Serious Infections

Serious and sometimes fatal infections due to bacterial, mycobacterial, invasive fungal, viral, protozoal, or other opportunistic pathogens have been reported in patients receiving TNF-blockers including SIMPONI. Among opportunistic infections, tuberculosis, histoplasmosis, aspergillosis, candidiasis, coccidioidomycosis, listeriosis, and pneumocystosis were the most commonly reported with TNF-blockers. Patients have frequently presented with disseminated rather than localized disease, and were often taking concomitant immunosuppressants such as methotrexate or corticosteroids. The concomitant use of a TNF-blocker and abatacept or anakinra was associated with a higher risk of serious infections; therefore, the concomitant use of SIMPONI and these biologic products is not recommended *[see Warning and Precautions (5.5, 5.6) and Drug Interactions (7.2)]*.

Treatment with SIMPONI should not be initiated in patients with an active infection, including clinically important localized infections. The risks and benefits of treatment should be considered prior to initiating SIMPONI in patients:

- with chronic or recurrent infection;
- who have been exposed to tuberculosis;
- with a history of an opportunistic infection;
- who have resided or traveled in areas of endemic tuberculosis or endemic mycoses, such as histoplasmosis, coccidioidomycosis, or blastomycosis; or
- with underlying conditions that may predispose them to infection.

Patients should be closely monitored for the development of signs and symptoms of infection during and after treatment with SIMPONI. SIMPONI should be discontinued if a patient develops a serious infection, an opportunistic infection, or sepsis. A patient who develops a new infection during treatment with SIMPONI should undergo a prompt and complete diagnostic workup appropriate for an immunocompromised patient, appropriate antimicrobial therapy should be initiated, and the patient should be closely monitored.

In controlled Phase 3 trials through Week 16 in patients with RA, PsA, and AS, serious infections were observed in 1.4% of SIMPONI-treated patients and 1.3% of control-treated patients. In the controlled Phase 3 trials through Week 16 in patients with RA, PsA, and AS, the incidence of serious infections per 100 patient-years of follow-up was 5.4 (95% CI: 4.0, 7.2) for the SIMPONI group and 5.3 (95% CI: 3.1, 8.7) for the placebo group. Serious infections observed in SIMPONI-treated patients included sepsis, pneumonia, cellulitis, abscess, tuberculosis, invasive fungal infections, and hepatitis B infection.

Tuberculosis

Cases of reactivation of tuberculosis or new tuberculosis infections have been observed in patients receiving TNF-blockers, including patients who have previously received treatment for latent or active tuberculosis. Patients should be evaluated for tuberculosis risk factors and tested for latent infection prior to initiating SIMPONI and periodically during therapy.

Treatment of latent tuberculosis infection prior to therapy with TNF-blockers has been shown to reduce the risk of tuberculosis reactivation during therapy. Induration of 5 mm or greater with tuberculin skin testing should be considered a positive test result when assessing if treatment for latent tuberculosis is needed prior to initiating SIMPONI, even for patients previously vaccinated with Bacille Calmette-Guerin (BCG).

Anti-tuberculosis therapy should also be considered prior to initiation of SIMPONI in patients with a past history of latent or active tuberculosis in whom an adequate course of treatment cannot be confirmed, and for patients with a negative test for latent tuberculosis but having risk factors for tuberculosis infection. Consultation with a physician with expertise in the treatment of tuberculosis is recommended to aid in the decision whether initiating anti-tuberculosis therapy is appropriate for an individual patient.

Patients should be closely monitored for the development of signs and symptoms of tuberculosis including patients who tested negative for latent tuberculosis infection prior to initiating therapy.

Tuberculosis should be strongly considered in patients who develop a new infection during SIMPONI treatment, especially in patients who have previously or recently traveled to countries with a high prevalence of tuberculosis, or who have had close contact with a person with active tuberculosis.

In the controlled and uncontrolled portions of the Phase 2 RA and Phase 3 RA, PsA, and AS trials, the incidence of active TB was 0.23 and 0 per 100 patient-years in 2347 SIMPONI-treated patients and 674 placebo-treated patients, respectively. Cases of TB included pulmonary and extra pulmonary TB. The overwhelming majority of the TB cases occurred in countries with a high incidence rate of TB.

Invasive Fungal Infections

For SIMPONI-treated patients who reside or travel in regions where mycoses are endemic, invasive fungal infection should be suspected if they develop a serious systemic illness. Appropriate empiric antifungal therapy should be considered while a diagnostic workup is being performed. Antigen and antibody testing for histoplasmosis may be negative in some patients with active infection. When feasible, the decision to administer empiric antifungal therapy in these patients should be made in consultation with a physician with expertise in the diagnosis and treatment of invasive fungal infections and should take into account both the risk for severe fungal infection and the risks of antifungal therapy.

Hepatitis B Virus Reactivation

The use of TNF-blockers including SIMPONI has been associated with reactivation of hepatitis B virus (HBV) in patients who are chronic hepatitis B carriers (i.e., surface antigen positive). In some instances, HBV reactivation occurring in conjunction with TNF-blocker therapy has been fatal. The majority of these reports have occurred in patients who received concomitant immunosuppressants.

Patients at risk for HBV infection should be evaluated for prior evidence of HBV infection before initiating TNF-blocker therapy. The risks and benefits of treatment should be considered prior to prescribing TNF-blockers, including SIMPONI, to patients who are carriers of HBV. Adequate data are not available on whether anti-viral therapy can reduce the risk of HBV reactivation in HBV carriers who are treated with TNF-blockers. Patients who are carriers of HBV and require treatment with TNF-blockers should be closely monitored for clinical and laboratory signs of active HBV infection throughout therapy and for several months following termination of therapy.

In patients who develop HBV reactivation, TNF-blockers should be stopped and antiviral therapy with appropriate supportive treatment should be initiated. The safety of resuming TNF-blockers after HBV reactivation has been controlled is not known. Therefore, prescribers should exercise caution when considering resumption of TNF-blockers in this situation and monitor patients closely.

5.2 Malignancies

The risks and benefits of TNF-blocker treatment including SIMPONI should be considered prior to initiating therapy in patients with a known malignancy other than a successfully treated non-melanoma skin cancer (NMSC) or when considering continuing a TNF-blocker in patients who develop a malignancy.

In the controlled portions of clinical trials of TNF-blockers including SIMPONI, more cases of lymphoma have been observed among patients receiving anti-TNF treatment compared with patients in the control groups. During the controlled portions of the Phase 2 trials in RA, and the Phase 3 trials in RA, PsA and AS, the incidence of lymphoma per 100 patient-years of follow-up was 0.21 (95% CI: 0.03, 0.77) in the combined SIMPONI group compared with an incidence of 0 (95% CI: 0, 0.96) in the placebo group. In the controlled and uncontrolled portions of these clinical trials in 2347 SIMPONI-treated patients with a median follow-up of 1.4 years, the incidence of lymphoma was 3.8-fold higher than expected in the general U.S. population according to the SEER database (adjusted for age, gender, and race).[1] Patients with RA and other chronic inflammatory diseases, particularly patients with highly active disease and/or chronic exposure to immunosuppressant therapies, may be at higher risk (up to several fold) than the general population for the development of lymphoma, even in the absence of TNF-blocking therapy.

During the controlled portions of the Phase 2 trial in RA, and the Phase 3 trials in RA, PsA and AS, the incidence of malignancies other than lymphoma per 100 patient-years of follow-up was not elevated in the combined SIMPONI group compared with the placebo group. In the controlled and uncontrolled portions of these trials, the incidence of malignancies, other than lymphoma, in SIMPONI-treated patients was similar to that expected in the general U.S. population according to the SEER database (adjusted for age, gender, and race).[1]

In controlled trials of other TNF-blockers in patients at higher risk for malignancies (e.g., patients with COPD, patients with Wegener's granulomatosis treated with concomitant cyclophosphamide) a greater portion of malignancies occurred in the TNF-blocker group compared to the controlled group. In an exploratory 1-year clinical trial evaluating the use of 50, 100 and 200 mg of SIMPONI in 309 patients with severe persistent asthma, 6 patients developed malignancies other than NMSC in the SIMPONI groups compared to none in the control group. Three of the 6 patients were in the 200 mg SIMPONI group.

5.3 Congestive Heart Failure

Cases of worsening congestive heart failure (CHF) and new onset CHF have been reported with TNF-blockers. In several exploratory trials of other TNF-blockers in the treatment of CHF, there were greater proportions of TNF-blocker treated patients who had CHF exacerbations requiring hospitalization or increased mortality. SIMPONI has not been studied in patients with a history of CHF and SIMPONI should be used with caution in patients with CHF. If a decision is made to administer SIMPONI to patients with CHF, these patients should be closely monitored during therapy, and SIMPONI should be discontinued if new or worsening symptoms of CHF appear.

5.4 Demyelinating Disorders

Use of TNF-blockers has been associated with cases of new onset or exacerbation of central nervous system (CNS) demyelinating disorders, including multiple sclerosis (MS). While no trials have been performed evaluating SIMPONI in the treatment of patients with MS, another TNF-blocker was associated with increased disease activity in patients with MS. Therefore, prescribers should exercise caution in considering the use of TNF-blockers including SIMPONI in patients with CNS demyelinating disorders including MS.

5.5 Use with Abatacept

In controlled trials, the concurrent administration of another TNF-blocker and abatacept was associated with a greater proportion of serious infections than the use of a TNF-blocker alone; and the combination therapy, compared to the use of a TNF-blocker alone, has not demonstrated improved clinical benefit in the treatment of RA. Therefore, the combination of TNF-blockers including SIMPONI and abatacept is not recommended [*see Drug Interactions (7.2)*].

5.6 Use with Anakinra

Concurrent administration of anakinra (an interleukin-1 antagonist) and another TNF-blocker, was associated with a greater portion of serious infections and neutropenia and no additional benefits compared with the TNF-blocker alone. Therefore, the combination of anakinra with TNF-blockers, including SIMPONI, is not recommended [*see Drug Interactions 7.2*].

5.7 Hematologic Cytopenias

There have been post-marketing reports of pancytopenia, leukopenia, neutropenia, aplastic anemia, and thrombocytopenia in patients receiving TNF-blockers. Although, there were no cases of severe cytopenias seen in the SIMPONI clinical trials, caution should be exercised when using TNF-blockers, including SIMPONI, in patients who have significant cytopenias.

5.8 Vaccinations

Patients treated with SIMPONI may receive vaccinations, except for live vaccines. No data are available on the response to live vaccination or the risk of infection, or transmission of infection after the administration of live vaccines to patients receiving SIMPONI. In the Phase 3 PsA study, after pneumococcal vaccination, a similar proportion of SIMPONI-treated and placebo-treated patients were able to mount an adequate immune response of at least a 2-fold increase in antibody titers to pneumococcal polysaccharide vaccine. In both SIMPONI-treated and placebo-treated patients, the proportions of patients with response to pneumococcal vaccine were lower among patients receiving MTX compared with patients not receiving MTX. The data suggest that SIMPONI does not suppress the humoral immune response to the pneumococcal vaccine.

6.0 ADVERSE REACTIONS

Because clinical trials are conducted under widely varying conditions, adverse reaction rates observed in the clinical trials of a drug cannot be directly compared to rates in the clinical trials of another drug and may not reflect the rates observed in clinical practice.

6.1 Clinical Studies Experience

The safety data described below are based on 5 pooled, randomized, double-blind, controlled Phase 3 trials in patients with RA, PsA, and AS (Studies RA-1, RA-2, RA-3, PsA, and AS) [*see Clinical Studies (14.1, 14.2 and 14.3)*]. These 5 trials included 639 control-treated patients and 1659 SIMPONI-treated patients including 1089 with RA, 292 with PsA, and 278 with AS. The proportion of patients who discontinued treatment due to adverse reactions in the controlled Phase 3 trials through Week 16 in RA, PsA and AS was 2% for SIMPONI-treated patients and 3% for placebo-treated patients. The most common adverse reactions leading to discontinuation of SIMPONI in the controlled Phase 3 trials through Week 16 were sepsis (0.2%), alanine aminotransferase increased (0.2%), and aspartate aminotransferase increased (0.2%).

The most serious adverse reactions were:

- Serious Infections [*see Warnings and Precautions (5.1)*]
- Malignancies [*see Warnings and Precautions (5.2)*]

Upper respiratory tract infection and nasopharyngitis were the most common adverse reactions reported in the combined Phase 3 RA, PsA and AS trials through Week 16, occurring in 7% and 6% of SIMPONI-treated patients as compared with 6% and 5% of control-treated patients, respectively.

Infections

In controlled Phase 3 trials through Week 16 in RA, PsA, and AS, infections were observed in 28% of SIMPONI-treated patients compared to 25% of control-treated patients [for Serious Infections, *see Warnings and Precautions (5.1)*].

Liver Enzyme Elevations

There have been reports of severe hepatic reactions including acute liver failure in patients receiving TNF-blockers. In controlled Phase 3 trials of SIMPONI in patients with RA, PsA, and AS through Week 16, ALT elevations ≥ 5 × ULN occurred in 0.2% of control-treated patients and 0.7% of SIMPONI-treated patients and ALT elevations ≥ 3 × ULN occurred in 2% of control-treated patients and 2% of SIMPONI-treated patients. Since many of the patients in the Phase 3 trials were also taking medications that cause liver enzyme elevations (e.g., NSAIDS, MTX), the relationship between golimumab and liver elevation is not clear.

Autoimmune Disorders and Autoantibodies

The use of TNF-blockers has been associated with the formation of autoantibodies and, rarely, with the development of a lupus-like syndrome. In the controlled Phase 3 trials in patients with RA, PsA, and AS through Week 14, there was no association of SIMPONI treatment and the development of newly positive anti-dsDNA antibodies.

Injection Site Reactions

In controlled Phase 3 trials through Week 16 in RA, PsA and AS, 6% of SIMPONI-treated patients had injection site reactions compared with 2% of control-treated patients. The majority of the injection site reactions were mild and the most frequent manifestation was injection site erythema. In controlled Phase 2 and 3 trials in RA, PsA, and AS, no patients treated with SIMPONI developed anaphylactic reactions.

Psoriasis: New-Onset and Exacerbations

Cases of new onset psoriasis, including pustular psoriasis and palmoplantar psoriasis, have been reported with the use of TNF-blockers, including SIMPONI. Cases of exacerbation of pre-existing psoriasis have also been reported with the use of TNF-blockers. Many of these patients were taking concomitant immunosuppressants (e.g., MTX, corticosteroids). Some of these patients required hospitalization. Most patients had improvement of their psoriasis following discontinuation of their TNF-blocker. Some patients have had recurrences of the psoriasis when they were re-challenged with a different TNF-blocker. Discontinuation of SIMPONI should be considered for severe cases and those that do not improve or that worsen despite topical treatments.

Immunogenicity

Antibodies to SIMPONI were detected in 57 (4%) of SIMPONI-treated patients across the Phase 3 RA, PsA and AS trials through Week 24. Similar rates were observed in each of the three indications. Patients who received SIMPONI with concomitant MTX had a lower proportion of antibodies to SIMPONI than patients who received SIMPONI without MTX (approximately 2% versus 7%, respectively). Of the patients with a positive antibody response to SIMPONI in the Phase 2 and 3 trials, most were determined to have neutralizing antibodies to golimumab as measured by a cell-based functional assay. The small number of patients positive for antibodies to SIMPONI limits the ability to draw definitive conclusions regarding the relationship between antibodies to golimumab and clinical efficacy or safety measures.

The data above reflect the percentage of patients whose test results were considered positive for antibodies to SIMPONI in an ELISA assay, and are highly dependent on the sensitivity and specificity of the assay. Additionally, the observed incidence of antibody positivity in an assay may be influenced by several factors including sample handling, timing of sample collection, concomitant medications, and underlying disease. For these reasons, comparison of the incidence of antibodies to SIMPONI with the incidence of antibodies to other products may be misleading.

Other Adverse Reactions

Table 1 summarizes the adverse drug reactions that occurred at a rate of at least 1% in the combined SIMPONI groups during the controlled period of the 5 pooled Phase 3 trials through Week 16 in patients with RA, PsA, and AS.

Table 1. Adverse Drug Reactions Reported by ≥ 1% of Patients in the Phase 3 Trials of RA, PsA, and AS through Week 16[a]

	Placebo ± DMARDs	SIMPONI ± DMARDs
Patients treated	639	1659
Adverse Reaction (Preferred Term)		
Upper respiratory tract infection	37 (6%)	120 (7%)
Nasopharyngitis	31 (5%)	91 (6%)
Alanine aminotransferase increased	18 (3%)	58 (4%)
Injection site erythema	6 (1%)	56 (3%)
Hypertension	9 (1%)	48 (3%)
Aspartate aminotransferase increased	10 (2%)	44 (3%)
Bronchitis	9 (1%)	31 (2%)
Dizziness	7 (1%)	32 (2%)
Sinusitis	7 (1%)	27 (2%)
Influenza	7 (1%)	25 (2%)
Pharyngitis	8 (1%)	22 (1%)
Rhinitis	4 (< 1%)	20 (1%)
Pyrexia	4 (< 1%)	20 (1%)
Oral herpes	2 (< 1%)	16 (1%)
Paraesthesia	2 (< 1%)	16 (1%)

a Patients may have taken concomitant MTX, sulfasalazine, hydroxychloroquine, low dose corticosteroids (≤ 10 mg of prednisone/day or equivalent), and/or NSAIDs during the trials).

7.0 DRUG INTERACTIONS

7.1 Methotrexate

For the treatment of RA, SIMPONI should be used with methotrexate (MTX) [*see Clinical Studies (14.1)*]. Since the presence or absence of concomitant MTX did not appear to influence the efficacy or safety of SIMPONI in the treatment of PsA or AS, SIMPONI can be used with or without MTX in the treatment of PsA and AS [*see Clinical Studies (14.1) and Clinical Pharmacology (12.3)*].

7.2 Biologic Products for RA, PsA, and/or AS

An increased risk of serious infections has been seen in clinical RA studies of other TNF-blockers used in combination with anakinra or abatacept, with no added benefit; therefore, use of SIMPONI with abatacept or anakinra is not recommended [*see Warnings and Precautions (5.5 and 5.6)*]. A higher rate of serious infections has also been observed in RA patients treated with rituximab who received subsequent treatment with a TNF-blocker. There is insufficient information to provide recommendations regarding the concomitant use of SIMPONI and other biologic products approved to treat RA, PsA, or AS.

7.3 Live Vaccines

Live vaccines should not be given concurrently with SIMPONI [*see Warnings and Precautions (5.8)*].

7.4 Cytochrome P450 Substrates

The formation of CYP450 enzymes may be suppressed by increased levels of cytokines (e.g., TNFα) during chronic inflammation. Therefore, it is expected that for a molecule that antagonizes cytokine activity, such as golimumab, the formation of CYP450 enzymes could be normalized. Upon initiation or discontinuation of SIMPONI in patients being treated with CYP450 substrates with a narrow therapeutic index, monitoring of the effect (e.g., warfarin) or drug concentration (e.g., cyclosporine or theophylline) is recommended and the individual dose of the drug product may be adjusted as needed.

8.0 USE IN SPECIFIC POPULATIONS

8.1 Pregnancy

Pregnancy Category B—There are no adequate and well-controlled studies of SIMPONI in pregnant women. Because animal reproduction and developmental studies are not always predictive of human response, it is not known whether SIMPONI can cause fetal harm when administered to a pregnant woman or can affect reproduction capacity. SIMPONI should be used during pregnancy only if clearly needed.

An embryofetal developmental toxicology study was performed in which pregnant cynomolgus monkeys were treated subcutaneously with golimumab during the first trimester with doses up to 50 mg/kg twice weekly (360 times greater than the maximum recommended human dose-MHRD) and has revealed no evidence of harm to maternal animals or fetuses. Umbilical cord blood samples collected at the end of the second trimester showed that fetuses were exposed to golimumab during gestation. In this study, *in utero* exposure to golimumab produced no developmental defects to the fetus.

A pre- and post-natal developmental study was performed in which pregnant cynomolgus monkeys were treated with golimumab during the second and third trimesters, and during lactation at doses up to 50 mg/kg twice weekly (860 times and 310 times greater than the maximal steady state human blood levels for maternal animals and neonates, respectively) and has revealed no evidence of harm to maternal animals or neonates. Golimumab was present in the neonatal serum from the time of birth and for up to six months postpartum. Exposure to golimumab during gestation and during the postnatal period caused no developmental defects in the infants.

8.3 Nursing Mothers

It is not known whether SIMPONI is excreted in human milk or absorbed systemically after ingestion. Because many drugs and immunoglobulins are excreted in human milk, and because of the potential for adverse reactions in nursing infants from SIMPONI, a decision should be made whether to discontinue nursing or to discontinue the drug, taking into account the importance of the drug to the mother.

In the pre- and post-natal development study in cynomolgus monkeys in which golimumab was administered subcutaneously during pregnancy and lactation, golimumab was detected in the breast milk at concentrations that were approximately 400-fold lower than the maternal serum concentrations.

8.4 Pediatric Use

Safety and effectiveness of SIMPONI in pediatric patients less than 18 years of age have not been established.

8.5 Geriatric Use

In the Phase 3 trials in RA, PsA, and AS, there were no overall differences in SAEs, serious infections, and AEs in SIMPONI-treated patients ages 65 or older (N = 155) compared with younger SIMPONI-treated patients. Because there is a higher incidence of infections in the geriatric population in general, caution should be used in treating geriatric patients with SIMPONI.

10.0 OVERDOSAGE

In a clinical study, 5 patients received protocol-directed single infusions of 10 mg/kg of intravenous SIMPONI without

Continued on next page

Simponi—Cont.

serious adverse reactions or other significant reactions. The highest weight patient was 100 kg, and therefore received a single intravenous infusion of 1000 mg of SIMPONI. There were no SIMPONI overdoses in the clinical studies.

11.0 DESCRIPTION

SIMPONI (golimumab) is a human IgG1κ monoclonal antibody specific for human tumor necrosis factor alpha (TNFα) that exhibits multiple glycoforms with molecular masses of approximately 150 to 151 kilodaltons. SIMPONI was created using genetically engineered mice immunized with human TNF, resulting in an antibody with human-derived antibody variable and constant regions. SIMPONI is produced by a recombinant cell line cultured by continuous perfusion and is purified by a series of steps that includes measures to inactivate and remove viruses.

The SIMPONI drug product is a sterile solution of the golimumab antibody supplied as either a single dose prefilled syringe (with a passive needle safety guard) or a single dose prefilled autoinjector. The Type 1 glass syringe has a coated stopper. The fixed stainless steel needle (5 bevel, 27G, half-inch) is covered with a needle shield to prevent leakage of the solution through the needle and to protect the needle during handling prior to administration. The needle shield is made of a dry natural rubber containing latex.

SIMPONI does not contain preservatives. The solution is clear to slightly opalescent, colorless to light yellow with a pH of approximately 5.5. SIMPONI is provided in one strength: 50 mg of the golimumab antibody in 0.5 mL of solution. Each 0.5 mL of SIMPONI contains 50 mg of the golimumab antibody, 0.44 mg of L-histidine and L-histidine monohydrochloride monohydrate, 20.5 mg of sorbitol, 0.08 mg of polysorbate 80, and Water for Injection.

12.0 CLINICAL PHARMACOLOGY

12.1 Mechanism of Action

Golimumab is a human monoclonal antibody that binds to both the soluble and transmembrane bioactive forms of human TNFα. This interaction prevents the binding of TNFα to its receptors, thereby inhibiting the biological activity of TNFα (a cytokine protein). There was no evidence of the golimumab antibody binding to other TNF superfamily ligands; in particular, the golimumab antibody did not bind or neutralize human lymphotoxin. Golimumab did not lyse human monocytes expressing transmembrane TNF in the presence of complement or effector cells.

Elevated TNFα levels in the blood, synovium, and joints have been implicated in the pathophysiology of several chronic inflammatory diseases such as rheumatoid arthritis, psoriatic arthritis, and ankylosing spondylitis. TNFα is an important mediator of the articular inflammation that is characteristic of these diseases. Golimumab modulated the *in vitro* biological effects mediated by TNF in several bioassays, including the expression of adhesion proteins responsible for leukocyte infiltration (E-selectin, ICAM-1 and VCAM-1) and the secretion of proinflammatory cytokines (IL-6, IL-8, G-CSF and GM-CSF).

12.2 Pharmacodynamics

In clinical studies, decreases in C-reactive protein (CRP), interleukin (IL)-6, matrix metalloproteinase 3 (MMP-3), intercellular adhesion molecule (ICAM)-1 and vascular endothelial growth factor (VEGF) were observed following SIMPONI administration in patients with RA, PsA, and AS.

12.3 Pharmacokinetics

Following subcutaneous (SC) administration of SIMPONI to healthy subjects and patients with active RA, the median time to reach maximum serum concentrations (T_{max}) ranged from 2 to 6 days. A SC injection of 50 mg SIMPONI to healthy subjects produced a mean maximum serum concentration (C_{max}) of approximately 2.5 μg/mL. SIMPONI exhibited dose-proportional pharmacokinetics (PK) in patients with active RA over the dose range of 0.1 to 10.0 mg/kg following a single intravenous (IV) dose. Following a single IV administration over the same dose range in patients with active RA, mean systemic clearance of SIMPONI was estimated to be 4.9 to 6.7 mL/day/kg, and mean volume of distribution ranged from 58 to 126 mL/kg. The volume of distribution for SIMPONI indicates that SIMPONI is distributed primarily in the circulatory system with limited extravascular distribution. Median terminal half-life values were estimated to be approximately 2 weeks in healthy subjects and patients with active RA, PsA or AS. By cross-study comparisons of mean AUC_{inf} values following an IV or SC administration of SIMPONI, the absolute bioavailability of SC SIMPONI was estimated to be approximately 53%.

When 50 mg SIMPONI was administered SC to patients with RA, PsA or AS every 4 weeks, serum concentrations appeared to reach steady state by Week 12. With concomitant use of methotrexate (MTX), treatment with 50 mg SIMPONI SC every 4 weeks resulted in a mean steady-state trough serum concentration of approximately 0.4–0.6 μg/mL in patients with active RA, approximately 0.5 μg/mL in patients with active PsA, and approximately 0.8 μg/mL in patients with active AS. Patients with RA, PsA and AS treated with SIMPONI 50 mg and MTX had approximately 52%, 36% and 21% higher mean steady-state trough concentrations of golimumab, respectively compared with those treated with SIMPONI 50 mg without MTX. The presence of MTX also decreased anti-golimumab antibody incidence from 7% to 2% [*see Adverse Reactions (6.1)*]. For RA, SIMPONI should be used with MTX. In the PsA and AS trials, the presence or absence of concomitant MTX did not appear to influence clinical efficacy and safety parameters [*see Drug Interactions (7.1) and Clinical Studies (14.1)*].

Population PK analyses indicated that concomitant use of NSAIDs, oral corticosteroids, or sulfasalazine did not influence the apparent clearance of SIMPONI.

Population PK analyses showed there was a trend toward higher apparent clearance of SIMPONI with increasing weight. However, across the PsA and AS populations, no meaningful differences in clinical efficacy were observed among the subgroups by weight quartile. The RA trial in MTX-experienced and TNF-blocker-naïve patients (Study RA-2) did show evidence of a reduction in clinical efficacy with increasing body weight, but this effect was observed for both tested doses of SIMPONI (50 mg and 100 mg). Therefore, there is no need to adjust the dosage of SIMPONI based on a patient's weight.

Population PK analyses suggested no PK differences between male and female patients after body weight adjustment in the RA and PsA trials. In the AS trial, female patients showed 13% higher apparent clearance than male patients after body weight adjustment. Subgroup analysis based on gender showed that both female and male patients achieved clinically significant response at the proposed clinical dose. Dosage adjustment based on gender is not needed.

Population PK analyses indicated that PK parameters of SIMPONI were not influenced by age in adult patients. Patients with age ≥ 65 years had apparent clearance of SIMPONI similar to patients with age < 65 years. No ethnicity-related PK differences were observed between Caucasians and Asians, and there were too few patients of other races to assess for PK differences.

Patients who developed anti-SIMPONI antibodies generally had lower steady-state serum trough concentrations of SIMPONI.

No formal study of the effect of renal or hepatic impairment on the PK of golimumab was conducted.

13.0 NONCLINICAL TOXICOLOGY

13.1 Carcinogenesis, Mutagenesis, Impairment of Fertility

Long-term animal studies of golimumab have not been conducted to evaluate its carcinogenic potential. Mutagenicity studies have not been conducted with golimumab. A fertility study conducted in mice using an analogous anti-mouse TNFα antibody showed no impairment of fertility.

14.0 CLINICAL STUDIES

14.1 Rheumatoid Arthritis

The efficacy and safety of SIMPONI were evaluated in 3 multicenter, randomized, double-blind, controlled trials (Studies RA-1, RA-2, and RA-3) in 1542 patients ≥ 18 years of age with moderately to severely active RA, diagnosed according to the American College of Rheumatology (ACR) criteria, for at least 3 months prior to administration of study agent. Patients were required to have at least 4 swollen and 4 tender joints. SIMPONI was administered subcutaneously at doses of 50 mg or 100 mg every 4 weeks. Double-blinded controlled efficacy data were collected and analyzed through Week 24. Patients were allowed to continue stable doses of concomitant low dose corticosteroids (equivalent to ≤ 10 mg of prednisone a day) and/or NSAIDs and patients may have received oral MTX during the trials.

Study RA-1 evaluated 461 patients who were previously treated (at least 8 to 12 weeks prior to administration of study agent) with one or more doses of a biologic TNF-blocker without a serious adverse reaction. Patients may have discontinued the biologic TNF-blocker for a variety of reasons. Patients were randomized to receive placebo (n = 155), SIMPONI 50 mg (n = 153), or SIMPONI 100 mg (n = 153). Patients were allowed to continue stable doses of concomitant MTX, sulfasalazine (SSZ), and/or hydroxychloroquine (HCQ) during the trial. The use of other DMARDs including cytotoxic agents or other biologics was prohibited.

Study RA-2 evaluated 444 patients who had active RA despite a stable dose of at least 15 mg/week of MTX and who had not been previously treated with a biologic TNF-blocker. Patients were randomized to receive background MTX (n = 133), SIMPONI 50 mg + background MTX (n = 89), SIMPONI 100 mg + background MTX (n = 89), or SIMPONI 100 mg monotherapy (n = 133). The use of other DMARDs including SSZ, HCQ, cytotoxic agents, or other biologics was prohibited.

Study RA-3 evaluated 637 patients with active RA who were MTX-naïve and had not previously been treated with a biologic TNF-blocker. Patients were randomized to receive MTX (n = 160), SIMPONI 50 mg + MTX (n = 159), SIMPONI 100 mg + MTX (n = 159), or SIMPONI 100 mg monotherapy (n = 159). For patients receiving MTX, MTX was administered at a dose of 10 mg/week beginning at Week 0 and increased to 20 mg/week by Week 8. The use of other DMARDs including SSZ, HCQ, cytotoxic agents, or other biologics was prohibited.

The primary endpoint in Study RA-1 and Study RA-2 was the percentage of patients achieving an ACR 20 response at Week 14 and the primary endpoint in Study RA-3 was the percentage of patients achieving an ACR 50 response at Week 24.

In Studies RA-1, RA-2, and RA-3, the median duration of RA disease was 9.4, 5.7, and 1.2 years; and 99%, 75%, and 54% of the patients used at least one DMARD in the past,

Table 2. Studies RA-1, RA-2, and RA-3 Proportion of Patients with an ACR Response[a]

	Study RA-1 Active RA previously treated with one or more doses of TNF-blockers		Study RA-2 Active RA, despite MTX		Study RA-3 Active RA, MTX Naïve	
	Placebo ± DMARDs[b]	SIMPONI 50 mg ± DMARDs[b]	Background MTX	SIMPONI 50 mg + Background MTX	MTX	SIMPONI 50 mg + MTX
N[c]	155	153	133	89	160	159
ACR 20						
Week 14	18%	35%	33%	55%	NA	NA
Week 24	17%	34%	28%	60%	49%	62%
ACR 50						
Week 14	6%	16%	10%	35%	NA	NA
Week 24	5%	18%	14%	37%	29%	40%
ACR 70						
Week 14	2%	10%	4%	13%	NA	NA
Week 24	3%	12%	5%	20%	16%	24%[d]

a Approximately 78% and 58% of the patients received concomitant low dose corticosteroids (equivalent to ≤ 10 mg of prednisone a day) and NSAIDs, respectively, during the 3 pooled RA trials.
b DMARDs in Study RA-1 included MTX, HCQ, and/or SSZ (about 68%, 8%, and 5% of patients received MTX, HCQ, and SSZ, respectively).
c N reflects randomized patients.
d Not significantly different from MTX monotherapy.
NA Not applicable, as data was not collected at Week 14 in Study RA-3.

respectively. Approximately 77% and 57% of patients received concomitant NSAIDs and low dose corticosteroids, respectively, in the 3 pooled RA trials.

Clinical Response

In the 3 RA trials, a greater percentage of patients treated with the combination of SIMPONI and MTX achieved ACR responses at Week 14 (Studies RA-1 and RA-2) and Week 24 (Studies RA-1, RA-2, and RA-3) versus patients treated with the MTX alone. There was no clear evidence of improved ACR response with the higher SIMPONI dose group (100 mg) compared to the lower SIMPONI dose group (50 mg). In Studies RA-2 and RA-3, the SIMPONI monotherapy groups were not statistically different from the MTX monotherapy groups in ACR responses. Table 2 shows the proportion of patients with the ACR response for the SIMPONI 50 mg and control groups in Studies RA-1, RA-2, and RA-3. In the subset of patients who received SIMPONI in combination with MTX in Study RA-1, the proportion of patients achieving ACR 20, 50 and 70 responses at week 14 were 40%, 18%, and 13%, respectively, in the SIMPONI 50 mg + MTX group (N = 103) compared with 17%, 6%, and 2%, respectively, in the placebo + MTX group (N = 107). Table 3 shows the percent improvement in the components of the ACR response criteria for the SIMPONI 50 mg + MTX and MTX groups in Study RA-2. The percent of patients achieving ACR 20 responses by visit for Study RA-2 is shown in Figure 1. ACR 20 responses were observed in 38% of patients in the SIMPONI 50 mg + MTX group at the first assessment (Week 4) after the initial SIMPONI administration.

[See table at top of previous page]

Table 3. Study RA-2—Median Percent Improvement from Baseline in the Individual ACR Components at Week 14[a]

	Background MTX	SIMPONI 50 mg + Background MTX
N[b]	133	89
Number of swollen joints (0–66)		
Baseline	12	13
Week 14	38%	62%
Number of tender joints (0–68)		
Baseline	21	26
Week 14	30%	60%
Patient's assessment of pain (0–10)		
Baseline	5.7	6.1
Week 14	18%	55%
Patient's global assessment of disease activity (0–10)		
Baseline	5.3	6.0
Week 14	15%	45%
Physician's global assessment of disease activity (0–10)		
Baseline	5.7	6.1
Week 14	35%	55%
HAQ score (0–3)		
Baseline	1.25	1.38
Week 14	10%	29%
CRP (mg/dl)		
Baseline	0.8	1.0
Week 14	2%	44%

Note: Baseline values are medians.
a In Study RA-2, about 70% and 85% of patients received concomitant low dose corticosteroids (equivalent to ≤ 10 mg of prednisone a day) and/or NSAIDs during the trials, respectively.
b N reflects randomized patients; actual number of patients evaluable for each endpoint may vary.

[See figure 1 at top of next column]

Physical Function Response in Patients with RA

In Studies RA-1 and RA-2, the SIMPONI 50 mg groups demonstrated a greater improvement compared to the control groups in the change in mean Health Assessment Ques-

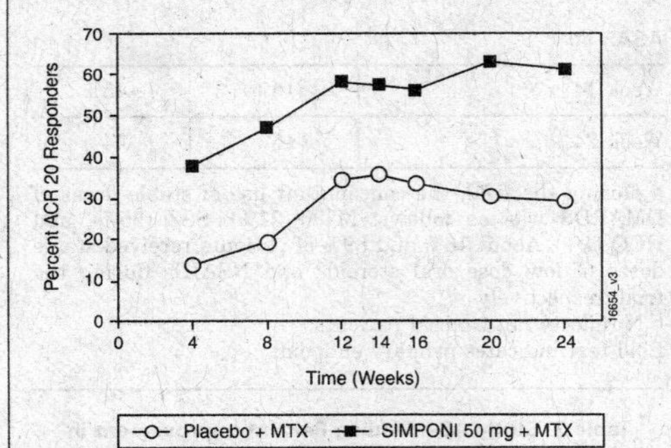

* The same patients may not have responded at each timepoint.

Figure 1. Study RA - 2 – Percent of Patients Achieving ACR 20 Response by Visit: Randomized Patients*

tionnaire Disability Index (HAQ-DI) score from baseline to Week 24: 0.25 vs. 0.05 in RA-1, 0.47 vs. 0.13 in RA-2, respectively. Also in Studies RA-1 and RA-2, the SIMPONI 50 mg groups compared to the control groups had a greater proportion of HAQ responders (change from baseline > 0.22) at Week 24: 44% vs. 28%, 65% vs. 35%, respectively.

14.2 Psoriatic Arthritis

The safety and efficacy of SIMPONI were evaluated in a multi-center, randomized, double-blind, placebo-controlled trial in 405 adult patients with moderately to severely active PsA (≥ 3 swollen joints and ≥ 3 tender joints) despite NSAID or DMARD therapy (Study PsA). Patients in this study had a diagnosis of PsA for at least 6 months with a qualifying psoriatic skin lesion of at least 2 cm in diameter. Previous treatment with a biologic TNF-blocker was not allowed. Patients were randomly assigned to placebo (n = 113), SIMPONI 50 mg (n = 146), or SIMPONI 100 mg (n = 146) given subcutaneously every 4 weeks. Patients were allowed to receive stable doses of concomitant MTX (≤ 25 mg/week), low dose oral corticosteroids (equivalent to ≤ 10 mg of prednisone a day), and/or NSAIDs during the trial. The use of other DMARDs including SSZ, HCQ, cytotoxic agents, or other biologics was prohibited. The primary endpoint was the percentage of patients achieving ACR 20 response at Week 14. Placebo-controlled efficacy data were collected and analyzed through Week 24.

Patients with each subtype of PsA were enrolled, including polyarticular arthritis with no rheumatoid nodules (43%), asymmetric peripheral arthritis (30%), distal interphalangeal (DIP) joint arthritis (15%), spondylitis with peripheral arthritis (11%), and arthritis mutilans (1%). The median duration of PsA disease was 5.1 years, 78% of patients received at least one DMARD in the past, and approximately 48% of patients received MTX, and 16% received low dose oral steroids.

Clinical Response in Patients with PsA

SIMPONI ± MTX, compared with placebo ± MTX, resulted in significant improvement in signs and symptoms as demonstrated by the proportion of patients with an ACR 20 response at Week 14 in Study PsA (see Table 4). There was no clear evidence of improved ACR response with the higher SIMPONI dose group (100 mg) compared to the lower SIMPONI dose group (50 mg). ACR responses observed in the SIMPONI-treated groups were similar in patients receiving and not receiving concomitant MTX. Similar ACR 20 responses at Week 14 were observed in patients with different PsA subtypes. However, the number of patients with arthritis mutilans was too small to allow meaningful assessment. SIMPONI 50 mg treatment also resulted in significantly greater improvement compared with placebo for each ACR component in Study PsA (Table 5). Treatment with SIMPONI resulted in improvement in enthesitis and skin manifestations in patients with PsA. However, the safety and efficacy of SIMPONI in the treatment of patients with plaque psoriasis has not been established.

The percent of patients achieving ACR 20 responses by visit for Study PsA is shown in Figure 2. ACR 20 responses were observed in 31% of patients in the SIMPONI 50 mg + MTX group at the first assessment (Week 4) after the initial SIMPONI administration.

Table 4. Study PsA - Proportion of Patients with ACR Responses

	Placebo ± MTX[a]	SIMPONI 50 mg ± MTX[a]
N[b]	113	146
ACR 20		
Week 14	**9 %**	**51 %**
Week 24	12 %	52 %
ACR 50		
Week 14	2 %	30 %
Week 24	4 %	32 %
ACR 70		
Week 14	1 %	12 %
Week 24	1 %	19 %

a In Study PsA, about 48%, 16%, and 72% of the patients received stable doses of MTX (≤ 25 mg/day), low dose corticosteroids (equivalent to ≤ 10 mg of prednisone a day), and NSAIDs, respectively.
b N reflects randomized patients.
Bold text indicates primary endpoint

Table 5. Study PsA - Percent Improvement in ACR Components at Week 14

	Placebo ± MTX[a]	SIMPONI 50 mg ± MTX[a]
N[b]	113	146
Number of swollen joints (0–66)		
Baseline	10.0	11.0
Week 14	8 %	60 %
Number of tender joints (0–68)		
Baseline	18.0	19.0
Week 14	0 %	54 %
Patient's assessment of pain (0–10)		
Baseline	5.4	5.8
Week 14	−1 %	48 %
Patient's global assessment of disease activity (0–10)		
Baseline	5.2	5.2
Week 14	2 %	49 %
Physician's global assessment of disease activity (0–10)		
Baseline	5.2	5.4
Week 14	7 %	59 %
HAQ score (0–10)		
Baseline	1.0	1.0
Week 14	0 %	28 %
CRP (mg/dL) (0–10)		
Baseline	0.6	0.6
Week 14	0 %	40 %

Note: Baseline are median values
a In Study PsA, about 48%, 16%, and 78% of the patients received stable doses of MTX (≤ 25 mg/day), low dose corticosteroids (equivalent to ≤ 10 mg of prednisone a day), and NSAIDs, respectively.
b N reflects randomized patients; actual number of patients evaluable for each endpoint may vary by timepoint

[See figure 2 at top of next column]

Physical Function Response in Patients with PsA

In Study PsA, SIMPONI 50 mg demonstrated a greater improvement compared to placebo in the change in mean Health Assessment Questionnaire Disability Index

Continued on next page

Simponi—Cont.

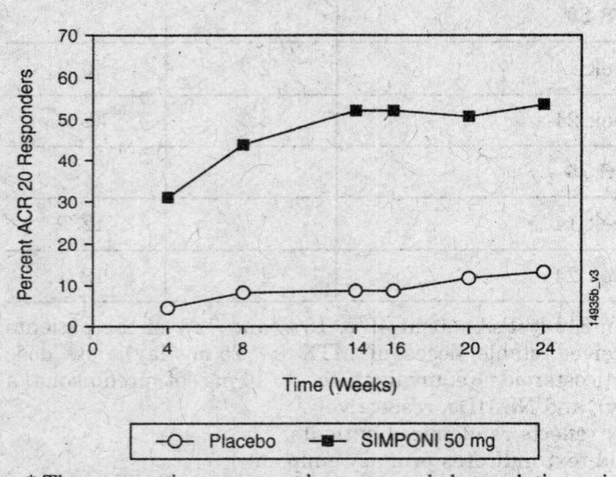

* The same patients may not have responded at each timepoint.

Figure 2. Study PsA – Percent of ACR 20 PsA Responders by Visit: Randomized Patients*

(HAQDI) score from baseline to Week 24 (0.33 and −0.01, respectively). In addition, the SIMPONI 50 mg group compared to the placebo group had a greater proportion of HAQ responders (≥ 0.3 change from baseline) at Week 24: 43% vs. 22%, respectively.

14.3 Ankylosing Spondylitis

The safety and efficacy of SIMPONI were evaluated in a multi-center, randomized, double-blind, placebo-controlled trial in 356 adult patients with active ankylosing spondylitis according to modified New York criteria for at least 3 months (Study AS). Patients had symptoms of active disease [defined as a Bath AS Disease Activity Index (BASDAI) ≥ 4 and VAS for total back pain of ≥ 4, on scales of 0 to 10 cm] despite current or previous NSAID therapy. Patients were excluded if they were previously treated with a biologic TNF-blocker or if they had complete ankylosis of the spine. Patients were randomly assigned to placebo (n = 78), SIMPONI 50 mg (n = 138), or SIMPONI 100 mg (n = 140) administered subcutaneously every 4 weeks. Patients were allowed to continue stable doses of concomitant MTX, sulfasalazine (SSZ), hydroxychloroquine (HCQ), low dose corticosteroids (equivalent to < 10 mg of prednisone a day), and/or NSAIDs during the trial. The use of other DMARDs including cytotoxic agents or other biologics was prohibited.

The primary endpoint was the percentage of patients achieving an ASsessment in Ankylosing Spondylitis (ASAS) 20 response at Week 14. Placebo-controlled efficacy data were collected and analyzed through Week 24.

In Study AS, the median duration of AS disease was 5.6 years, median duration of inflammatory back pain was 12 years, 83% were HLA-B27 positive, 24% had prior joint surgery or procedure, and 55% received at least one DMARD in the past. During the trial, the use of concomitant DMARDs and/or NSAIDs was as follows: MTX (20%), SSZ (26%), HCQ (1%), low dose oral steroids (16%), and NSAIDs (90%).

Clinical Response in Patients with AS

In Study AS, SIMPONI ± DMARDs treatment, compared with placebo ± DMARDs, resulted in a significant improvement in signs and symptoms as demonstrated by the proportion of patients with an ASAS 20 response at Week 14 (see Table 6). There was no clear evidence of improved ASAS response with the higher SIMPONI dose group (100 mg) compared to the lower SIMPONI dose group (50 mg). Table 7 shows the percent improvement in the components of the ASAS response criteria for the SIMPONI 50 mg ± DMARDs and placebo ± DMARDs groups in Study AS.

The percent of patients achieving ASAS 20 responses by visit for Study AS is shown in Figure 3. ASAS 20 responses were observed in 48% of patients in the SIMPONI 50 mg + MTX group at the first assessment (Week 4) after the initial SIMPONI administration.

Table 6. Study AS—Proportion of ASAS Responders at Weeks 14 and 24

	Placebo ± DMARDs[a]	SIMPONI 50 mg ± DMARDs[a]
N[b]	78	138
Responders, % of patients		
ASAS 20		
Week 14	**22%**	**59%**
Week 24	23%	56%
ASAS 40		
Week 14	15%	45%
Week 24	15%	44%

a During the trial, the concomitant use of stable doses of DMARDS was as follows: MTX (21%), SSZ (25%), and HCQ (1%). About 16% and 89% of patients received stable doses of low dose oral steroids and NSAIDs during the trial, respectively.
b N reflects randomized patients.
Bold text indicates primary endpoint

Table 7. Study AS—Median Percent Improvement in ASAS Components at Week 14

	Placebo ± DMARDs[a]	SIMPONI 50 mg ± DMARDs[a]
N[b]	78	138
ASAS components		
Patient global assessment (0–10)		
Baseline	7.2	7.0
Week 14	13%	47%
Total back pain (0–10)		
Baseline	7.6	7.5
Week 14	9%	50%
BASFI (0–10)[c]		
Baseline	4.9	5.0
Week 14	−3%	37%
Inflammation (0–10)[d]		
Baseline	7.1	7.1
Week 14	6%	59%

a During the trial, the concomitant use of stable doses of DMARDS was as follows: MTX (21%), SSZ (25%), and HCQ (1%). About 16% and 89% of patients received stable doses of low dose oral steroids and NSAIDs during the trial, respectively.
b N reflects randomized patients
c BASFI is Bath Ankylosing Spondylitis Functional Index
d Inflammation is the mean of two patient-reported stiffness self-assessments in the Bath AS Disease Activity Index (BASDAI)

* The same patients may not have responded at each timepoint.

Figure 3. Study AS – Percent of AS Patients Achieving ASAS 20 Response by Visit: Randomized Patients*

15.0 REFERENCES

1. SEER [database online]. US Population Data – 1969–2004. Bethesda, MD: National Cancer Institute. Release date: January 3, 2007. Available at: http//seer.cancer.gov/popdata/.

16.0 HOW SUPPLIED/STORAGE AND HANDLING

Each SIMPONI prefilled autoinjector or prefilled syringe is packaged in a light-blocking, cardboard outer carton. SIMPONI is available in packs of 1 prefilled syringe NDC 57894-070-01 or 1 prefilled SmartJect autoinjector NDC 57894-070-02.

Prefilled SmartJect Autoinjector

Each single dose SmartJect autoinjector contains a prefilled glass syringe (27 gauge ½ inch) providing 50 mg of SIMPONI per 0.5 mL of solution.

Prefilled Syringe

Each single dose prefilled glass syringe (27 gauge ½ inch) contains 50 mg of SIMPONI per 0.5 mL of solution.

Storage and Stability

SIMPONI must be refrigerated at 2ºC to 8ºC (36ºF to 46ºF) and protected from light. Keep the product in the original carton to protect from light until the time of use. Do not freeze. Do not shake. Do not use SIMPONI beyond the expiration date (EXP) on the carton or the expiration date on the prefilled syringe (observed through the viewing window) or the prefilled SmartJect autoinjector.

17.0 PATIENT COUNSELING INFORMATION

See Medication Guide (17.3)

17.1 Patient Counseling

Patients should be advised of the potential benefits and risks of SIMPONI. Physicians should instruct their patients to read the Medication Guide before starting SIMPONI therapy and to read it each time the prescription is renewed.

Infections

Inform patients that SIMPONI may lower the ability of their immune system to fight infections. Instruct the patient of the importance of contacting their doctor if they develop any symptoms of infection, including tuberculosis, invasive fungal infections, and hepatitis B reactivation.

Malignancies

Patients should be counseled about the risk of lymphoma and other malignancies while receiving SIMPONI.

Allergic Reactions

Advise latex-sensitive patients that the needle cover on the prefilled syringe as well as the prefilled syringe in the prefilled SmartJect autoinjector contains dry natural rubber (a derivative of latex).

Other Medical Conditions

Advise patients to report any signs of new or worsening medical conditions such as congestive heart failure, demyelinating disorders, autoimmune diseases, liver disease, cytopenias, or psoriasis.

17.2 Instruction on Injection Technique

The first self-injection should be performed under the supervision of a qualified healthcare professional. If a patient or caregiver is to administer SIMPONI, he/she should be instructed in injection techniques and their ability to inject subcutaneously should be assessed to ensure the proper administration of SIMPONI [*see Medication Guide (17.3)*].

Prior to use, remove the prefilled syringe or the prefilled SmartJect autoinjector from the refrigerator and allow SIMPONI to sit at room temperature outside of the carton for 30 minutes and out of the reach of children.

Do not warm SIMPONI in any other way. For example, do not warm SIMPONI in a microwave or in hot water.

Do not remove the prefilled syringe needle cover or SmartJect autoinjector cap while allowing SIMPONI to reach room temperature. Remove these immediately before injection.

Do not pull the autoinjector away from the skin until you hear a first "click" sound and then a second "click" sound (the injection is finished and the needle is pulled back). It usually takes about 3 to 6 seconds but may take up to 15 seconds for you to hear the second "click" after the first "click". If the autoinjector is pulled away from the skin before the injection is completed, a full dose of SIMPONI may not be administered.

A puncture-resistant container for disposal of needles and syringes should be used. Patients or caregivers should be instructed in the technique of proper syringe and needle disposal, and be advised not to reuse these items.

17.3 Medication Guide

Rx Only

MEDICATION GUIDE

SIMPONI™ (SIM-po-nee)

(golimumab)

Read the Medication Guide that comes with SIMPONI before you start taking it and each time you get a refill. There may be new information. This Medication Guide does not take the place of talking with your doctor about your medical condition or treatment. It is important to remain under your doctor's care while using SIMPONI.

What is the most important information I should know about SIMPONI?

SIMPONI is a medicine that affects your immune system. SIMPONI can lower the ability of your immune system to fight infections. Some people have serious infections while taking SIMPONI, including tuberculosis (TB), and infections caused by bacteria, fungi, or viruses that spread throughout their body. Some people have died from these serious infections.

- Your doctor should test you for TB before starting SIMPONI.

- Your doctor should monitor you closely for signs and symptoms of TB during treatment with SIMPONI.

You should not start taking SIMPONI if you have any kind of infection unless your doctor says it is okay.

Before starting SIMPONI, tell your doctor if you:

- think you have an infection or have symptoms of an infection such as:
  - fever, sweat, or chills
  - muscle aches
  - cough
  - shortness of breath
  - blood in phlegm
  - weight loss
  - warm, red, or painful skin or sores on your body
  - diarrhea or stomach pain
  - burning when you urinate or urinate more often than normal
  - feel very tired
- are being treated for an infection
- get a lot of infections or have infections that keep coming back
- have diabetes, HIV, or a weak immune system. People with these conditions have a higher chance for infections.
- have TB, or have been in close contact with someone with TB
- live, have lived, or traveled to certain parts of the country (such as the Ohio and Mississippi River valleys and the Southwest) where there is an increased chance for getting certain kinds of fungal infections (histoplasmosis, coccidioidomycosis, blastomycosis). These infections may happen or become more severe if you use SIMPONI. Ask your doctor, if you do not know if you have lived in an area where these infections are common.
- have or have had hepatitis B
- use the medicine Orencia (abatacept), Kineret (anakinra), or Rituxan (rituximab)

After starting SIMPONI, call your doctor right away if you have any symptoms of an infection. SIMPONI can make you more likely to get infections or make worse any infection that you have.

What is SIMPONI?

SIMPONI is a prescription medicine called a Tumor Necrosis Factor (TNF) blocker. SIMPONI is used in adults:

- with the medicine methotrexate to treat moderately to severely active rheumatoid arthritis (RA)
- to treat active psoriatic arthritis (PsA) alone or with methotrexate
- to treat active ankylosing spondylitis (AS)

You may continue to use other medicines that help treat your condition while taking SIMPONI, such as nonsteroidal anti-inflammatory drugs (NSAIDs) and prescription steroids, as recommended by your doctor.

What should I tell my doctor before starting treatment with SIMPONI?

SIMPONI may not be right for you. Before starting SIMPONI, tell your doctor about all your medical conditions, including if you:

- have an infection (see "What is the most important information I should know about SIMPONI?").
- have or have had lymphoma or any other type of cancer.
- have or had heart failure.
- have or have had a condition that affects your nervous system, such as multiple sclerosis.
- have recently received or are scheduled to receive a vaccine. People taking SIMPONI should not receive live vaccines. People taking SIMPONI can receive non-live vaccines.
- are allergic to rubber or latex. The needle cover on the prefilled syringe and SmartJect autoinjector contains dry natural rubber.
- are pregnant or planning to become pregnant. It is not known if SIMPONI will harm your unborn baby.
- are breastfeeding. You and your doctor should decide if you will take SIMPONI or breastfeed. You should not do both without talking to your doctor first.

Tell your doctor about all the medicines you take, including prescription and non-prescription medicines, vitamins, and herbal supplements. Especially, tell your doctor if you use:

- ORENCIA (abatacept), KINERET (anakinra), or RITUXAN (rituximab). You should not take SIMPONI while you are also taking ORENCIA or KINERET. Your doctor may not want to give you SIMPONI if you have received RITUXAN recently.
- Another TNF-blocker medicine. You should not take SIMPONI while you are also taking REMICADE (infliximab), HUMIRA (adalimumab), ENBREL (etanercept), or CIMZIA (certolizumab pegol).

Ask your doctor if you are not sure if your medicine is one listed above.

Keep a list of all your medications with you to show your doctor and pharmacist each time you get a new medicine.

How should I use SIMPONI?

- SIMPONI is given as an injection under the skin (subcutaneous injection or SC).
- SIMPONI should be injected one time each month.
- If your doctor decides that you or a caregiver may be able to give your injections of SIMPONI at home, you should receive training on the right way to prepare and inject SIMPONI. Do not try to inject SIMPONI yourself until you have been shown the right way to give the injections by your doctor or nurse.
- Use SIMPONI exactly as prescribed by your doctor.
- SIMPONI comes in a prefilled syringe or SmartJect™ autoinjector. Your doctor will prescribe the type that is best for you.
- See the detailed ***Patient Instructions for Use*** at the end of this Medication Guide for instructions about the right way to prepare and give your SIMPONI injections at home.
- Do not miss any doses of SIMPONI. If you forget to use SIMPONI, inject your dose as soon as you remember. Then, take your next dose at your regular scheduled time. In case you are not sure when to inject SIMPONI, call your doctor or pharmacist.

What are the possible side effects with SIMPONI?

SIMPONI can cause serious side effects including:

Serious Infections

(See **"What is the most important information I should know about SIMPONI?"**).

Hepatitis B infection in people who carry the virus in their blood.

- If you are a carrier of the hepatitis B virus (a virus that affects the liver), the virus can become active while you use SIMPONI. Your doctor may do blood tests before you start treatment with SIMPONI and while you are using SIMPONI. Tell your doctor if you have any of the following symptoms of a possible hepatitis B infection:
  - feel very tired
  - skin or eyes look yellow
  - little or no appetite
  - vomiting
  - muscle aches
  - dark urine
  - clay-colored bowel movements
  - fevers
  - chills
  - stomach discomfort
  - skin rash

Cancer

- People with inflammatory diseases including rheumatoid arthritis, psoriatic arthritis, or ankylosing spondylitis, especially those with very active disease, may be more likely to get lymphoma.
- If you use SIMPONI or other TNF-blockers, your risk of getting lymphoma or other cancers may increase.

Heart failure, including new heart failure or worsening of heart failure that you already have. New or worse heart failure can happen in people who use TNF-blocker medicines like SIMPONI.

- If you have heart failure, your condition should be watched closely while you take SIMPONI.
- Call your doctor right away if you get new or worsening symptoms of heart failure while taking SIMPONI (such as shortness of breath or swelling of your lower legs or feet).

Nervous System Problems

Rarely, people using TNF-blocker medicine have nervous system problems such as multiple sclerosis.

- Tell your doctor right away if you get any of these symptoms:
  - vision changes
  - weakness in your arms or legs
  - numbness or tingling in any part of your body

Liver Problems

Liver problems can happen in people who use TNF-blocker medicines, including SIMPONI. These problems can lead to liver failure and death. Call your doctor right away if you have any of these symptoms:

- feel very tired
- skin or eyes look yellow
- poor appetite or vomiting
- pain on the right side of your stomach (abdomen)

Blood Problems

Low blood counts have been seen with other TNF-blockers. Your body may not make enough blood cells that help fight infections or help stop bleeding. Symptoms include fever, bruising or bleeding easily, or looking pale. Your doctor will check your blood counts before and during treatment with SIMPONI.

Common side effects with SIMPONI include:

- upper respiratory tract infection
- nausea
- abnormal liver tests
- redness at the site of injection
- high blood pressure
- bronchitis
- dizziness
- sinus infection (sinusitis)
- flu
- runny nose
- fever
- cold sores
- numbness or tingling

Other side effects with SIMPONI include:

- **Immune System Problems.** Rarely, people using TNF-blocker medicines have developed symptoms that are like the symptoms of Lupus. Tell your doctor if you have any of these symptoms:
  - a rash on your cheeks or other parts of the body
  - sensitivity to the sun
  - new joint or muscle pains
  - becoming very tired
  - chest pain or shortness of breath
  - swelling of the feet, ankles, and/or legs
- **Psoriasis.** Some people using TNF-blocker medicines including SIMPONI had new psoriasis or worsening of psoriasis that they already had. Symptoms of psoriasis include: red scaly patches or raised bumps that are filled with pus on the skin. Psoriasis may go away or get better after stopping SIMPONI in some people.
- **Allergic Reactions.** Allergic reactions can happen in people who use TNF-blocker medicines. Call your doctor right away if you have any of these symptoms of an allergic reaction:
  - hives
  - swollen face
  - breathing trouble
  - chest pain

These are not all of the side effects with SIMPONI. Tell your doctor about any side effect that bothers you or does not go away. Call your doctor for medical advice about side effects. You may report side effects to the FDA at 1-800-FDA-1088.

How do I store SIMPONI?

- Refrigerate SIMPONI at 36°F to 46°F (2°C to 8°C).
- Do not freeze SIMPONI.
- Keep SIMPONI in the carton to protect it from light when not being used.
- Do not shake SIMPONI.

Keep SIMPONI and all medicines out of the reach of children.

General Information about SIMPONI

- Medicines are sometimes prescribed for purposes other than those listed in the Medication Guide. Do not use SIMPONI for a condition for which it was not prescribed.
- Do not give SIMPONI to other people, even if they have the same condition that you have. It may harm them.
- This Medication Guide summarizes the most important information about SIMPONI. If you would like more information, talk to your doctor. You can ask your doctor or pharmacist for information about SIMPONI that is written for health professionals. For more information go to www.simponi.com or call 1-800-457-6399.

What are the ingredients in SIMPONI?

Active ingredient: golimumab.

Inactive ingredients: L-histidine, L-histidine monohydrochloride monohydrate, sorbitol, polysorbate 80, and water for injection. SIMPONI does not contain preservatives.

Patient Instructions for Use

SIMPONI™ (SIM-po-nee)
(golimumab)
SmartJect™ Autoinjector

If your doctor decides that you or a caregiver may be able to give your injections of SIMPONI at home, you should receive training on the right way to prepare and inject SIMPONI. **Do not** try to inject SIMPONI yourself until you have been shown the right way to give the injections by your doctor or nurse.

It is important to read, understand, and follow these instructions so that you inject SIMPONI the right way. Call your doctor if you or your caregiver has any questions about the right way to inject SIMPONI.

Important information about your SmartJect autoinjector:

- When the button on the SmartJect autoinjector is pressed to give the dose of SIMPONI you will hear a loud 'click' sound. It is very important that you practice injecting SIMPONI with your doctor or nurse so that you are not startled by this click when you start giving the injections to yourself at home.
- If you pull the SmartJect autoinjector away from the skin before the injection is completed, you may not get your full dose of medicine and may lose some of the medicine.

Do not:

- shake the SmartJect autoinjector at any time
- remove the SmartJect autoinjector cap until you get to that step

Step 1: Gather and inspect the supplies for your injection

You will need these supplies for an injection of SIMPONI. See Figure 1.

- 1 alcohol swab
- 1 cotton ball or gauze
- 1 SIMPONI prefilled SmartJect autoinjector
- sharps container for autoinjector disposal

Continued on next page

Simponi—Cont.

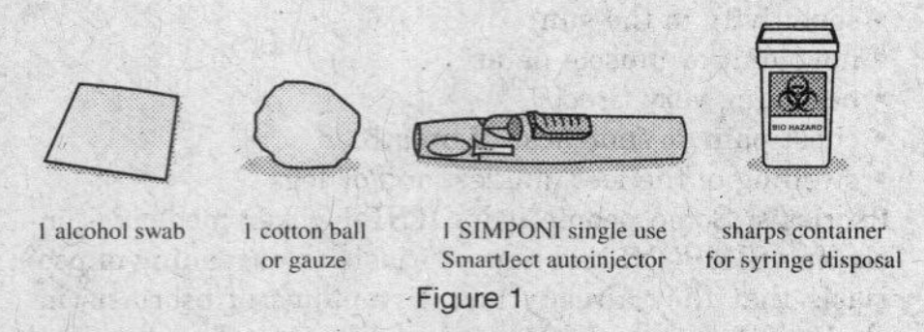

Figure 1

The figure below shows what the SmartJect autoinjector looks like. See Figure 2.

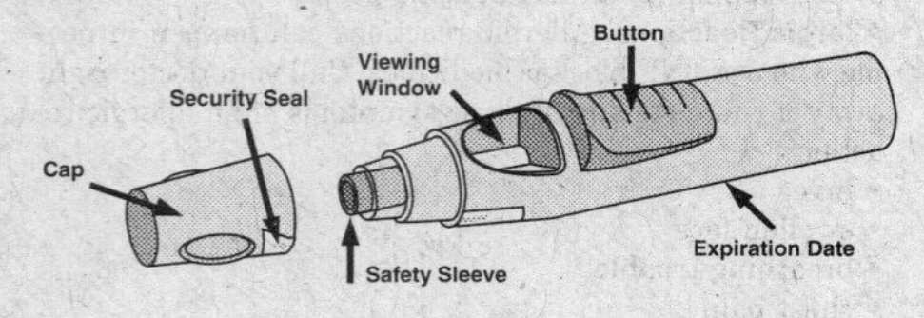

Figure 2

1.1 Check Expiration Date

- Check the expiration date ("EXP") on the SmartJect autoinjector.
- You can also check the expiration date printed on the carton.
- If the expiration date has passed, do not use the SmartJect autoinjector. Call your doctor or pharmacist, or call 1-800-457-6399 for help.

1.2 Check Security Seal

- Check the security seal around the cap of the SmartJect autoinjector. If the security seal is broken, do not use the SmartJect autoinjector.

1.3 Wait 30 minutes

- To ensure proper injection, allow the autoinjector to sit at room temperature outside the carton for 30 minutes and out of the reach of children.

Do not warm the SmartJect autoinjector in any other way (For example, **do not** warm it in a microwave or in hot water).

Do not remove the SmartJect autoinjector cap while allowing it to reach room temperature.

1.4 Check the Liquid in the SmartJect autoinjector

- Look through the viewing window of the SmartJect autoinjector. See Figure 3. Make sure that the liquid in the prefilled syringe is clear and colorless to slightly yellow in color. You may see a small amount of tiny particles that are white, or that you can see through. Do not inject the liquid if it is cloudy or discolored, or has large particles in it.
- You may also notice an air bubble. This is normal. See Figure 3.

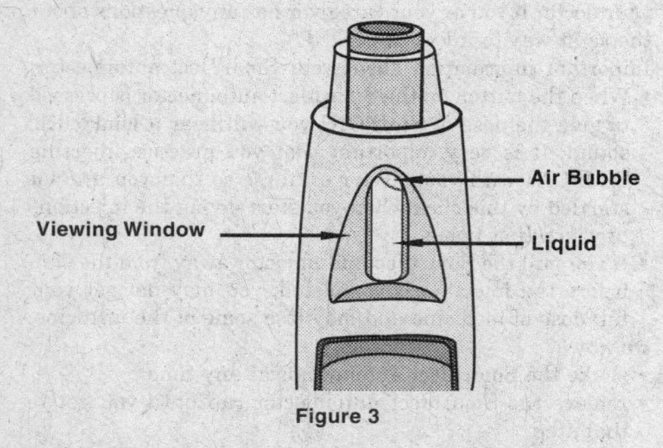

Figure 3

Step 2: Choose and prepare the injection site

2.1 Choose the Injection Site

- The recommended injection site is the front of your middle thighs. See Figure 4.

[See figure 4 at top of next column]

- You can also use the lower part of the abdomen below the navel (belly button), except for the two-inch area directly around the navel. See Figure 5.
- If a caregiver is giving you the injection, the outer area of the upper arms may also be used. See Figure 5.

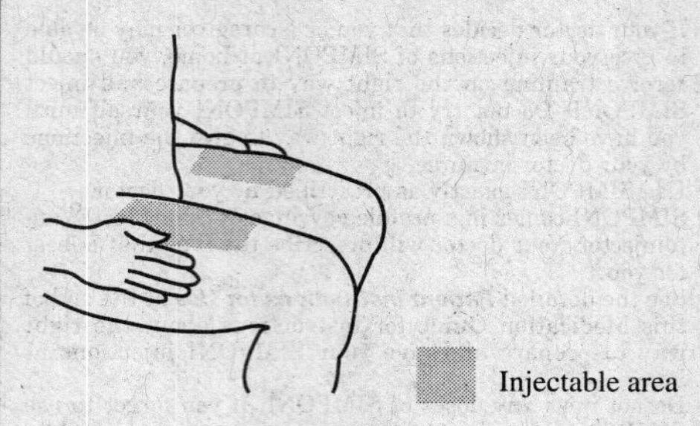

Figure 4

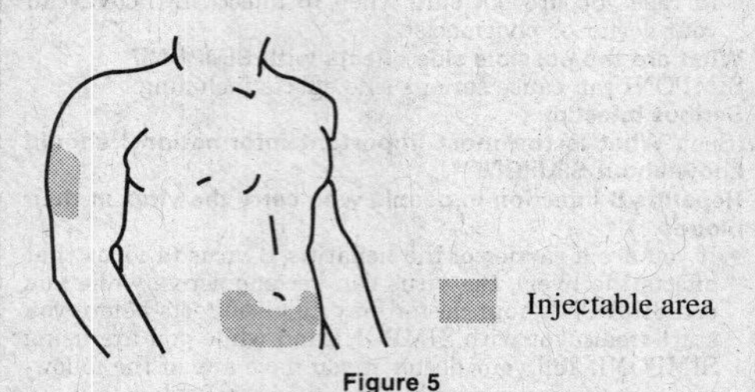

Figure 5

- **Do not** inject into areas where the skin is tender, bruised, red, scaly, or hard. Avoid areas with scars or stretch marks.

2.2 Prepare the Injection Site

- Wash your hands well with soap and warm water.
- Wipe the injection site with an alcohol swab.
- **Do not** touch this area again before giving the injection. Allow the skin to dry before injecting.
- **Do not** fan or blow on the clean area.

Step 3: Injecting SIMPONI using the single dose SmartJect autoinjector

3.1 Remove the Cap

- Do not remove the cap until you are ready to inject SIMPONI. Inject SIMPONI within 5 minutes after the cap has been removed.
- When you are ready to inject, twist the cap slightly to break the security seal. See Figure 6.

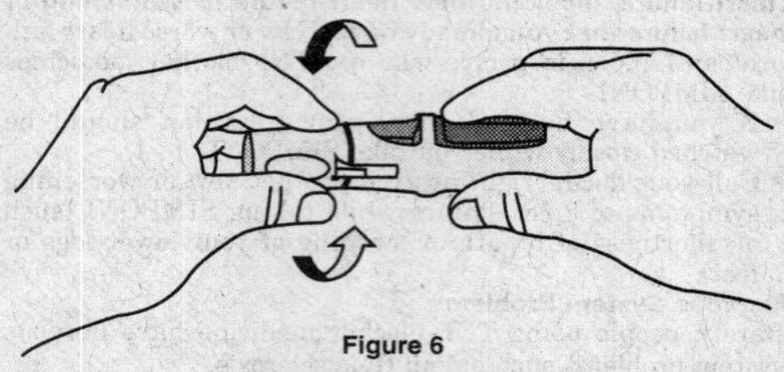

Figure 6

- Pull the cap off and throw it in the trash right away. See Figure 7.

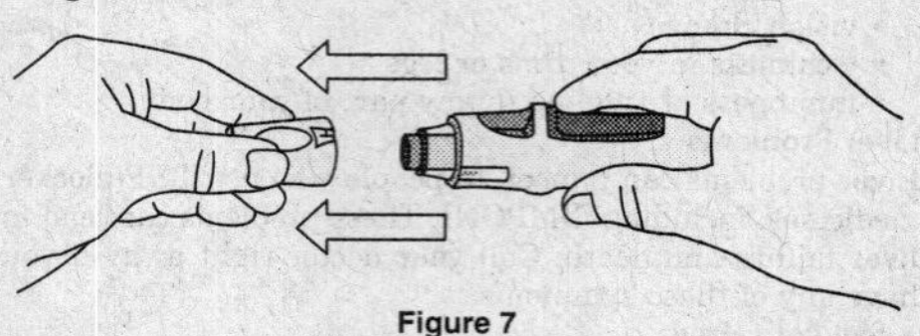

Figure 7

- **Do not** put the cap back on because it may damage the needle inside the SmartJect autoinjector.
- **Do not** use your SmartJect autoinjector if it is dropped without the cap in place.

3.2 Push the SmartJect autoinjector against the skin

- Hold the SmartJect autoinjector comfortably in your hand.
- Do not press the button. Push the open end of the SmartJect autoinjector firmly against the skin at **90-degree angle**. See Figure 8.

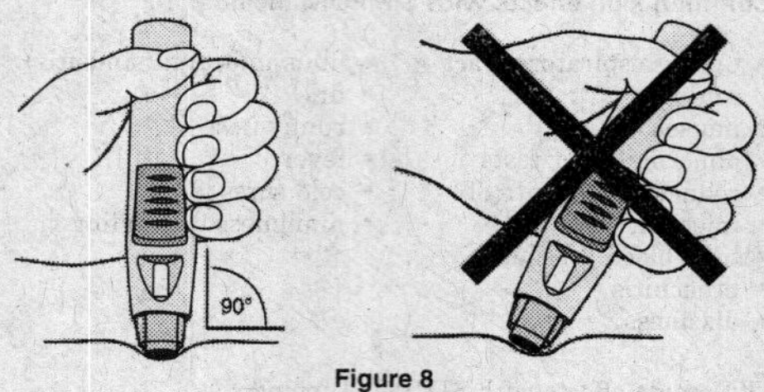

Figure 8

- **Use your** free hand to pinch and hold the skin at the injection site. This may make injecting easier.

3.3 Press button to inject

- Continue to hold the SmartJect autoinjector firmly against the skin, and press the button with your fingers (see Figure 9) or thumb (see Figure 10). You will not be able to push in the button unless the SmartJect autoinjector is pushed firmly against your skin.

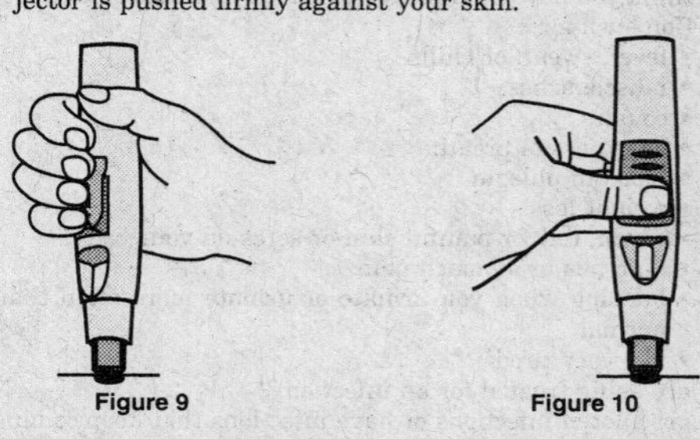

Figure 9 Figure 10

- After the button is pressed, it will stay pressed in so you do not need to keep pressure on it. See Figure 11.

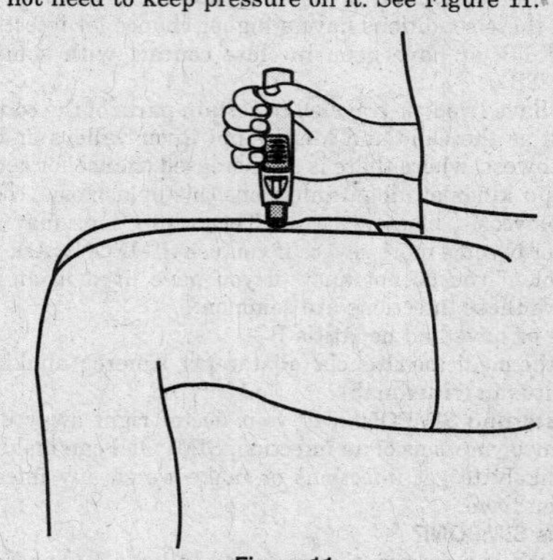

Figure 11

- You will hear a loud 'click' sound. This means that the injection has started. Do not pull the SmartJect autoinjector away from your skin. If you pull the SmartJect autoinjector away from the skin, you may not get your full dose of medicine. See Figure 12.
- **Do not** lift the SmartJect autoinjector yet.

3.4 Wait for Second "Click"

- Keep holding the SmartJect autoinjector against your skin until you hear the second 'click' sound. It usually takes about 3 to 6 seconds, but may take up to 15 seconds for you to hear the second 'click' sound. See Figure 13.
- The second 'click' sound means that the injection is finished and the needle has pulled back (retracted) into the SmartJect autoinjector.
- Lift the SmartJect autoinjector from the injection site. See Figure 14.
- If you have hearing problems, count for 15 seconds from the time you pressed the button and then lift the SmartJect autoinjector from the injection site.

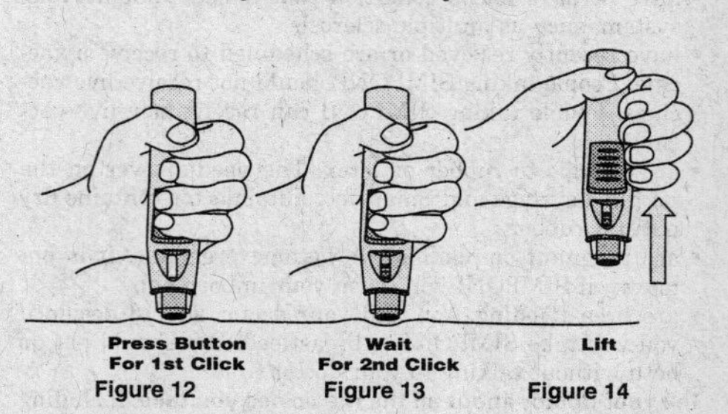

Figure 12 Figure 13 Figure 14

Step 4: After the injection

4.1 Check the Viewing Window

- After you finish injecting, check the viewing window to see the yellow indicator. See Figure 15. This means the SmartJect autoinjector has worked the right way.

[See figure 15 at top of next column]

- If you do not see the yellow indicator in the viewing window, call 1-800-457-6399 for help.

4.2 Dispose of the used SmartJect autoinjector

- Place the used SmartJect autoinjector into a closable puncture-resistant container. You may use a sharps container (such as a red biohazard container), a hard plastic container (such as a detergent bottle), or a metal container (such as an empty coffee can). See Figure 16.

[See figure 16 at top of next column]

- Ask your doctor for instructions on the right way to throw away (dispose of) the container. There may be local or state laws about how you should throw away used needles and syringes.

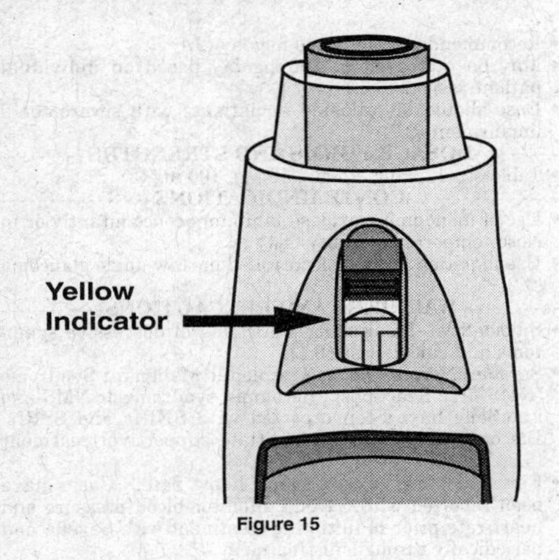

Figure 15

Figure 16

• Do not throw away your used SmartJect autoinjector in household trash. Do not recycle.

4.3 Use Cotton Ball or Gauze

• There may be a small amount of blood or liquid at the injection site, which is normal.
• You can press a cotton ball or gauze over the injection site for 10 seconds. Do not rub the injection site.
• You may cover the injection site with a small adhesive bandage, if needed.

Patient Instructions for Use
SIMPONI™
Prefilled Syringe

If your doctor decides that you or a caregiver may be able to give your injections of SIMPONI at home, you should receive training on the right way to prepare and inject SIMPONI. **Do not** try to inject SIMPONI yourself until you have been shown the right way to give the injections by your doctor or nurse.

It is important to read, understand, and follow these instructions so that you inject SIMPONI the right way. Call your doctor if you or your caregiver has any questions about the right way to inject SIMPONI.

Important information about your prefilled syringe:

• Always hold the prefilled syringe by the body of the syringe.

Do not:

• pull back on the plunger at any time.
• shake the SIMPONI prefilled syringe. This may damage the medicine.
• remove the needle cover from the prefilled syringe until you get to that step.
• touch the needle guard activation clips to prevent covering the needle with the needle guard too soon (See Figure 2).
• use SIMPONI if it has been frozen or if it has been kept at a room temperature that is too warm. See the Medication Guide section: "How should I store SIMPONI?"
• use your SIMPONI prefilled syringe if it looks damaged.

Step 1: Gather the supplies for your injection

You will need these supplies for an injection of SIMPONI. See Figure 1.

• 1 alcohol swab
• 1 cotton ball or gauze
• 1 SIMPONI prefilled syringe
• sharps container for syringe disposal

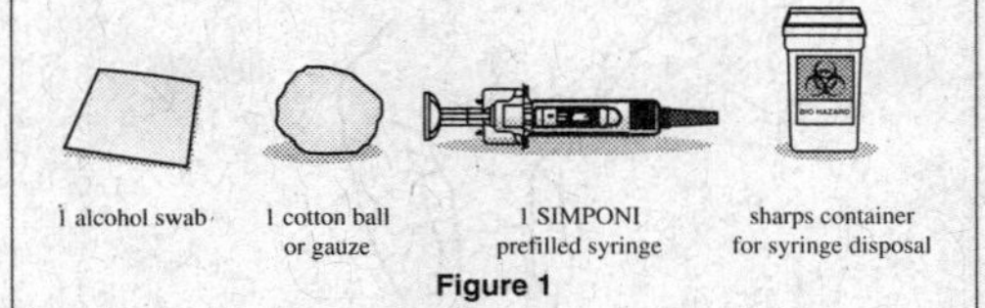

Figure 1

The diagram below shows what the prefilled syringe looks like. See Figure 2.

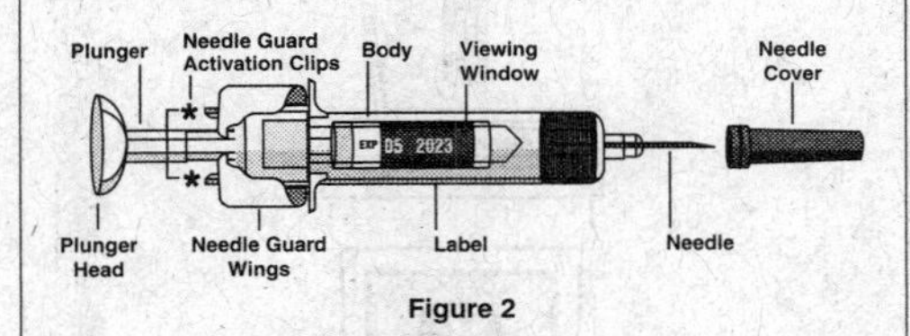

Figure 2

Step 2: Get ready to use your prefilled syringe

2.1 Check the Expiration Date

• Look for the expiration date printed on the back panel of the SIMPONI carton.
• **If the expiration date has passed, do not use the prefilled syringe. Call your doctor** or pharmacist or call 1-800-457-6399 for help.

2.2 Wait 30 minutes

• To ensure proper injection, allow the prefilled syringe to sit at room temperature outside of the carton for 30 minutes and out of the reach of children.

• **Do not** warm the prefilled syringe in any other way (For example, **do not** warm it in a microwave or in hot water).
• **Do not** remove the prefilled syringe needle cover while allowing it to reach room temperature.

2.3 Check the Liquid in the Prefilled Syringe

• Hold your SIMPONI prefilled syringe by the body with the covered needle pointing down. See Figure 3.

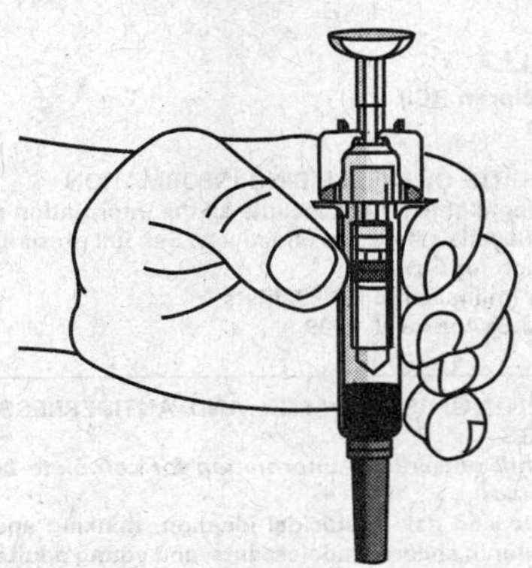

Figure 3

• Look at the liquid through the viewing window of the prefilled syringe. Make sure that the liquid in the prefilled syringe is clear and colorless to slightly yellow in color. You may see a small amount of tiny particles that are white, or that you can see through. Do not inject the liquid if it is cloudy or discolored, or has large particles in it.
• You may also see an air bubble. This is normal.

Step 3: Choose and prepare the injection site

3.1 Choose the Injection Site

• The recommended injection site is the front of your middle thighs. See Figure 4.

[See figure 4 at top of next column]

• You can also use the lower part of the abdomen below the navel (belly button), except for the two-inch area directly around the navel. See Figure 5.
• If a caregiver is giving you the injection, the outer area of the upper arms may also be used. See Figure 5.

[See figure 5 at top of next column]

Do not inject into areas where the skin is tender, bruised, red, scaly, or hard. Avoid areas with scars or stretch marks.

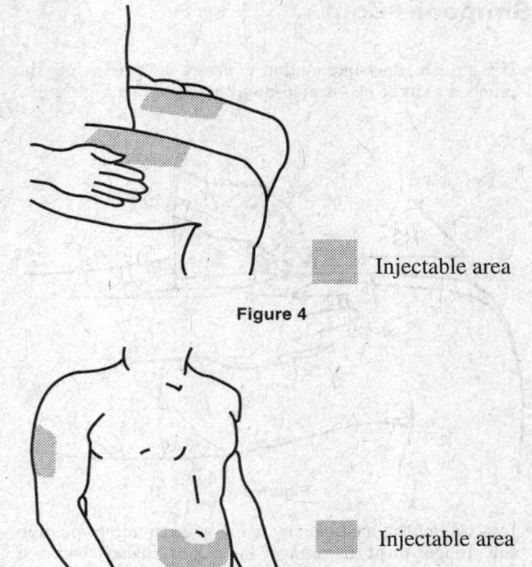

Figure 4

Injectable area

Figure 5

3.2 Prepare the Injection Site

• Wash your hands well with soap and warm water.
• Wipe the injection site with an alcohol swab.
• **Do not** touch this area again before giving the injection. Let your skin dry before injecting.
• **Do not** fan or blow on the clean area.

Step 4: Inject SIMPONI

Do not remove the needle cover until you are ready to inject SIMPONI. Inject SIMPONI within 5 minutes after you remove the needle cover.

4.1 Remove the Needle Cover

• **Do not** touch the plunger while removing the needle cover.
• Hold the body of the prefilled syringe with one hand, and pull the needle cover straight off. See Figure 6.

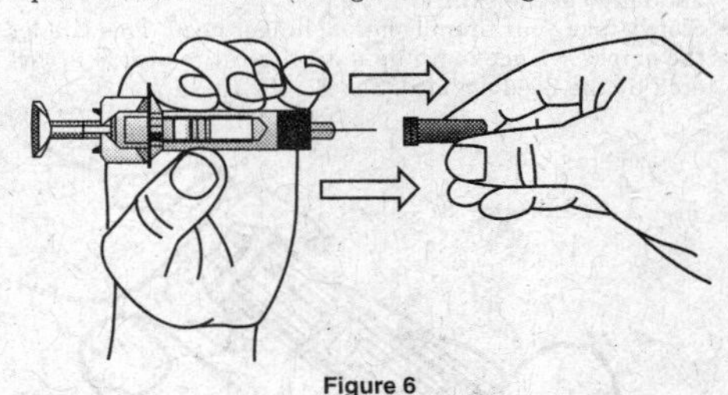

Figure 6

• Put the needle cover in the trash.
• You may see an air bubble in the prefilled syringe. This is normal.
• You may also see a drop of liquid at the end of the needle. This is normal.
• **Do not** touch the needle or let it touch any surface.
• **Do not** use the prefilled syringe if it is dropped without the needle cover in place.

4.2 Position the prefilled syringe and inject SIMPONI

• Hold the body of the prefilled syringe in one hand between the thumb and index fingers. See Figure 7.

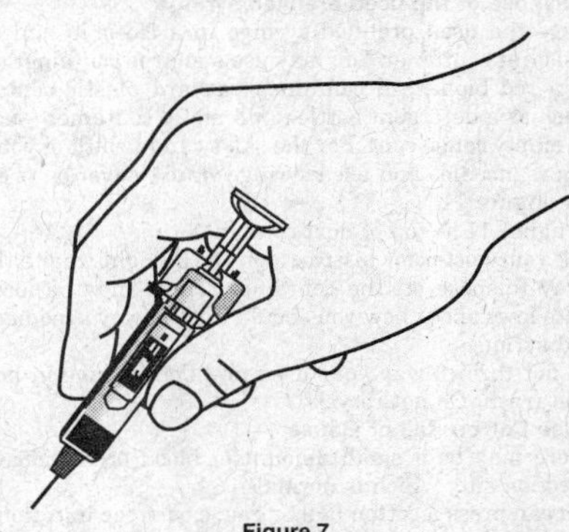

Figure 7

• **Do not** pull back on the plunger at any time.
• Use the other hand to gently pinch the area of skin that you previously cleaned. Hold firmly.

Continued on next page

Simponi—Cont.

- Use a quick, dart-like motion to insert the needle into the pinched skin at about a **45-degree angle**. See Figure 8.

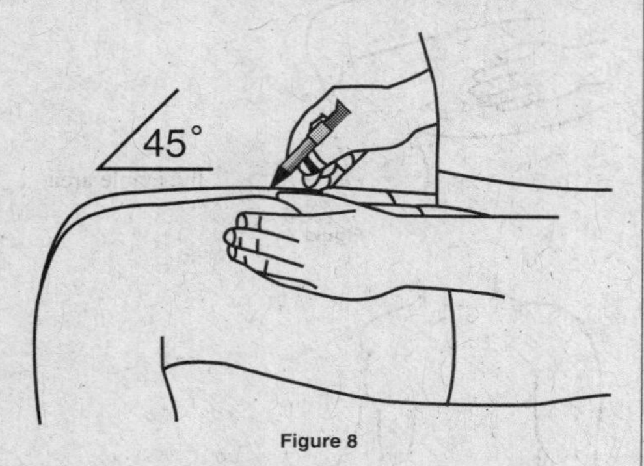

Figure 8

- Inject all of the medicine by using your thumb to push in the plunger until the plunger head is completely between the needle guard wings. See Figure 9.

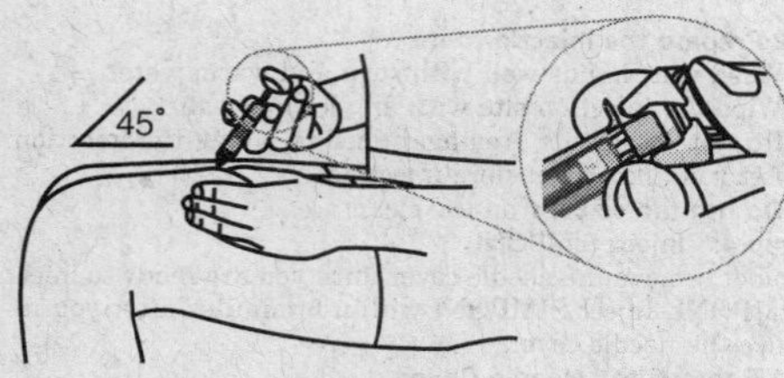

Figure 9

- When the plunger is pushed as far as it will go, keep pressure on the plunger head. Take the needle out of the skin and let go of the skin.
- Slowly take your thumb off the plunger head. This will let the empty syringe move up until the entire needle is covered by the needle guard. See Figure 10.

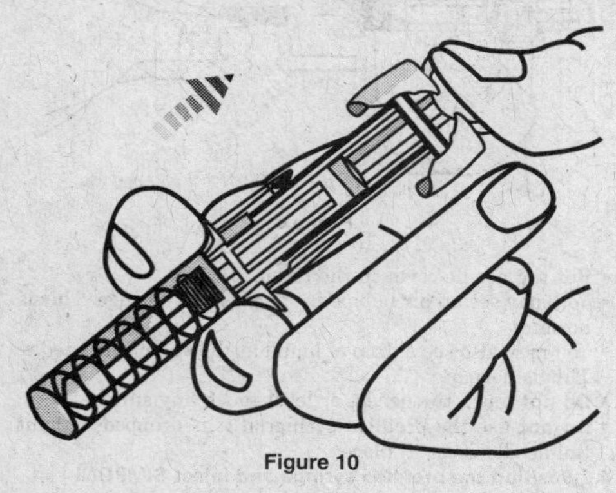

Figure 10

Step 5: After the injection

5.1 Dispose of the used prefilled syringe

- Place the used prefilled syringe in a closable puncture-resistant container. You may use a sharps container (such as a red biohazard container), a hard plastic container (such as a detergent bottle), or a metal container (such as an empty coffee can). For the safety and health of you and others, needles and used syringes **must never** be re-used. See Figure 11.

[See figure 11 at top of next column]

- Ask your doctor for instructions on the right way to throw away (dispose of) the container. There may be local or state laws about how you should throw away used needles and syringes.
- Do not throw away your used prefilled syringe in household trash. Do not recycle.

5.2 Use Cotton Ball or Gauze

- There may be a small amount of blood or liquid at the injection site, which is normal.
- You can press a cotton ball or gauze over the injection site and hold for 10 seconds. Do not rub the injection site.
- You may cover the injection site with a small adhesive bandage, if needed.

Manufactured by:
Centocor Ortho Biotech Inc.
Horsham, PA 19044
US License No. 1821

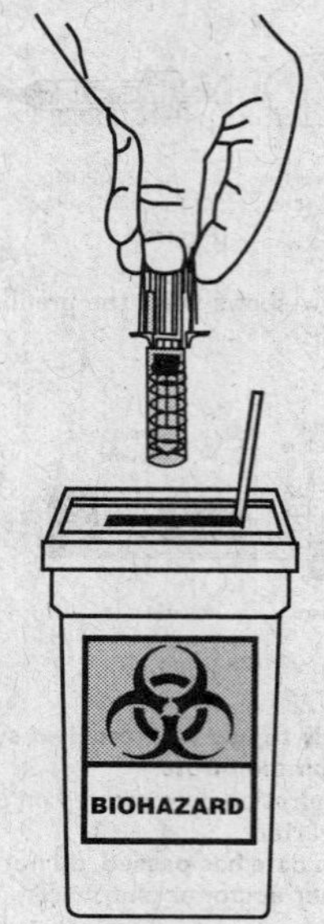

Figure 11

Revised: 4/2009
This Medication Guide has been approved by the U.S. Food and Drug Administration.

Forest Pharmaceuticals, Inc.

(Subsidiary of Forest Laboratories, Inc.)
13600 SHORELINE DRIVE
ST. LOUIS, MO 63045

Direct Inquiries to:
Professional Affairs Department
13600 Shoreline Drive
St. Louis, MO 63045
(800) 678-1605

SAVELLA ℞

(milnacipran HCl)
Tablets

HIGHLIGHTS OF PRESCRIBING INFORMATION
These highlights do not include all the information needed to use Savella safely and effectively. See full prescribing information for Savella.
Savella (milnacipran HCl) Tablets
Initial U.S. Approval: 2009

> **WARNING: SUICIDALITY AND ANTIDEPRESSANT DRUGS**
> ***See full prescribing information for complete boxed warning.***
> - **Increased risk of suicidal ideation, thinking and behavior in children, adolescents, and young adults taking antidepressants for major depressive disorder (MDD) and other psychiatric disorders. Savella is not approved for use in pediatric patients (5.1)**

------------RECENT MAJOR CHANGES------------

Warnings and Precautions, Serotonin Syndrome or Neuroleptic Malignant Syndrome (NMS)-like Reactions (5.2) 06/2009

------------INDICATIONS AND USAGE------------

Savella™ is a selective serotonin and norepinephrine reuptake inhibitor (SNRI) indicated for the management of fibromyalgia (1)

Savella is not approved for use in pediatric patients (5.1)

---------DOSAGE AND ADMINISTRATION---------

- Administer Savella in two divided doses per day (2.1)
- Begin dosing at 12.5 mg on the first day and increase to 100 mg/day over a 1-week period (2.1):
  Day 1: 12.5 mg once
  Days 2-3: 25 mg/day (12.5 mg twice daily)
  Days 4-7: 50 mg/day (25 mg twice daily)
  After Day 7: 100 mg/day (50 mg twice daily)
- Recommended dose is 100 mg/day (2.1)
- May be increased to 200 mg/day based on individual patient response (2.1)
- Dose should be adjusted in patients with severe renal impairment (2.2)

--------DOSAGE FORMS AND STRENGTHS--------

- Tablets: 12.5 mg, 25 mg, 50 mg, 100 mg (3)

---------------CONTRAINDICATIONS---------------

- Use of monoamine oxidase inhibitors concomitantly or in close temporal proximity (4.1)
- Use in patients with uncontrolled narrow-angle glaucoma (4.2)

----------WARNINGS AND PRECAUTIONS----------

- Suicidality: Monitor for worsening of depressive symptoms and suicide risk (5.1)
- Serotonin Syndrome or Neuroleptic Malignant Syndrome (NMS)-like Reactions: Serotonin syndrome or NMS-like reactions have been reported with SNRIs and SSRIs. Discontinue Savella and initiate supportive treatment (5.2, 7)
- Elevated blood pressure and heart rate: Cases have been reported with Savella. Monitor blood pressure and heart rate prior to initiating treatment with Savella and periodically throughout treatment (5.3, 5.4)
- Seizures: Cases have been reported with Savella therapy. Prescribe Savella with care in patients with a history of seizure disorder (5.5)
- Hepatotoxicity: More patients treated with Savella than with placebo experienced mild elevations of ALT and AST. Rarely, fulminant hepatitis has been reported in patients treated with Savella. Avoid concomitant use of Savella in patients with substantial alcohol use or chronic liver disease (5.6)
- Discontinuation: Withdrawal symptoms have been reported in patients when discontinuing treatment with Savella. A gradual dose reduction is recommended (5.7)
- Abnormal Bleeding: Savella may increase the risk of bleeding events. Caution patients about the risk of bleeding associated with the concomitant use of Savella and NSAIDs, aspirin, or other drugs that affect coagulation (5.9)
- Male patients with a history of obstructive uropathies may experience higher rates of genitourinary adverse events (5.11)

---------------ADVERSE REACTIONS---------------

The most frequently occurring adverse reactions (≥ 5% and greater than placebo) were nausea, headache, constipation, dizziness, insomnia, hot flush, hyperhidrosis, vomiting, palpitations, heart rate increased, dry mouth, and hypertension (6.3)

To report SUSPECTED ADVERSE REACTIONS, contact Forest Pharmaceuticals, Inc., at (800) 678-1605 or FDA at 1-800-FDA-1088 or www.fda.gov/medwatch.

---------------DRUG INTERACTIONS---------------

- Savella is unlikely to be involved in clinically significant pharmacokinetic drug interactions (7)
- Pharmacodynamic interactions of Savella with other drugs can occur (7)

---------USE IN SPECIFIC POPULATIONS---------

- Pregnancy and nursing mothers: Use only if the potential benefit justifies the potential risk to the fetus or child. (8.1, 8.3)

See 17 for PATIENT COUNSELING INFORMATION and Medication Guide.
Revised: July 2009

FULL PRESCRIBING INFORMATION: CONTENTS*

*Sections or subsections omitted from the full prescribing information are not listed.

FULL PRESCRIBING INFORMATION

WARNING: SUICIDALITY AND ANTIDEPRESSANT DRUGS Savella is a selective serotonin and norepinephrine reuptake inhibitor (SNRI), similar to some drugs used for the treatment of depression and other psychiatric disorders. Antidepressants increased the risk compared to placebo of suicidal thinking and behavior (suicidality) in children, adolescents, and young adults in short-term studies of major depressive disorder (MDD) and other psychiatric disorders. Anyone considering the use of such drugs in a child, adolescent, or young adult must balance this risk with the clinical need. Short-term studies did not show an increase in the risk of suicidality with antidepressants compared to placebo in adults beyond age 24; there was a reduction in risk with antidepressants compared to placebo in adults aged 65 and older. Depression and certain other psychiatric disorders are themselves associated with increases in the risk of suicide. Patients of all ages who are started on Savella should be monitored appropriately and observed closely for clinical worsening, suicidality, or unusual changes in behavior. Families and caregivers should be advised of the need for close observation and communication with the prescriber. Savella is not approved for use in the treatment of major depressive disorder. Savella is not approved for use in pediatric patients [*see Warnings and Precautions (5.1), Use in Specific Populations (8.4)*]

1 INDICATIONS AND USAGE

Savella is indicated for the management of fibromyalgia.
Savella is not approved for use in pediatric patients [*see Use in Specific Populations (8.4)*].

2 DOSAGE AND ADMINISTRATION

Savella is given orally with or without food.
Taking Savella with food may improve the tolerability of the drug.

2.1 Recommended Dosing

The recommended dose of Savella is 100 mg/day (50 mg twice daily).
Dosing should be titrated according to the following schedule:
Day 1: 12.5 mg once
Days 2-3: 25 mg/day (12.5 mg twice daily)
Days 4-7: 50 mg/day (25 mg twice daily)
After Day 7: 100 mg/day (50 mg twice daily)
Based on individual patient response, the dose may be increased to 200 mg/day (100 mg twice daily).
Doses above 200 mg/day have not been studied.
Savella should be tapered and not abruptly discontinued after extended use [*see Discontinuing Savella (2.4) and Warnings and Precautions (5.7)*]

2.2 Patients with Renal Insufficiency

No dosage adjustment is necessary in patients with mild renal impairment.
Savella should be used with caution in patients with moderate renal impairment.
For patients with severe renal impairment (indicated by an estimated creatinine clearance of 5-29 mL/min), the maintenance dose should be reduced by 50% to 50 mg/day (25 mg twice daily).
Based on individual patient response, the dose may be increased to 100 mg/day (50 mg twice daily).
Savella is not recommended for patients with end-stage renal disease.

2.3 Patients with Hepatic Insufficiency

No dosage adjustment is necessary for patients with hepatic impairment.
As with any drug, caution should be exercised in patients with severe hepatic impairment.

2.4 Discontinuing Savella

Withdrawal symptoms have been observed in clinical trials following discontinuation of milnacipran, as with other serotonin and norepinephrine re-uptake inhibitors (SNRIs) and selective serotonin re-uptake inhibitors (SSRIs). Patients should be monitored for these symptoms when discontinuing treatment. Savella should be tapered and not abruptly discontinued after extended use [*see Warnings and Precautions (5.7)*].

2.5 Switching patients to or from a Monoamine Oxidase Inhibitor (MAOI)

At least 14 days should elapse between discontinuation of a MAOI and initiation of therapy with Savella. In addition, at least 5 days should be allowed after stopping Savella before starting a MAOI [*see Contraindications (4.1)*].

3 DOSAGE FORMS AND STRENGTHS

Film-coated, immediate release tablets in four strengths: 12.5 mg, 25 mg, 50 mg, and 100 mg of milnacipran hydrochloride.
12.5 mg tablets are round, blue, "F" on one side, "L" on the reverse;
25 mg tablets are round, white, "FL" on one side, "25" on the reverse;
50 mg tablets are oval, white, "FL" on one side, "50" on the reverse;
100 mg tablets are oval, pink, "FL" on one side, "100" on the reverse
[*see Description (11) and How Supplied/Storage and Handling (16)*].

4 CONTRAINDICATIONS

4.1 Monoamine Oxidase Inhibitors

Concomitant use of Savella in patients taking monoamine oxidase inhibitors (MAOIs) is contraindicated. In patients receiving a serotonin reuptake inhibitor in combination with a monoamine oxidase inhibitor (MAOI), there have been reports of serious, sometimes fatal, reactions including hyperthermia, rigidity, myoclonus, autonomic instability with possible rapid fluctuations of vital signs, and mental status changes that include extreme agitation progressing to delirium and coma. These reactions have also been reported in patients who have recently discontinued serotonin reuptake inhibitors and have been started on an MAOI. Some cases presented with features resembling neuroleptic malignant syndrome. The effects of combined use of Savella and MAOIs have not been evaluated in humans. Therefore, it is recommended that Savella should not be used in combination with an MAOI, or within 14 days of discontinuing treatment with an MAOI. Similarly, at least 5 days should be allowed after stopping Savella before starting an MAOI [see *Dosage and Administration (2.5), Warnings and Precautions (5.2)*].

4.2 Uncontrolled Narrow-Angle Glaucoma

In clinical trials, Savella was associated with an increased risk of mydriasis. Mydriasis has been reported with other dual reuptake inhibitors of norepinephrine and serotonin; therefore, do not use Savella in patients with uncontrolled narrow-angle glaucoma.

5 WARNINGS AND PRECAUTIONS

5.1 Suicide Risk

Savella is a selective serotonin and norepinephrine reuptake inhibitor (SNRI), similar to some drugs used for the treatment of depression and other psychiatric disorders.
Patients, both adult and pediatric, with depression or other psychiatric disorders may experience worsening of their depression and/or the emergence of suicidal ideation and behavior (suicidality) or unusual changes in behavior, whether or not they are taking these medications, and this risk may persist until significant remission occurs. Suicide is a known risk of depression and certain other psychiatric disorders, and these disorders themselves are the strongest predictors of suicide. There has been a long-standing concern, however, that antidepressants, including drugs that inhibit the reuptake of norepinephrine and/or serotonin, may have a role in inducing worsening of depression and the emergence of suicidality in certain patients during the early phases of treatment.
In the placebo-controlled clinical trials of adults with fibromyalgia, among the patients who had a history of depression at treatment initiation, the incidence of suicidal ideation was 0.5% in patients treated with placebo, 0% in patients treated with Savella 100 mg/day, and 1.3% in patients treated with Savella 200 mg/day. No suicides occurred in the short-term or longer-term (up to 1 year) fibromyalgia trials.
Pooled analyses of short-term placebo-controlled trials of drugs used to treat depression (SSRIs and others) showed that these drugs increase the risk of suicidal thinking and behavior (suicidality) in children, adolescents, and young adults (ages 18-24) with major depressive disorder (MDD) and other psychiatric disorders. Short-term studies did not show an increase in the risk of suicidality with these drugs compared to placebo in adults beyond age 24; there was a reduction in suicidality risk with antidepressants compared to placebo in adults age 65 and older.
The pooled analyses of placebo-controlled trials in children and adolescents with MDD, obsessive compulsive disorder (OCD), or other psychiatric disorders included a total of 24 short-term trials of 9 drugs used to treat depression in over 4400 patients. The pooled analyses of placebo-controlled trials in adults with MDD or other psychiatric disorders included a total of 295 short-term trials (median duration of 2 months) of 11 antidepressant drugs in over 77,000 patients. There was considerable variation in risk of suicidality among drugs, but a tendency toward an increase in the younger patients for almost all drugs studied. There were differences in absolute risk of suicidality across the different indications, with the highest incidence in MDD. The risk of differences (drug versus placebo), however, were relatively stable within age strata and across indications. These risk differences (drug-placebo difference in the number of cases of suicidality per 1000 patients treated) are provided in Table 1.

Table 1. Risk Differences (Drug – Placebo) in the number of Cases of Suicidality, per 1000 patients treated

Age Range	Drug-Placebo Difference in Number of Cases of Suicidality per 1000 Patients Treated
< 18	14 additional cases
18-24	5 additional cases
	Decreases Compared to Placebo
25-64	1 fewer case
≥ 65	6 fewer cases

No suicides occurred in any of the pediatric trials. There were suicides in the adult trials, but the number was not sufficient to reach any conclusion about drug effect on suicide.
It is unknown whether the suicidality risk extends to longer-term use, i.e., beyond several months. However, there is substantial evidence from placebo-controlled maintenance trials in adults with depression that the use of antidepressants can delay the recurrence of depression.
All patients being treated with drugs inhibiting the reuptake of norepinephrine and/or serotonin for any indication should be monitored appropriately and observed closely for clinical worsening, suicidality, and unusual changes in behavior, especially during the initial few months of a course of drug therapy, or at times of dose changes, either increases or decreases.
The following symptoms, anxiety, agitation, panic attacks, insomnia, irritability, hostility, aggressiveness, impulsivity, akathisia (psychomotor restlessness), hypomania, mania, have been reported in adult and pediatric patients being treated with drugs inhibiting the reuptake of norepinephrine and/or serotonin for major depressive disorder as well as for other indications, both psychiatric and nonpsychiatric. Although a causal link between the emergence of such symptoms and either the worsening of depression and/or the emergence of suicidal impulses has not been established, there is concern that such symptoms may represent precursors to emerging suicidality.
Consideration should be given to changing the therapeutic regimen, including possibly discontinuing the medication, in patients who may experience worsening depressive symptoms, or who are experiencing emergent suicidality or symptoms that might be precursors to worsening depression or suicidality, especially if these symptoms are severe or abrupt in onset, or were not part of the patient's presenting symptoms.

Continued on next page

Savella—Cont.

If the decision has been made to discontinue treatment due to worsening depressive symptoms or emergent suicidality, medication should be tapered, as rapidly as is feasible, but with recognition that abrupt discontinuation can produce withdrawal symptoms *[see Dosage and Administration—Recommended Dosing (2.1), Dosage—Discontinuing Savella (2.4), and Warnings and Precautions—Discontinuation of Treatment with Savella (5.7)].*

Families and caregivers of patients being treated with drugs inhibiting the reuptake of norepinephrine and/or serotonin for major depressive disorder or other indications, both psychiatric and nonpsychiatric, should be alerted about the need to monitor patients for the emergence of agitation, irritability, unusual changes in behavior, and the other symptoms described above, as well as the emergence of suicidality, and to report such symptoms immediately to health care providers. Such monitoring should include daily observation by families and caregivers. Prescriptions for Savella should be written for the smallest quantity of tablets consistent with good patient management, in order to reduce the risk of overdose.

5.2 Serotonin Syndrome or Neuroleptic Malignant Syndrome (NMS)-Like Reactions

The development of a potentially life-threatening serotonin syndrome or Neuroleptic Malignant Syndrome (NMS)-like reactions have been reported with SNRIs and SSRIs alone, including Savella, but particularly with concomitant use of serotonergic drugs (including triptans), with drugs which impair metabolism of serotonin (including MAOIs) or with antipsychotics or other dopamine antagonists.. Serotonin syndrome symptoms may include mental status changes (e.g., agitation, hallucinations, coma), autonomic instability (e.g., tachycardia, labile blood pressure, hyperthermia), neuromuscular aberrations (e.g., hyperreflexia, incoordination) and/or gastrointestinal symptoms (e.g., nausea, vomiting, diarrhea) *[see Drug Interactions (7)].* Serotonin syndrome, in its most severe form can resemble neuroleptic malignant syndrome, which includes hyperthermia, muscle rigidity, autonomic instability with possible rapid fluctuation of vital signs, and mental status changes. Patients should be monitored for the emergence of serotonin syndrome or NMS-like signs and symptoms.

The concomitant use of Savella with MAOIs is contraindicated *[see Contraindications (4.1)].* If concomitant treatment of Savella with a 5-hydroxytryptamine receptor agonist (triptan) is clinically warranted, careful observation of the patient is advised, particularly during treatment initiation and dose increases *[see Drug Interactions (7)].*

The concomitant use of Savella with serotonin precursors (such as tryptophan) is not recommended *[see Drug Interactions (7)].* Treatment with Savella and any concomitant serotonergic or antidopaminergic agents, including antipsychotics, should be discontinued immediately if the above events occur and supportive symptomatic treatment should be initiated.

5.3 Effects on Blood Pressure

Inhibition of the reuptake of norepinephrine (NE) and serotonin (5-HT) can lead to cardiovascular effects. SNRIs, including Savella, have been associated with reports of increase in blood pressure.

In a double-blind, placebo-controlled clinical pharmacology study in healthy subjects designed to evaluate the effects of milnacipran on various parameters, including blood pressure at supratherapeutic doses, there was evidence of mean increases in supine blood pressure at doses up to 300 mg twice daily (600 mg/day). At the highest 300 mg twice daily dose, the mean increase in systolic blood pressure was up to 8.1 mm Hg for the placebo group and up to 10.0 mm Hg for the Savella treated group over the 12 hour steady state dosing interval. The corresponding mean increase in diastolic blood pressure over this interval was up to 4.6 mm Hg for placebo and up to 11.5 mm Hg for the Savella treated group.

In the 3-month placebo-controlled fibromyalgia clinical trials, Savella treatment was associated with mean increases of up to 3.1 mm Hg in systolic blood pressure (SBP) and diastolic blood pressure (DBP) *[see Adverse Reactions (6.3)].*

In the placebo-controlled trials, among fibromyalgia patients who were non-hypertensive at baseline, approximately twice as many patients in the Savella treatment arms became hypertensive at the end of the study (SBP ≥ 140 mmHg or DBP ≥ 90 mmHg) compared with the placebo patients: 7.2% of patients in the placebo arm versus 19.5% of patients treated with Savella 100 mg/day and 16.6% of patients treated with Savella 200 mg/day. Among patients who met systolic criteria for pre-hypertension at baseline (SBP 120-139 mmHg), more patients became hypertensive at the end of the study in the Savella treatment arms than placebo: 9% of patients in the placebo arm versus 14% in both the Savella 100 mg/day and the Savella 200 mg/day treatment arms.

Among fibromyalgia patients who were hypertensive at baseline, more patients in the Savella treatment arms had a >15 mmHg increase in SBP than placebo at the end of the study: 1% of patients in the placebo arm versus 7% in the Savella 100 mg/day and 2% in the Savella 200 mg/day treatment arms. Similarly, more patients who were hypertensive at baseline and were treated with Savella had DBP increases > 10 mmHg than placebo at the end of study: 3% of patients in the placebo arm versus 8% in the Savella 100 mg/day and 6% in the Savella 200 mg/day treatment arms.

Sustained increases in SBP (increase of ≥ 15 mmHg on three consecutive post-baseline visits) occurred in 2% of placebo patients versus 9% of patients receiving Savella 100 mg/day and 6% of patients receiving Savella 200 mg/day. Sustained increases in DBP (increase of ≥ 10 mmHg on 3 consecutive post-baseline visits) occurred in 4% of patients receiving placebo versus 13% of patients receiving Savella 100 mg/day and 10% of patients receiving Savella 200 mg/day.

Sustained increases in blood pressure could have adverse consequences. Cases of elevated blood pressure requiring immediate treatment have been reported.

Concomitant use of Savella with drugs that increase blood pressure and pulse has not been evaluated and such combinations should be used with caution *[see Drug Interactions (7)].*

Effects of Savella on blood pressure in patients with significant hypertension or cardiac disease have not been systematically evaluated. Savella should be used with caution in these patients.

Blood pressure should be measured prior to initiating treatment and periodically measured throughout Savella treatment. Pre-existing hypertension and other cardiovascular disease should be treated before starting therapy with Savella. For patients who experience a sustained increase in blood pressure while receiving Savella, either dose reduction or discontinuation should be considered.

5.4 Effects on Heart Rate

SNRIs have been associated with reports of increase in heart rate.

In clinical trials, relative to placebo, Savella treatment was associated with mean increases in pulse rate of approximately 7 to 8 beats per minute *[see Adverse Reactions (6.2, 6.3)].*

Increases in pulse ≥ 20 bpm occurred more frequently in Savella-treated patients when compared to placebo: 0.3% in the placebo arm versus 8% in the Savella 100 mg/day and 8% in the 200 mg/day treatment arms. The effect of Savella on heart rate did not appear to increase with increasing dose.

Savella has not been systematically evaluated in patients with a cardiac rhythm disorder.

Heart rate should be measured prior to initiating treatment and periodically measured throughout Savella treatment. Pre-existing tachyarrhythmias and other cardiac disease should be treated before starting therapy with Savella. For patients who experience a sustained increase in heart rate while receiving Savella, either dose reduction or discontinuation should be considered.

5.5 Seizures

Savella has not been systematically evaluated in patients with a seizure disorder. In clinical trials evaluating Savella in patients with fibromyalgia, seizures/convulsions have not been reported. However, seizures have been reported infrequently in patients treated with Savella for disorders other than fibromyalgia. Savella should be prescribed with care in patients with a history of a seizure disorder.

5.6 Hepatotoxicity

In the placebo-controlled fibromyalgia trials, increases in the number of patients treated with Savella with mild elevations of ALT or AST (1-3 times the upper limit of normal, ULN) were observed. Increases in ALT were more frequently observed in the patients treated with Savella 100 mg/day (6%) and Savella 200 mg/day (7%), compared to the patients treated with placebo (3%). One patient receiving Savella 100 mg/day (0.2%) had an increase in ALT greater than 5 times the upper limit of normal but did not exceed 10 times the upper limit of normal. Increases in AST were more frequently observed in the patients treated with Savella 100 mg/day (3%) and Savella 200 mg/day (5%) compared to the patients treated with placebo (2%).

The increases of bilirubin observed in the fibromyalgia clinical trials were not clinically significant. No case met the criteria of elevated ALT > 3× ULN and associated with an increase in bilirubin ≥ 2× ULN.

There have been cases of increased liver enzymes and reports of severe liver injury, including fulminant hepatitis with milnacipran from foreign postmarketing experience. In the cases of severe liver injury there were significant underlying clinical conditions and/or the use of multiple concomitant medications. Because of underreporting, it is impossible to provide an accurate estimate of the true incidence of these reactions.

Savella should be discontinued in patients who develop jaundice or other evidence of liver dysfunction. Treatment with Savella should not be resumed unless another cause can be established.

Savella should ordinarily not be prescribed to patients with substantial alcohol use or evidence of chronic liver disease.

5.7 Discontinuation of Treatment with Savella

Withdrawal symptoms have been observed in clinical trials following discontinuation of milnacipran, as with other SNRIs and SSRIs.

During marketing of milnacipran, and other SNRIs and SSRIs, there have been spontaneous reports of adverse events indicative of withdrawal and physical dependence occurring upon discontinuation of these drugs, particularly when discontinuation is abrupt. The adverse events include the following: dysphoric mood, irritability, agitation, dizziness, sensory disturbances (e.g., paresthesias such as electric shock sensations), anxiety, confusion, headache, lethargy, emotional lability, insomnia, hypomania, tinnitus, and seizures. Although these events are generally self-limiting, some have been reported to be severe.

Patients should be monitored for these symptoms when discontinuing treatment with Savella. Savella should be tapered and not abruptly discontinued after extended use. If intolerable symptoms occur following a decrease in the dose or upon discontinuation of treatment, then resuming the previously prescribed dose may be considered. Subsequently, the physician may continue decreasing the dose but at a more gradual rate *[see Dosage and Administration (2.4)].*

5.8 Hyponatremia

Hyponatremia may occur as a result of treatment with SSRIs and SNRIs, including Savella. In many cases, this hyponatremia appears to be the result of the syndrome of inappropriate antidiuretic hormone secretion (SIADH). Cases with serum sodium lower than 110 mmol/L have been reported. Elderly patients may be at greater risk of developing hyponatremia with SNRIs, SSRIs, or Savella. Also, patients taking diuretics or who are otherwise volume-depleted may be at greater risk *[see Geriatric Use (8.5)].* Discontinuation of Savella should be considered in patients with symptomatic hyponatremia.

Signs and symptoms of hyponatremia include headache, difficulty concentrating, memory impairment, confusion, weakness, and unsteadiness, which may lead to falls. Signs and symptoms associated with more severe and/or acute cases have included hallucination, syncope, seizure, coma, respiratory arrest, and death.

5.9 Abnormal Bleeding

SSRIs and SNRIs, including Savella, may increase the risk of bleeding events. Concomitant use of aspirin, nonsteroidal anti-inflammatory drugs, warfarin, and other anticoagulants may add to this risk. Case reports and epidemiological studies (case-control and cohort design) have demonstrated an association between use of drugs that interfere with serotonin reuptake and the occurrence of gastrointestinal bleeding. Bleeding events related to SSRIs and SNRIs use have ranged from ecchymoses, hematomas, epistaxis, and petechiae to life-threatening hemorrhages.

Patients should be cautioned about the risk of bleeding associated with the concomitant use of Savella and NSAIDs, aspirin, or other drugs that affect coagulation.

5.10 Activation of Mania

No activation of mania or hypomania was reported in the clinical trials evaluating effects of Savella in patients with fibromyalgia. However those clinical trials excluded patients with current major depressive episode. Activation of mania and hypomania have been reported in patients with mood disorders who were treated with other similar drugs for major depressive disorder. As with these other agents, Savella should be used cautiously in patients with a history of mania.

5.11 Patients with a History of Dysuria

Because of their noradrenergic effect, SNRIs including Savella, can affect urethral resistance and micturition. In the controlled fibromyalgia trials, dysuria occurred more frequently in patients treated with Savella (1%) than in placebo-treated patients (0.5%). Caution is advised in use of Savella in patients with a history of dysuria, notably in male patients with prostatic hypertrophy, prostatitis, and other lower urinary tract obstructive disorders. Male patients are more prone to genitourinary adverse effects, such as dysuria or urinary retention, and may experience testicular pain or ejaculation disorders.

5.12 Controlled Narrow-Angle Glaucoma

Mydriasis has been reported in association with SNRIs and Savella; therefore, Savella should be used cautiously in patients with controlled narrow-angle glaucoma.

Do not use Savella in patients with Uncontrolled Narrow-Angle Glaucoma *[see Contraindications (4.2)].*

5.13 Concomitant Use with Alcohol

In clinical trials, more patients treated with Savella developed elevated transaminases than did placebo treated pa-

Table 2. Treatment-Emergent Adverse Reaction Incidence in Placebo Controlled Trials in Fibromyalgia Patients (Events Occurring in at Least 2% of All Savella-Treated Patients and Occurring More Frequently in Either Savella Treatment Group Than in the Placebo Treatment Group)

System Organ Class–Preferred Term	Savella 100 mg/day (n = 623) %	Savella 200 mg/day (n = 934) %	All Savella (n = 1557) %	Placebo (n = 652) %
Cardiac Disorders				
Palpitations	8	7	7	2
Tachycardia	3	2	2	1
Eye Disorders				
Vision blurred	1	2	2	1
Gastrointestinal Disorders				
Nausea	35	39	37	20
Constipation	16	15	16	4
Vomiting	6	7	7	2
Dry mouth	5	5	5	2
Abdominal pain	3	3	3	2
General Disorders				
Chest pain	3	2	2	2
Chills	1	2	2	0
Chest discomfort	2	1	1	1
Infections				
Upper respiratory tract infection	7	6	6	6
Investigations				
Heart rate increased	5	6	6	1
Blood pressure increased	3	3	3	1
Metabolism and Nutrition Disorders				
Decreased appetite	1	2	2	0
Nervous System Disorders				
Headache	19	17	18	14
Dizziness	11	10	10	6
Migraine	6	4	5	3
Paresthesia	2	3	2	2
Tremor	2	2	2	1
Hypoesthesia	1	2	1	1
Tension headache	2	1	1	1
Psychiatric Disorders				
Insomnia	12	12	12	10
Anxiety	5	3	4	4
Respiratory Disorders				
Dyspnea	2	2	2	1
Skin Disorders				
Hyperhidrosis	8	9	9	2
Rash	3	4	3	2
Pruritus	3	2	2	2
Vascular Disorders				
Hot flush	11	12	12	2
Hypertension	7	4	5	2
Flushing	2	3	3	1

tients [see *Warnings and Precautions (5.6)*]. Because it is possible that milnacipran may aggravate pre-existing liver disease, Savella should not be prescribed to patients with substantial alcohol use or evidence of chronic liver disease.

6 ADVERSE REACTIONS

6.1 Clinical Trial Data Sources

Savella was evaluated in three double-blind placebo-controlled trials involving 2209 fibromyalgia patients (1557 patients treated with Savella and 652 patients treated with placebo) for a treatment period up to 29 weeks.

The stated frequencies of adverse reactions represent the proportion of individuals who experienced, at least once, a treatment-emergent adverse reaction of the type listed. A reaction was considered treatment emergent if it occurred for the first time or worsened while receiving therapy following baseline evaluation.

Because clinical trials are conducted under widely varying conditions, adverse reaction rates observed in the clinical trials of a drug cannot be directly compared to rates in the clinical trials of another drug and may not reflect the rates observed in practice.

6.2 Adverse Reactions Leading to Discontinuation

In placebo-controlled trials in patients with fibromyalgia, 23% of patients treated with Savella 100 mg/day, 26% of patients treated with Savella 200 mg/day discontinued prematurely due to adverse reactions, compared to 12% of patients treated with placebo. The adverse reactions that led to withdrawal in ≥ 1% of patients in the Savella treatment group and with an incidence rate greater than that in the placebo treatment group were nausea (milnacipran 6%, placebo 1%), palpitations (milnacipran 3%, placebo 1%), headache (milnacipran 2%, placebo 0%), constipation (milnacipran 1%, placebo 0%), heart rate increased (milnacipran 1%, placebo 0%), hyperhidrosis (milnacipran 1%, placebo 0%), vomiting (milnacipran 1%, placebo 0%), and dizziness (milnacipran 1% and placebo 0.5%). Discontinuation due to adverse reactions was generally more common among patients treated with Savella 200 mg/day compared to Savella 100 mg/day.

6.3 Most Common Adverse Reactions

In the placebo-controlled fibromyalgia patient trials the most frequently occurring adverse reaction in clinical trials was nausea. The most common adverse reactions (incidence ≥ 5% and twice placebo) in patients treated with Savella were constipation, hot flush, hyperhidrosis, vomiting, palpitations, heart rate increased, dry mouth, and hypertension.

Table 2 lists all adverse reactions that occurred in at least 2% of patients treated with Savella at either 100 or 200 mg/day and at an incidence greater than that of placebo.

[See table 2 above]

6.4 Weight Changes

In placebo-controlled fibromyalgia clinical trials, patients treated with Savella for up to 3 months experienced a mean weight loss of approximately 0.8 kg in both the Savella 100 mg/day and the Savella 200 mg/day treatment groups, compared with a mean weight loss of approximately 0.2 kg in placebo-treated patients.

6.5 Genitourinary Adverse Reactions in Males

In the placebo-controlled fibromyalgia studies, the following treatment-emergent adverse reactions related to the genitourinary system were observed in at least 2% of male patients treated with Savella, and occurred at a rate greater than in placebo-treated male patients: dysuria, ejaculation disorder, erectile dysfunction, ejaculation failure, libido decreased, prostatitis, scrotal pain, testicular pain, testicular swelling, urinary hesitation, urinary retention, urethral pain, and urine flow decreased.

6.6 Other Adverse Reactions Observed During Clinical Trials of Savella in Fibromyalgia

Following is a list of frequent (those occurring on one or more occasions in at least 1/100 patients) treatment-emergent adverse reactions reported from 1824 fibromyalgia patients treated with Savella for periods up to 68 weeks. The listing does not include those events already listed in Table 2, those events for which a drug cause was remote, those events which were so general as to be uninformative, and those events reported only once which did not have a substantial probability of being acutely life threatening.

Adverse reactions are categorized by body system and listed in order of decreasing frequency. Adverse reactions of major clinical importance are described in the *Warnings and Precautions* section (5).

Gastrointestinal Disorders—diarrhea, dyspepsia, gastroesophageal reflux disease, flatulence, abdominal distension

General Disorders—fatigue, peripheral edema, irritability, pyrexia

Infections—urinary tract infection, cystitis

Injury, Poisoning, and Procedural Complications—contusion, fall

Investigations—weight decreased or increased

Metabolism and Nutrition Disorders—hypercholesterolemia

Nervous System Disorders—somnolence, dysgeusia

Psychiatric Disorders—depression, stress

Skin Disorders—night sweats

Continued on next page

Savella—Cont.

6.7 Postmarketing Spontaneous Reports
The following additional adverse reactions have been identified from spontaneous reports of Savella received worldwide. These adverse reactions have been chosen for inclusion because of a combination of seriousness, frequency of reporting, or potential causal connection to Savella. However, because these adverse reactions were reported voluntarily from a population of uncertain size, it is not always possible to reliably estimate their frequency or establish a causal relationship to drug exposure. These events include:
Blood and Lymphatic System Disorders—leukopenia, neutropenia, thrombocytopenia
Cardiac Disorders—supraventricular tachycardia
Eye Disorders—accommodation disorder
Endocrine Disorders—hyperprolactinemia
Hepatobiliary Disorders—hepatitis
Metabolism and Nutrition Disorders—anorexia, hyponatremia
Musculoskeletal and Connective Tissue Disorders—rhabdomyolysis
Nervous System Disorders—convulsions (including grand mal), loss of consciousness, Parkinsonism
Psychiatric Disorders—delirium, hallucination
Renal and Urinary Disorders—acute renal failure, urinary retention
Reproductive System and Breast Disorders—galactorrhea
Skin Disorders—erythema multiforme, Stevens Johnson syndrome
Vascular Disorders—hypertensive crisis

7 DRUG INTERACTIONS

Milnacipran undergoes minimal CYP450 related metabolism, with the majority of the dose excreted unchanged in urine (55%), and has a low binding to plasma proteins (13%). In vitro and in vivo studies showed that Savella is unlikely to be involved in clinically significant pharmacokinetic drug interactions *[see Pharmacokinetics in Special Populations (12.4)]*.

Clinically Important Interactions with Other Drugs

Lithium: Serotonin syndrome may occur when lithium is co-administered with Savella and with other drugs that impair metabolism of serotonin [*see Warnings and Precautions – Serotonin Syndrome or Neuroleptic Malignant Syndrome (NMS)-Like Reactions (5.2)*].

Epinephrine and norepinephrine: Savella inhibits the reuptake of norepinephrine. Therefore concomitant use of Savella with epinephrine and norepinephrine may be associated with paroxysmal hypertension and possible arrhythmia [*see Warnings and Precautions – Effects on Blood Pressure (5.3)* and *Effects on Heart Rate (5.4)*]

Serotonergic Drugs: Co-administration of Savella with other inhibitors of serotonin re-uptake may result in hypertension and coronary artery vasoconstriction, through additive serotonergic effects *[see Warnings and Precautions (5.2)]*.

Digoxin: Use of Savella concomitantly with digoxin may be associated with potentiation of adverse hemodynamic effects. Postural hypotension and tachycardia have been reported in combination therapy with intravenously administered digoxin (1 mg). Co-administration of Savella and intravenous digoxin should be avoided [*see Warnings and Precautions (5.3, 5.4)*]

Clonidine: Because Savella inhibits norepinephrine reuptake, co-administration with clonidine may inhibit clonidine's anti-hypertensive effect.

Clomipramine: In a drug-drug interaction study, an increase in euphoria and postural hypotension was observed in patients who switched from clomipramine to Savella.

CNS-active drugs: Given the primary CNS effects of Savella, caution should be used when it is taken in combination with other centrally acting drugs, including those with a similar mechanism of action.

Monoamine Oxidase Inhibitors (MAOIs): [see Contraindications (4.1)].

8 USE IN SPECIFIC POPULATIONS

8.1 Pregnancy
Pregnancy Category C
Milnacipran increased the incidence of dead fetuses in utero in rats at doses of 5 mg/kg/day (0.25 times the MRHD on a mg/m^2 basis). Administration of milnacipran to mice and rabbits during the period of organogenesis did not result in embryotoxicity or teratogenicity at doses up to 125 mg/kg/day in mice (3 times the maximum recommended human dose [MRHD] of 200 mg/day on a mg/m^2 basis) and up to 60 mg/kg/day in rabbits (6 times the MRHD of 200 mg/day on a mg/m^2 basis). In rabbits, the incidence of the skeletal variation, extra single rib, was increased following administration of milnacipran at 15 mg/kg/day during the period of organogenesis.

There are no adequate and well-controlled studies in pregnant women. Savella should be used during pregnancy only if the potential benefit justifies the potential risk to the fetus.

Nonteratogenic Effects
Neonates exposed to dual reuptake inhibitors of serotonin and norepinephrine, or selective serotonin reuptake inhibitors late in the third trimester have developed complications requiring prolonged hospitalization, respiratory support, and tube feeding. Such complications can arise immediately upon delivery. Reported clinical findings have included respiratory distress, cyanosis, apnea, seizures, temperature instability, feeding difficulty, vomiting, hypoglycemia, hypotonia, hypertonia, hyperreflexia, tremor, jitteriness, irritability, and constant crying. These features are consistent with either a direct toxic effect of these classes of drugs or, possibly, a drug discontinuation syndrome. It should be noted that, in some cases, the clinical picture is consistent with serotonin syndrome [*see Warnings and Precautions (5.2)*].

In rats, a decrease in pup body weight and viability on postpartum day 4 were observed when milnacipran, at a dose of 5 mg/kg/day (approximately 0.2 times the MRHD on a mg/m^2 basis), was administered orally to rats during late gestation. The no-effect dose for maternal and offspring toxicity was 2.5 mg/kg/day (approximately 0.1 times the MRHD on a mg/m^2 basis).

8.2 Labor and Delivery
The effect of milnacipran on labor and delivery is unknown. The use of Savella during labor and delivery is not recommended.

8.3 Nursing Mothers
There are no adequate and well-controlled studies in nursing mothers. It is not known if milnacipran is excreted in human milk. Studies in animals have shown that milnacipran or its metabolites are excreted in breast milk. Because many drugs are excreted in human milk and because of the potential for serious adverse reactions in nursing infants from milnacipran, a decision should be made whether to discontinue the drug, taking into account the importance of the drug to the mother. Because the safety of Savella in infants is not known, nursing while on Savella is not recommended.

8.4 Pediatric Use
Safety and effectiveness of Savella in a fibromyalgia pediatric population below the age of 17 have not been established [*see Box Warning* and *Warnings and Precautions (5.1)*]. The use of Savella is not recommended in pediatric patients.

8.5 Geriatric Use
In controlled clinical studies of Savella, 402 patients were 60 years or older, and no overall differences in safety and efficacy were observed between these patients and younger patients. In view of the predominant excretion of unchanged milnacipran via kidneys and the expected decrease in renal function with age renal function should be considered prior to use of Savella in the elderly [*see Dosage and Administration (2.2)*].

SNRIs, SSRIs, and Savella, have been associated with cases of clinically significant hyponatremia in elderly patients, who may be at greater risk for this adverse event [*see Warnings and Precautions (5.8)*].

9 DRUG ABUSE AND DEPENDENCE

9.1 Controlled Substance
Milnacipran is not a controlled substance.

9.2 Abuse
Milnacipran did not produce behavioral signs indicative of abuse potential in animal or human studies.

9.3 Dependence
Milnacipran produces physical dependence, as evidenced by the emergence of withdrawal symptoms following drug discontinuation, similar to other SNRIs and SSRIs. These withdrawal symptoms can be severe. Thus, Savella should be tapered and not abruptly discontinued after extended use. [see *Section 5.7 Discontinuation of Treatment with Savella*].

10 OVERDOSAGE

There is limited clinical experience with Savella overdose in humans. In clinical trials, cases of acute ingestions up to 1000 mg, alone or in combination with other drugs, were reported with none being fatal.

In postmarketing experience, fatal outcomes have been reported for acute overdoses primarily involving multiple drugs but also with Savella only. The most common signs and symptoms included increased blood pressure, cardio-respiratory arrest, changes in the level of consciousness (ranging from somnolence to coma), confusional state, dizziness, and increased hepatic enzymes.

Management of Overdose
There is no specific antidote to Savella, but if serotonin syndrome ensues, specific treatment (such as with cyproheptadine and/or temperature control) may be considered. In case of acute overdose, treatment should consist of those general measures employed in the management of overdose with any drug.

An adequate airway, oxygenation, and ventilation should be assured and cardiac rhythm and vital signs should be monitored. Induction of emesis is not recommended. Gastric lavage with a large-bore orogastric tube with appropriate airway protection, if needed, may be indicated if performed soon after ingestion or in symptomatic patients. Because there is no specific antidote for Savella, symptomatic care and treatment with gastric lavage and activated charcoal should be considered as soon as possible for patients who experience a Savella overdose.

Due to the large volume of distribution of this drug, forced diuresis, dialysis, hemoperfusion, and exchange transfusion are unlikely to be beneficial.

In managing overdose, the possibility of multiple drug involvement should be considered. The physician should consider contacting a poison control center for additional information on the treatment of any overdose. Telephone numbers for certified poison control centers are listed in the *Physicians' Desk Reference* (PDR).

11 DESCRIPTION

Milnacipran hydrochloride is a selective norepinephrine and serotonin reuptake inhibitor; it inhibits norepinephrine uptake with greater potency than serotonin. It is a racemic mixture with the chemical name: (±)-[1R(S),2S(R)]-2-(aminomethyl)-N,N-diethyl-1-phenylcyclopropanecarboxamide hydrochloride. The structural formula is:

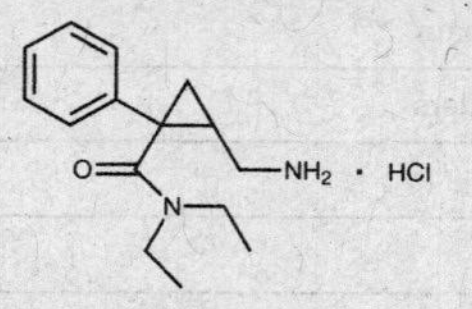

Milnacipran hydrochloride is a white to off-white crystalline powder with a melting point of 179°C. It is freely soluble in water, methanol, ethanol, chloroform, and methylene chloride and sparingly soluble in diethyl ether. It has an empirical formula of C15H23ClN2O and a molecular weight of 282.8 g/mol.

Savella is available for oral administration as film-coated tablets containing 12.5 mg, 25 mg, 50 mg, and 100 mg milnacipran hydrochloride. Each tablet also contains dibasic calcium phosphate, povidone, carboxymethylcellulose calcium, colloidal silicon dioxide, magnesium stearate, and talc as inactive ingredients. Additionally, the following inactive ingredients are also present as components of the film coat:
12.5 mg:
FD&C Blue #2 Aluminum Lake, hypromellose, polyethylene glycol, titanium dioxide
25 mg:
Hypromellose, polyethylene glycol, titanium dioxide
50 mg:
Hypromellose, polyethylene glycol, titanium dioxide
100 mg:
FD&C Red #40 Aluminum Lake, hypromellose, polyethylene glycol, titanium dioxide

12 CLINICAL PHARMACOLOGY

12.1 Mechanism of Action
The exact mechanism of the central pain inhibitory action of milnacipran and its ability to improve the symptoms of fibromyalgia in humans are unknown. Preclinical studies have shown that milnacipran is a potent inhibitor of neuronal norepinephrine and serotonin reuptake; milnacipran inhibits norepinephrine uptake with approximately 3-fold higher potency in vitro than serotonin without directly affecting the uptake of dopamine or other neurotransmitters. Milnacipran has no significant affinity for serotonergic (5-HT1-7), α- and β-adrenergic, muscarinic (M1-5), histamine (H1-4), dopamine (D1-5), opiate, benzodiazepine, and γ-aminobutyric acid (GABA) receptors in vitro. Pharmacologic activity at these receptors is hypothesized to be associated with the various anticholinergic, sedative, and cardiovascular effects seen with other psychotropic drugs. Milnacipran has no significant affinity for Ca++, K+, Na+ and Cl− channels and does not inhibit the activity of human monoamine oxidases (MAO-A and MAO-B) or acetylcholinesterase.

12.2 Pharmacodynamics
Cardiovascular Electrophysiology—The effect of Savella on the QTcF interval was measured in a double-blind placebo- and positive-controlled parallel study in 88 healthy subjects using 600 mg/day Savella (3 to 6 times the recommended therapeutic dose for fibromyalgia). After baseline and placebo adjustment, the maximum mean QTcF change was 8 ms (2-sided 90% CI, 3-12 ms). This increase is not considered to be clinically significant.

12.3 Pharmacokinetics
Milnacipran is well absorbed after oral administration with an absolute bioavailability of approximately 85% to 90%. The exposure to milnacipran increased proportionally within the therapeutic dose range. It is excreted predomi-

nantly unchanged in urine (55%) and has a terminal elimination half-life of about 6 to 8 hours. Steady-state levels are reached within 36 to 48 hours and can be predicted from single-dose data. The active enantiomer, *d*-milnacipran, has a longer elimination half-life (8-10 hours) than the *l*-enantiomer (4-6 hours). There is no interconversion between the enantiomers.

Absorption and Distribution

Savella is absorbed following oral administration with maximum concentrations (Cmax) reached within 2 to 4 hours post dose. Absorption of Savella is not affected by food. The absolute bioavailability is approximately 85% to 90%. The mean volume of distribution of milnacipran following a single intravenous dose to healthy subjects is approximately 400 L. Plasma protein binding is 13%.

Metabolism and Elimination

Milnacipran and its metabolites are eliminated primarily by renal excretion. Following oral administration of ^{14}C-milnacipran hydrochloride, approximately 55% of the dose was excreted in urine as unchanged milnacipran (24% as *l*-milnacipran and 31% as *d*-milnacipran). The *l*-milnacipran carbamoyl-O-glucuronide was the major metabolite excreted in urine and accounted for approximately 17% of the dose; approximately 2% of the dose was excreted in urine as *d*-milnacipran carbamoyl-O-glucuronide. Approximately 8% of the dose was excreted in urine as the N-desethyl milnacipran metabolite.

12.4 Pharmacokinetics in Special Populations

Renal Impairment—Milnacipran pharmacokinetics were evaluated following single oral administration of 50 mg Savella to subjects with mild (creatinine clearance [CLcr] 50-80 mL/min), moderate (CLcr 30-49 mL/min), and severe (CLcr 5-29 mL/min) renal impairment and to healthy subjects (CLcr > 80 mL/min). The mean AUC0-∞ increased by 16%, 52%, and 199%, and terminal elimination half-life increased by 38%, 41%, and 122% in subjects with mild, moderate, and severe renal impairment, respectively, compared with healthy subjects.

No dosage adjustment is necessary for patients with mild renal impairment. Caution should be exercised in patients with moderate renal impairment. Dose adjustment is necessary in severe renal impairment patients. [*see Dosage and Administration (2.2)*].

Hepatic Impairment—Milnacipran pharmacokinetics were evaluated following single oral administration of 50 mg Savella to subjects with mild (Child-Pugh A), moderate (Child-Pugh B), and severe (Child-Pugh C) hepatic impairment and to healthy subjects. AUC0-∞ and T½ were similar in healthy subjects and subjects with mild and moderate hepatic impairment. However, subjects with severe hepatic impairment had a 31% higher AUC0-∞ and a 55% higher T½ than healthy subjects. Caution should be exercised in patients with severe hepatic impairment.

Elderly—Cmax and AUC parameters of milnacipran were about 30% higher in elderly (> 65 years) subjects compared with young subjects due to age-related decreases in renal function. No dosage adjustment is necessary based on age unless renal function is severely impaired [*see Dosage and Administration (2.2)*].

Gender—Cmax and AUC parameters of milnacipran were about 20% higher in female subjects compared with male subjects. Dosage adjustment based on gender is not necessary.

Drug-Drug Interactions

In Vitro Studies

In general, milnacipran, at concentrations that were at least 25 times those attained in clinical trials, did not inhibit human CYP1A2, CYP2A6, CYP2C9, CYP2C19, CYP2D6, CYP2E1, and CYP3A4 or induce human CYP1A2, CYP2B6, CYP2C8, CYP2C9, CYP2C19, and CYP3A4/5 enzyme systems, indicating a low potential of interactions with drugs metabolized by these enzymes.

In vitro studies have shown that the biotransformation rate of milnacipran by human hepatic microsomes and hepatocytes was low. A low biotransformation was also observed following incubation of milnacipran with cDNA-expressed human CYP1A2, CYP2A6, CYP2B6, CYP2C9, CYP2C19, CYP2D6, CYP2E1, and CYP3A4 isozymes.

In Vivo Studies

The drug interaction studies described in this section were conducted in healthy adult subjects.

Carbamazepine—There were no clinically significant changes in the pharmacokinetics of milnacipran following coadministration of Savella (100 mg/day) and carbamazepine (200 mg twice a day). No changes were observed in the pharmacokinetics of carbamazepine or its epoxide metabolite due to coadministration with Savella.

Clomipramine—Switch from clomipramine (75 mg once a day) to milnacipran (100 mg/day) without a washout period did not lead to clinically significant changes in the pharmacokinetics of milnacipran. Because an increase in adverse events (eg, euphoria and postural hypotension) was observed after switching from clomipramine to milnacipran, monitoring of patients during treatment switch is recommended.

Digoxin—There was no pharmacokinetic interaction between Savella (200 mg/day) and digoxin (0.2 mg/day Lanoxicaps) following multiple-dose administration to healthy subjects.

Fluoxetine—Switch from fluoxetine (20 mg once a day), a strong inhibitor of CYP2D6 and a moderate inhibitor of CYP2C19, to milnacipran (100 mg/day) without a washout period did not affect the pharmacokinetics of milnacipran.

Lithium—Multiple doses of Savella (100 mg/day) did not affect the pharmacokinetics of lithium.

Lorazepam—There was no pharmacokinetic interaction between a single dose of Savella (50 mg) and lorazepam (1.5 mg).

Warfarin—Steady-state milnacipran (200 mg/day) did not affect the pharmacokinetics of R-warfarin and S-warfarin or the pharmacodynamics (as assessed by measurement of prothrombin INR) of a single dose of 25 mg warfarin. The pharmacokinetics of Savella were not altered by warfarin.

13 NONCLINICAL TOXICOLOGY

13.1 Carcinogenesis, Mutagenesis, and Impairment of Fertility

Carcinogenesis

Dietary administration of milnacipran to rats at doses of 50 mg/kg/day (2 times the MRHD on a mg/m^2 basis) for 2 years caused a statistically significant increase in the incidence of thyroid C-cell adenomas and combined adenomas and carcinomas in males. A carcinogenicity study was conducted in Tg.rasH2 mice for 6 months at oral gavage doses of up to 125 mg/kg/day. Milnacipran did not induce tumors in Tg.rasH2 mice at any dose tested.

Mutagenesis

Milnacipran was not mutagenic in the in vitro bacterial reverse mutation assay (Ames test) or in the L5178Y TK +/- mouse lymphoma forward mutation assay. Milnacipran was also not clastogenic in an in vitro chromosomal aberration test in human lymphocytes or in the in vivo mouse micronucleus assay.

Impairment of Fertility

Although administration of milnacipran to male and female rats had no statistically significant effect on mating or fertility at doses up to 80 mg/kg/day (4 times the MRHD on an mg/m^2 basis) there was an apparent dose-related decrease in the fertility index at clinically relevant doses based on body surface area.

13.2 Animal Toxicology and Pharmacology

Hepatic Effects

Chronic administration (2-years) of milnacipran to rats at 15 mg/kg (0.6 times the MRHD on an mg/m^2 basis) and higher doses showed increased incidences of centrilobular vacuolation of the liver in male rats and eosinophilic foci in male and female rats in the absence of any change in hepatic enzymes. The clinical significance of the finding is not known. Chronic (1-year) administration in the primate at doses up to 25 mg/kg (2 times the MRHD on a mg/m^2 basis) did not demonstrate similar evidence of hepatic changes.

Ocular Effects

Chronic (2-years) administration of milnacipran to rats at 15 mg/kg (0.6 times the MRHD on a mg/m^2 basis) and higher doses showed increased incidence of keratitis of the eye. One year studies in the rat and primate did not show this response.

14 CLINICAL STUDIES

Management of Fibromyalgia

The efficacy of Savella for the management of fibromyalgia was established in two double-blind, placebo-controlled, multicenter studies in adult patients (18-74 years of age). Enrolled patients met the American College of Rheumatology (ACR) criteria for fibromyalgia (a history of widespread pain for 3 months and pain present at 11 or more of the 18 specific tender point sites). Approximately 35% of patients had a history of depression. Study 1 was six months in duration and Study 2 was three months in duration.

A larger proportion of patients treated with Savella than with placebo experienced a simultaneous reduction in pain from baseline of at least 30% (VAS) and also rated themselves as much improved or very much improved based on the patient global assessment (PGIC). In addition, a larger proportion of patients treated with Savella met the criteria for treatment response, as measured by the composite endpoint that concurrently evaluated improvement in pain (VAS), physical function (SF-36 PCS), and patient global assessment (PGIC), in fibromyalgia as compared to placebo.

Study 1: This 6-month study compared total daily doses of Savella 100 mg and 200 mg to placebo. Patients were enrolled with a minimum mean baseline pain score of ≥ 50 mm on a 100 mm visual analog scale (VAS) ranging from 0 ("no pain") to 100 ("worst possible pain"). The mean baseline pain score in this trial was 69. The efficacy results for Study 1 are summarized in Figure 1.

Figure 1 shows the proportion of patients achieving various degrees of improvement in pain from baseline to the 3-month time point and who concurrently rated themselves globally improved (PGIC score of 1 or 2). Patients who did not complete the 3-month assessment were assigned 0% improvement. More patients in the Savella treatment arms experienced at least a 30% reduction in pain from baseline (VAS) and considered themselves globally improved (PGIC) than did patients in the placebo arm. Treatment with Savella 200 mg/day did not confer greater benefit than treatment with Savella 100 mg/day.

Figure 1: Patients Achieving Various Levels of Pain Relief with Concurrent Ratings of Being Much or Very Much Improved on the PGIC — Study 1

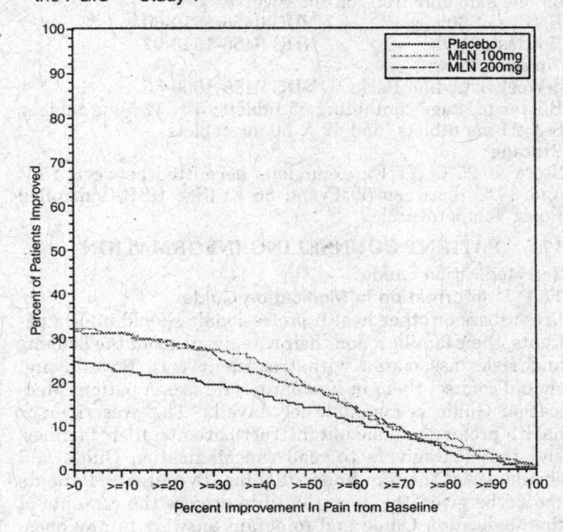

Study 2: This 3-month study compared total daily doses of Savella 100 mg and 200 mg to placebo. Patients were enrolled with a minimum mean baseline pain score of ≥ 40 mm on a 100-mm VAS ranging from 0 ("no pain") to 100 ("worst possible pain"). The mean baseline pain score in this trial was 65. The efficacy results for Study 2 are summarized in Figure 2.

Figure 2 shows the proportion of patients achieving various degrees of improvement in pain from baseline to the 3-month time point and who concurrently rated themselves globally improved (PGIC score of 1 or 2). Patients who did not complete the 3-month assessment were assigned 0% improvement. More patients in the Savella treatment arms experienced at least a 30% reduction in pain from baseline (VAS) and considered themselves globally improved (PGIC) than did patients in the placebo arm. Treatment with Savella 200 mg/day did not confer greater benefit than treatment with Savella 100 mg/day.

Figure 2: Patients Achieving Various Levels of Pain Relief with Concurrent Ratings of Being Much or Very Much Improved on the PGIC — Study 2

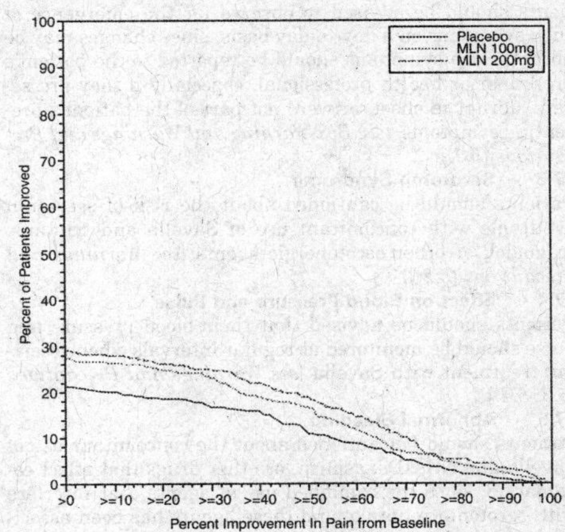

In both studies, some patients who rated themselves as globally "much" or "very much" improved experienced a decrease in pain as early as week 1 of treatment with a stable dose of Savella that persisted throughout these studies.

16 HOW SUPPLIED/STORAGE AND HANDLING

12.5-mg tablets:

Blue, round, film-coated tablets, debossed with "F" on one side and "L" on the reverse

Bottles of 60: NDC 0456-1512-60

25-mg tablets:

White, round, film-coated tablets, debossed with "FL" on one side and "25" on the reverse

Bottles of 60: NDC 0456-1525-60

Bottles of 180: NDC 0456-1525-01

Continued on next page

Savella—Cont.

50-mg tablets:
White, oval-shaped, film-coated tablets, debossed with "FL" on one side and "50" on the reverse
Bottles of 60: NDC 0456-1550-60
Bottles of 180: NDC 0456-1550-01
100-mg tablets:
Pink, oval-shaped film-coated tablets, debossed with "FL" on one side and "100" on the reverse
Bottles of 60: NDC 0456-1510-60
Bottles of 180: NDC 0456-1510-01
Titration Pack:
4-Week Titration Pack: NDC 0456-1500-55
Blister package containing 55 tablets: 5 × 12.5-mg tablets, 8 × 25-mg tablets, and 42 × 50 mg tablets.
Storage
Store at 25°C (77°F); excursions permitted between 15°C and 30°C (between 59°F and 86°F) [See USP Controlled Room Temperature].

17 PATIENT COUNSELING INFORMATION

See Medication Guide

17.1 Information in Medication Guide
Prescribers or other health professionals should inform patients, their families, and their caregivers about the benefits and risks associated with treatment with Savella and should counsel them in its appropriate use. A patient Medication Guide is available for Savella. The prescriber or health professional should instruct patients, their families, and their caregivers to read the Medication Guide and should assist them in understanding its contents. Patients should be given the opportunity to discuss the contents of the Medication Guide and to obtain answers to any questions they may have. The complete text of the Medication Guide is reprinted at the end of this document.
Patients should be advised of the following issues and asked to alert their prescriber if these occur while taking Savella:

17.2 Suicide Risk
Patients and their families and caregivers should be advised that Savella is a selective norepinephrine and serotonin reuptake inhibitor and therefore belongs to the same class of drugs as antidepressants. Patients, their families and their caregivers should be advised that patients with depression may be at increased risk for clinical worsening and/or suicidal ideation if they stop taking anti-depressant medication, change the dose, or start a new medication.
Patients, their families and their caregivers should be encouraged to be alert to the emergence of anxiety, agitation, panic attacks, insomnia, irritability, hostility, aggressiveness, impulsivity, akathisia (psychomotor restlessness), hypomania or other unusual changes in behavior, worsening of depression, and suicidal ideation, especially early during treatment with Savella or other drugs that inhibit the reuptake of norepinephrine and/or serotonin, and when the dose is adjusted up or down. Families and caregivers of patients should be advised to observe for the emergence of such symptoms on a day-to-day basis, since changes may be abrupt. Such symptoms should be reported to the patient's prescriber or health professional, especially if they are severe, abrupt in onset, or were not part of the patient's presenting symptoms. [*see Box Warning and Warnings and Precautions (5.1)*].

17.3 Serotonin Syndrome
Patients should be cautioned about the risk of serotonin syndrome with concomitant use of Savella and triptans, tramadol, or other serotonergic agents [*see Warnings and Precautions (5.2)*].

17.4 Effect on Blood Pressure and Pulse
Patients should be advised that their blood pressure and pulse should be monitored at regular intervals when receiving treatment with Savella [*see Warnings and Precautions (5.3, 5.4)*].

17.5 Abnormal Bleeding
Patients should be cautioned about the concomitant use of Savella and NSAIDs, aspirin, or other drugs that affect coagulation, since the combined use of agents that interfere with serotonin reuptake and these agents has been associated with an increased risk of abnormal bleeding [*see Warnings and Precautions (5.9)*].

17.6 Ability to Drive and Use Machinery
Savella might diminish mental and physical capacities necessary to perform certain tasks such as operating machinery, including motor vehicles. Patients should be cautioned about operating machinery or driving motor vehicles until they are reasonably certain that Savella treatment does not affect their ability to engage in such activities.

17.7 Alcohol
Patients should be advised to avoid consumption of alcohol while taking Savella [*see Warnings and Precautions (5.6, 5.13)*].

17.8 Discontinuation
Patients should be advised that withdrawal symptoms can occur when discontinuing treatment with Savella, particularly when discontinuation is abrupt. [*see Warnings and Precautions (5.7)*]

17.9 Pregnancy
Patients should be advised to notify their physician if they become pregnant or intend to become pregnant during Savella therapy [*see Use in Specific Populations (8.1)*].

17.10 Nursing
Patients should be advised to notify their physician if they are breast-feeding [*see Use in Specific Populations (8.3)*].

17.11 FDA-Approved Medication Guide

MEDICATION GUIDE
Savella (Sa-vel-la) Tablets
(milnacipran HCl)

Antidepressant Medicines, Depression and other serious Mental Illnesses, and Suicidal Thoughts or Actions
Savella is not used to treat depression, but it acts like medicines that are used to treat depression (antidepressants) and other psychiatric disorders.
Read the Medication Guide that comes with you or your family member's antidepressant medicine. This Medication Guide is only about the risk of suicidal thoughts or actions with antidepressant medicines. **Talk to your or your family member's healthcare provider about:**
- **all risks and benefits of treatment with antidepressant medicines**
- **all treatment choices for depression or other serious mental illness**

What is the most important information I should know about antidepressant medicines, depression and other serious mental illnesses, and suicidal thoughts or actions?
1. **Antidepressant medicines may increase suicidal thoughts or actions in some children, teenagers, and young adults within the first few months of treatment.**
2. **Depression and other serious mental illnesses are the most important causes of suicidal thoughts and actions. Some people may have a particularly high risk of having suicidal thoughts or actions.** These include people who have (or have a family history of) bipolar illness (also called manic-depressive illness) or suicidal thoughts or actions.
3. **How can I watch for and try to prevent suicidal thoughts and actions in myself or a family member?**
   - Pay close attention to any changes, especially sudden changes, in mood, behaviors, thoughts, or feelings. This is very important when an antidepressant medicine is started or when the dose is changed.
   - Call the healthcare provider right away to report new or sudden changes in mood, behavior, thoughts, or feelings.
   - Keep all follow-up visits with the healthcare provider as scheduled. Call the healthcare provider between visits as needed, especially if you have concerns about symptoms.

Call a healthcare provider right away if you or your family member has any of the following symptoms, especially if they are new, worse, or worry you:
- thoughts about suicide or dying
- attempts to commit suicide
- new or worse depression
- new or worse anxiety
- feeling very agitated or restless
- panic attacks
- new or worse irritability
- acting aggressive, being angry, or violent
- acting on dangerous impulses
- an extreme increase in activity and talking (mania)
- other unusual changes in behavior or mood
- trouble sleeping (insomnia)

What else do I need to know about antidepressant medicines?
- **Never stop an antidepressant medicine without first talking to a healthcare provider.** Stopping an antidepressant medicine suddenly can cause other symptoms.
- **Antidepressants are medicines used to treat depression and other illnesses.** It is important to discuss all the risks of treating depression and also the risks of not treating it. Patients and their families or other caregivers should discuss all treatment choices with the healthcare provider, not just the use of antidepressants.
- **Antidepressant medicines have other side effects.** Talk to the healthcare provider about the side effects of the medicine prescribed for you or your family member.
- **Antidepressant medicines can interact with other medicines.** Know all of the medicines that you or your family member takes. Keep a list of all medicines to show the healthcare provider. Do not start new medicines without first checking with your healthcare provider.
- **Not all antidepressant medicines prescribed for children are FDA approved for use in children.** Talk to your child's healthcare provider for more information. Call your doctor for medical advice about side effects. You may report side effects to FDA at 1-800-FDA-1088.

This Medication Guide has been approved by the U.S. Food and Drug Administration.
Manufactured for:
Forest Pharmaceuticals, Inc.
Manufactured by:
Forest Laboratories, Inc.
Licensed from Pierre Fabre Medicament and Cypress Bioscience, Inc.
Revised: March 2009

GlaxoSmithKline

FIVE MOORE DRIVE
RESEARCH TRIANGLE PARK, NC 27709

For all inquiries, including adverse event and quality assurance reporting, contact the GSK Response Center at 1-888-825-5249.
For updates to the product information listed below, also consult www.gsk.com.

LAMICTAL® XR™ ℞

[la-mĭk' tal]
(lamotrigine)
Extended-Release Tablets

HIGHLIGHTS OF PRESCRIBING INFORMATION
These highlights do not include all the information needed to use LAMICTAL XR safely and effectively. See full prescribing information for LAMICTAL XR.
LAMICTAL XR (lamotrigine) Extended-Release Tablets
Initial U.S. Approval: 1994

WARNING: SERIOUS SKIN RASHES
See full prescribing information for complete boxed warning.
Cases of life-threatening serious rashes, including Stevens-Johnson syndrome, toxic-epidermal necrolysis, and/or rash-related death, have been caused by lamotrigine. The rate of serious rash is greater in pediatric patients than in adults. Additional factors that may increase the risk of rash include (5.1):
- **coadministration with valproate**
- **exceeding recommended initial dose of LAMICTAL XR**
- **exceeding recommended dose escalation of LAMICTAL XR**

Benign rashes are also caused by lamotrigine; however, it is not possible to predict which rashes will prove to be serious or life-threatening. LAMICTAL XR should be discontinued at the first sign of rash unless the rash is clearly not drug-related. (5.1)

--------------INDICATIONS AND USAGE--------------
LAMICTAL XR is an antiepileptic drug (AED) indicated as adjunctive therapy for partial onset seizures with or without secondary generalization in patients ≥13 years of age. (1.1)

---------DOSAGE AND ADMINISTRATION---------
- Doses are administered once daily. Dose escalation and maintenance doses are based on concomitant medications. (2.1, 2.2)
- To avoid an increased risk of rash, the recommended initial dose and subsequent dose escalations should not be exceeded. LAMICTAL XR Patient Titration Kits are available for the first 5 weeks of treatment. (2.1, 16)
- For patients being converted from immediate-release lamotrigine to LAMICTAL XR, the initial dose of LAMICTAL XR should match the total daily dose of the immediate-release lamotrigine. Patients should be closely monitored for seizure control after conversion to LAMICTAL XR. (2.3)
- Do not restart LAMICTAL XR in patients who discontinued due to rash unless the potential benefits clearly outweigh the risks. (2.1, 5.1)
- Adjustments to maintenance doses will in most cases be required in patients starting or stopping estrogen-containing oral contraceptives. (2.1, 5.7)
- LAMICTAL XR should be discontinued over a period of at least 2 weeks (approximately 50% reduction per week). (2.1, 5.8)

--------DOSAGE FORMS AND STRENGTHS--------
Extended-Release Tablets: 25 mg, 50 mg, 100 mg, and 200 mg. (3.1, 16)

-----------------CONTRAINDICATIONS-----------------
Hypersensitivity to the drug or its ingredients. (Boxed Warning, 4)

-----------WARNINGS AND PRECAUTIONS-----------
- Life-threatening serious rash, and/or rash-related death, may result. (Boxed Warning, 5.1)
- Hypersensitivity reaction may be fatal or life-threatening. Early signs of hypersensitivity (e.g., fever, lymphadenopathy) may present without rash; if signs present, patient should be evaluated immediately.

- LAMICTAL XR should be discontinued if alternate etiology for hypersensitivity signs is not found. (5.2)
- Acute multiorgan failure has resulted (some cases fatal). (5.3)
- Blood dyscrasias (e.g., neutropenia, thrombocytopenia, pancytopenia) may result, either with or without an associated hypersensitivity syndrome. (5.4)
- Suicidal behavior and ideation. (5.5)
- Medication errors involving LAMICTAL have occurred. In particular, the names LAMICTAL or lamotrigine can be confused with the names of other commonly used medications. Medication errors may also occur between the different formulations of LAMICTAL. (3.2, 5.6, 16, 17.9)

ADVERSE REACTIONS

- Most common adverse reactions (treatment difference ≥4%, LAMICTAL XR - Placebo) are dizziness, tremor/intention tremor, cerebellar coordination/balance disorder, nausea, asthenic conditions (asthenia, fatigue, malaise), vertigo/positional vertigo, and diplopia. (6.1)

To report SUSPECTED ADVERSE REACTIONS, contact GlaxoSmithKline at 1-888-825-5249 or FDA at 1-800-FDA-1088 or www.fda.gov/medwatch.

DRUG INTERACTIONS

- Valproate increases lamotrigine concentrations more than 2-fold. (7, 12.3)
- Carbamazepine, phenytoin, phenobarbital, and primidone decrease lamotrigine concentrations by approximately 40%. (7, 12.3)
- Oral estrogen-containing contraceptives and rifampin also decrease lamotrigine concentrations by approximately 50%. (7, 12.3)

USE IN SPECIFIC POPULATIONS

- Pediatric use: Safety and effectiveness in patients below the age of 13 have not been established. (8.4)
- Effectiveness of lamotrigine, used as adjunctive treatment for partial seizures, was not demonstrated in a small randomized, double-blind, placebo-controlled, withdrawal study in very young pediatric patients (1 to 24 months). (8.4)
- Hepatic impairment: Dosage adjustments required. (2.1)
- Healthcare professionals can enroll patients in the Lamotrigine Pregnancy Registry (1-800-336-2176). Patients can enroll themselves in the North American Antiepileptic Drug Pregnancy Registry (1-888-233-2334). (8.1)

See 17 for PATIENT COUNSELING INFORMATION and Medication Guide.

Revised: May 2009
LXR:1PI

FULL PRESCRIBING INFORMATION: CONTENTS*

*Sections or subsections omitted from the full prescribing information are not listed.

FULL PRESCRIBING INFORMATION

WARNING: SERIOUS SKIN RASHES

LAMICTAL® XR™ can cause serious rashes requiring hospitalization and discontinuation of treatment. The incidence of these rashes, which have included Stevens-Johnson syndrome, is approximately 0.8% (8 per 1,000) in pediatric patients (2 to 16 years of age) receiving the immediate-release formulation of LAMICTAL as adjunctive therapy for epilepsy and 0.3% (3 per 1,000) in adults on adjunctive therapy for epilepsy. In a prospectively followed cohort of 1,983 pediatric patients (2 to 16 years of age) with epilepsy taking the adjunctive immediate-release formulation of LAMICTAL, there was 1 rash-related death. LAMICTAL XR is not approved for patients under the age of 13 years. In worldwide postmarketing experience, rare cases of toxic epidermal necrolysis and/or rash-related death have been reported in adult and pediatric patients, but their numbers are too few to permit a precise estimate of the rate.

The risk of serious rash caused by treatment with LAMICTAL XR is not expected to differ from that with the immediate-release formulation of LAMICTAL. However, the relatively limited treatment experience with LAMICTAL XR makes it difficult to characterize the frequency and risk of serious rashes caused by treatment with LAMICTAL XR.

Other than age, there are as yet no factors identified that are known to predict the risk of occurrence or the severity of rash caused by LAMICTAL XR. There are suggestions, yet to be proven, that the risk of rash may also be increased by (1) coadministration of LAMICTAL XR with valproate (includes valproic acid and divalproex sodium), (2) exceeding the recommended initial dose of LAMICTAL XR, or (3) exceeding the recommended dose escalation for LAMICTAL XR. However, cases have occurred in the absence of these factors.

Nearly all cases of life-threatening rashes caused by the immediate-release formulation of LAMICTAL have occurred within 2 to 8 weeks of treatment initiation. However, isolated cases have occurred after prolonged treatment (e.g., 6 months). Accordingly, duration of therapy cannot be relied upon as means to predict the potential risk heralded by the first appearance of a rash. Although benign rashes are also caused by LAMICTAL XR, it is not possible to predict reliably which rashes will prove to be serious or life-threatening. Accordingly, LAMICTAL XR should ordinarily be discontinued at the first sign of rash, unless the rash is clearly not drug-related. Discontinuation of treatment may not prevent a rash from becoming life-threatening or permanently disabling or disfiguring *[see Warnings and Precautions (5.1)]*.

1 INDICATIONS AND USAGE

LAMICTAL XR is indicated as adjunctive therapy for partial onset seizures with or without secondary generalization in patients ≥13 years of age.

Safety and effectiveness of LAMICTAL XR for use in patients below the age of 13 have not been established.

2 DOSAGE AND ADMINISTRATION

LAMICTAL XR Extended-Release Tablets are taken once daily, with or without food. Tablets must be swallowed whole and must not be chewed, crushed, or divided.

2.1 General Dosing Considerations

Rash: There are suggestions, yet to be proven, that the risk of severe, potentially life-threatening rash may be increased by (1) coadministration of LAMICTAL XR with valproate, (2) exceeding the recommended initial dose of LAMICTAL XR, or (3) exceeding the recommended dose escalation for LAMICTAL XR. However, cases have occurred in the absence of these factors *[see Boxed Warning]*. Therefore, it is important that the dosing recommendations be followed closely.

The risk of nonserious rash may be increased when the recommended initial dose and/or the rate of dose escalation of LAMICTAL XR is exceeded and in patients with a history of allergy or rash to other AEDs.

LAMICTAL XR Patient Titration Kits provide LAMICTAL XR at doses consistent with the recommended titration schedule for the first 5 weeks of treatment, based upon concomitant medications for patients with partial onset seizures and are intended to help reduce the potential for rash. The use of LAMICTAL XR Patient Titration Kits is recommended for appropriate patients who are starting or restarting LAMICTAL XR *[see How Supplied/Storage and Handling (16)]*.

It is recommended that LAMICTAL XR not be restarted in patients who discontinued due to rash associated with prior treatment with lamotrigine, unless the potential benefits clearly outweigh the risks. If the decision is made to restart a patient who has discontinued LAMICTAL XR, the need to restart with the initial dosing recommendations should be assessed. The greater the interval of time since the previous dose, the greater consideration should be given to restarting with the initial dosing recommendations. If a patient has discontinued lamotrigine for a period of more than 5 half-lives, it is recommended that initial dosing recommendations and guidelines be followed. The half-life of lamotrigine is affected by other concomitant medications *[see Clinical Pharmacology (12.3)]*.

LAMICTAL XR Added to Drugs Known to Induce or Inhibit Glucuronidation: Drugs other than those listed in the Clinical Pharmacology section *[see Clinical Pharmacology (12.3)]* have not been systematically evaluated in combination with lamotrigine. Because lamotrigine is metabolized predominantly by glucuronic acid conjugation, drugs that are known to induce or inhibit glucuronidation may affect the apparent clearance of lamotrigine and doses of LAMICTAL XR may require adjustment based on clinical response.

Target Plasma Levels: A therapeutic plasma concentration range has not been established for lamotrigine. Dosing of LAMICTAL XR should be based on therapeutic response *[see Clinical Pharmacology (12.3)]*.

Women Taking Estrogen-Containing Oral Contraceptives: *Starting LAMICTAL XR in Women Taking Estrogen-Containing Oral Contraceptives:* Although estrogen-containing oral contraceptives have been shown to increase the clearance of lamotrigine *[see Clinical Pharmacology (12.3)]*, no adjustments to the recommended dose-escalation guidelines for LAMICTAL XR should be necessary solely based on the use of estrogen-containing oral contraceptives. Therefore, dose escalation should follow the recommended guidelines for initiating adjunctive therapy with LAMICTAL XR based on the concomitant AED or other concomitant medications (see Table 1). See below for adjustments to maintenance doses of LAMICTAL XR in women taking estrogen-containing oral contraceptives.

Adjustments to the Maintenance Dose of LAMICTAL XR In Women Taking Estrogen-Containing Oral Contraceptives:

(1) Taking Estrogen-Containing Oral Contraceptives: For women not taking carbamazepine, phenytoin, phenobarbital, primidone, or other drugs such as rifampin that induce lamotrigine glucuronidation *[see Drug Interactions, (7)]*, the maintenance dose of LAMICTAL XR will in most cases need to be increased, by as much as 2-fold over the recommended target maintenance dose, in order to maintain a consistent lamotrigine plasma level *[see Clinical Pharmacology (12.3)]*.

(2) Starting Estrogen-Containing Oral Contraceptives: In women taking a stable dose of LAMICTAL XR and not taking carbamazepine, phenytoin, phenobarbital, primidone, or other drugs such as rifampin that induce lamotrigine glucuronidation *[see Drug Interactions, (7), Clinical Pharmacology (12.3)]*, the maintenance dose will in most cases need to be increased by as much as 2-fold in or-

Continued on next page

Product information on these pages is effective as of June 2009. Further information is available at 1-888-825-5249 or www.gsk.com.

Lamictal XR—Cont.

der to maintain a consistent lamotrigine plasma level. The dose increases should begin at the same time that the oral contraceptive is introduced and continue, based on clinical response, no more rapidly than 50 to 100 mg/day every week. Dose increases should not exceed the recommended rate (see Table 1) unless lamotrigine plasma levels or clinical response support larger increases. Gradual transient increases in lamotrigine plasma levels may occur during the week of inactive hormonal preparation ("pill-free" week), and these increases will be greater if dose increases are made in the days before or during the week of inactive hormonal preparation. Increased lamotrigine plasma levels could result in additional adverse reactions, such as dizziness, ataxia, and diplopia. If adverse reactions attributable to LAMICTAL XR consistently occur during the "pill-free" week, dose adjustments to the overall maintenance dose may be necessary. Dose adjustments limited to the "pill-free" week are not recommended. For women taking LAMICTAL XR in addition to carbamazepine, phenytoin, phenobarbital, primidone, or other drugs such as rifampin that induce lamotrigine glucuronidation *[see Drug Interactions, (7), Clinical Pharmacology (12.3)]*, no adjustment should be necessary to the dose of LAMICTAL XR.

(3) Stopping Estrogen-Containing Oral Contraceptives: For women not taking carbamazepine, phenytoin, phenobarbital, primidone, or other drugs such as rifampin that induce lamotrigine glucuronidation *[see Drug Interactions, (7), Clinical Pharmacology (12.3)]*, the maintenance dose of LAMICTAL XR will in most cases need to be decreased by as much as 50% in order to maintain a consistent lamotrigine plasma level. The decrease in dose of LAMICTAL XR should not exceed 25% of the total daily dose per week over a 2-week period, unless clinical response or lamotrigine plasma levels indicate otherwise *[see Clinical Pharmacology (12.3)]*. For women taking LAMICTAL XR in addition to carbamazepine, phenytoin, phenobarbital, primidone, or other drugs such as rifampin that induce lamotrigine glucuronidation *[see Drug Interactions, (7), Clinical Pharmacology (12.3)]*, no adjustment to the dose of LAMICTAL XR should be necessary.

Women and Other Hormonal Contraceptive Preparations or Hormone Replacement Therapy: The effect of other hormonal contraceptive preparations or hormone replacement therapy on the pharmacokinetics of lamotrigine has not been systematically evaluated. It has been reported that ethinylestradiol, not progestogens, increased the clearance of lamotrigine up to 2-fold, and the progestin-only pills had no effect on lamotrigine plasma levels. Therefore, adjustments to the dosage of LAMICTAL XR in the presence of progestogens alone will likely not be needed.

Patients With Hepatic Impairment: Experience in patients with hepatic impairment is limited. Based on a clinical pharmacology study in 24 patients with mild, moderate, and severe liver impairment *[see Use in Specific Populations (8.6), Clinical Pharmacology (12.3)]*, the following general recommendations can be made. No dosage adjustment is needed in patients with mild liver impairment. Initial, escalation, and maintenance doses should generally be reduced by approximately 25% in patients with moderate and severe liver impairment without ascites and 50% in patients with severe liver impairment with ascites. Escalation and maintenance doses may be adjusted according to clinical response.

Patients With Renal Impairment: Initial doses of LAMICTAL XR should be based on patients' concomitant medications (see Table 1); reduced maintenance doses may be effective for patients with significant renal impairment *[see Use in Specific Populations (8.7), Clinical Pharmacology (12.3)]*. Few patients with severe renal impairment have been evaluated during chronic treatment with immediate-release lamotrigine. Because there is inadequate experience in this population, LAMICTAL XR should be used with caution in these patients.

Discontinuation Strategy: For patients receiving LAMICTAL XR in combination with other AEDs, a re-evaluation of all AEDs in the regimen should be considered if a change in seizure control or an appearance or worsening of adverse reactions is observed.

If a decision is made to discontinue therapy with LAMICTAL XR, a step-wise reduction of dose over at least 2 weeks (approximately 50% per week) is recommended unless safety concerns require a more rapid withdrawal *[see Warnings and Precautions (5.8)]*.

Discontinuing carbamazepine, phenytoin, phenobarbital, primidone, or other drugs such as rifampin that induce lamotrigine glucuronidation should prolong the half-life of lamotrigine; discontinuing valproate should shorten the half-life of lamotrigine.

2.2 Partial Onset Seizures

This section provides specific dosing recommendations for patients ≥13 years of age. Specific dosing recommendations are provided depending upon concomitant AED or other concomitant medications.

[See table 1 below]

2.3 Conversion From Immediate-Release Lamotrigine Tablets to LAMICTAL XR

Patients may be converted directly from immediate-release lamotrigine to LAMICTAL XR Extended-Release Tablets. The initial dose of LAMICTAL XR should match the total daily dose of immediate-release lamotrigine. However, some subjects on concomitant enzyme-inducing agents may have lower plasma levels of lamotrigine on conversion and should be monitored *[see Clinical Pharmacology (12.3)]*.

Following conversion to LAMICTAL XR, all patients (but especially those on an enzyme-inducing AED) should be closely monitored for seizure control. Depending on the therapeutic response after conversion, the total daily dose may need to be adjusted within the recommended dosing instructions (Table 1).

3 DOSAGE FORMS AND STRENGTHS

3.1 Extended-Release Tablets

25 mg, yellow with white center, round, biconvex, film-coated tablets printed with "LAMICTAL" and "XR 25."

50 mg, green with white center, round, biconvex, film-coated tablets printed with "LAMICTAL" and "XR 50."

100 mg, orange with white center, round, biconvex, film-coated tablets printed with "LAMICTAL" and "XR 100."

200 mg, blue with white center, round, biconvex, film-coated tablets printed with "LAMICTAL" and "XR 200."

3.2 Potential Medication Errors

Patients should be strongly advised to visually inspect their tablets to verify that they are receiving LAMICTAL XR, as opposed to other medications, and that they are receiving the correct formulation of LAMICTAL each time they fill their prescription. Depictions of the LAMICTAL XR tablets can be found in the Medication Guide *[see Patient Counseling Information (17.10)]*.

4 CONTRAINDICATIONS

LAMICTAL XR is contraindicated in patients who have demonstrated hypersensitivity to the drug or its ingredients *[see Boxed Warning, Warnings and Precautions (5.1), (5.2)]*.

5 WARNINGS AND PRECAUTIONS

5.1 Serious Skin Rashes *[see Boxed Warning]*

The risk of serious rash caused by treatment with LAMICTAL XR is not expected to differ from that with the immediate-release formulation of LAMICTAL *[see Boxed Warning]*. However, the relatively limited treatment experience with LAMICTAL XR makes it difficult to characterize the frequency and risk of serious rashes caused by treatment with LAMICTAL XR.

Pediatric Population: The incidence of serious rash associated with hospitalization and discontinuation of the immediate-release formulation of LAMICTAL in a prospectively followed cohort of pediatric patients (2 to 16 years of age) with epilepsy receiving adjunctive therapy with immediate-release lamotrigine was approximately 0.8% (16 of 1,983). When 14 of these cases were reviewed by 3 expert dermatologists, there was considerable disagreement as to their proper classification. To illustrate, one dermatologist considered none of the cases to be Stevens-Johnson syndrome; another assigned 7 of the 14 to this diagnosis. There was 1 rash-related death in this 1,983-patient cohort. Additionally, there have been rare cases of toxic epidermal necrolysis with and without permanent sequelae and/or death in US and foreign postmarketing experience.

There is evidence that the inclusion of valproate in a multidrug regimen increases the risk of serious, potentially life-threatening rash in pediatric patients. In pediatric patients who used valproate concomitantly, 1.2% (6 of 482) experienced a serious rash compared with 0.6% (6 of 952) patients not taking valproate.

LAMICTAL XR is not approved in patients under the age of 13 years.

Adult Population: Serious rash associated with hospitalization and discontinuation of the immediate-release formulation of LAMICTAL occurred in 0.3% (11 of 3,348) of adult patients who received the immediate-release formulation of LAMICTAL in premarketing clinical trials of epilepsy. In worldwide postmarketing experience, rare cases of rash-related death have been reported, but their numbers are too few to permit a precise estimate of the rate.

Among the rashes leading to hospitalization were Stevens-Johnson syndrome, toxic epidermal necrolysis, angioedema, and a rash associated with a variable number of the following systemic manifestations: fever, lymphadenopathy, facial swelling, and hematologic and hepatologic abnormalities.

There is evidence that the inclusion of valproate in a multidrug regimen increases the risk of serious, potentially life-threatening rash in adults. Specifically, of 584 patients administered the immediate-release formulation of LAMICTAL with valproate in epilepsy clinical trials, 6 (1%) were hospitalized in association with rash; in contrast, 4 (0.16%) of 2,398 clinical trial patients and volunteers administered the immediate-release formulation of LAMICTAL in the absence of valproate were hospitalized.

Patients With History of Allergy or Rash to Other AEDs: The risk of nonserious rash may be increased when the recommended initial dose and/or the rate of dose escalation of LAMICTAL is exceeded and in patients with a history of allergy or rash to other AEDs.

5.2 Hypersensitivity Reactions

Hypersensitivity reactions, some fatal or life-threatening, have also occurred. Some of these reactions have included clinical features of multiorgan failure/dysfunction, including hepatic abnormalities and evidence of disseminated intravascular coagulation. It is important to note that early manifestations of hypersensitivity (e.g., fever, lymphadenopathy) may be present even though a rash is not evident. If such signs or symptoms are present, the patient should be evaluated immediately. LAMICTAL XR should be discontinued if an alternative etiology for the signs or symptoms cannot be established.

Prior to initiation of treatment with LAMICTAL XR, the patient should be instructed that a rash or other signs or symptoms of hypersensitivity (e.g., fever, lymphadenopathy) may herald a serious medical event and that the patient should report any such occurrence to a physician immediately.

5.3 Acute Multiorgan Failure

Multiorgan failure, which in some cases has been fatal or irreversible, has been observed in patients receiving the immediate-release formulation of LAMICTAL. Fatalities associated with multiorgan failure and various degrees of hepatic failure have been reported in 2 of 3,796 adult patients and 4 of 2,435 pediatric patients who received the immediate-release formulation of LAMICTAL in epilepsy

Table 1. Escalation Regimen for LAMICTAL XR in Patients ≥13 Years of Age

	For Patients TAKING Valproate*	For Patients NOT TAKING Carbamazepine, Phenytoin, Phenobarbital, Primidone†, or Valproate*	For Patients TAKING Carbamazepine, Phenytoin, Phenobarbital, or Primidone† and NOT TAKING Valproate*
Weeks 1 and 2	25 mg every *other* day	25 mg every day	50 mg every day
Weeks 3 and 4	25 mg every day	50 mg every day	100 mg every day
Week 5	50 mg every day	100 mg every day	200 mg every day
Week 6	100 mg every day	150 mg every day	300 mg every day
Week 7	150 mg every day	200 mg every day	400 mg every day
Maintenance Range (Week 8 and onward)	200 to 250 mg every day‡	300 to 400 mg every day‡	400 to 600 mg every day‡

* Valproate has been shown to inhibit glucuronidation and decrease the apparent clearance of lamotrigine *[see Drug Interactions (7), Clinical Pharmacology (12.3)]*.

† These drugs induce glucuronidation and increase clearance *[see Drug Interactions (7), Clinical Pharmacology (12.3)]*. Other drugs which have similar effects include estrogen-containing oral contraceptives *[see Drug Interactions (7), Clinical Pharmacology (12.3)]*. Dosing recommendations for oral contraceptives can be found in General Dosing Considerations *[see Dosage and Administration (2.1)]*. Patients on rifampin, or other drugs that induce glucuronidation and increase clearance, should follow the same dosing titration/maintenance regimen as that used with anticonvulsants that have this effect.

‡ Dose increases at week 8 or later should not exceed 100 mg daily at weekly intervals.

clinical trials. Rare fatalities from multiorgan failure have been reported in compassionate plea and postmarketing use. The majority of these deaths occurred in association with other serious medical events, including status epilepticus and overwhelming sepsis, and hantavirus, making it difficult to identify the initial cause.
Additionally, 3 patients (a 45-year-old woman, a 3.5-year-old boy, and an 11-year-old girl) developed multiorgan dysfunction and disseminated intravascular coagulation 9 to 14 days after the immediate-release formulation of LAMICTAL was added to their AED regimens. Rash and elevated transaminases were also present in all patients and rhabdomyolysis was noted in 2 patients. Both pediatric patients were receiving concomitant therapy with valproate, while the adult patient was being treated with carbamazepine and clonazepam. All patients subsequently recovered with supportive care after treatment with the immediate-release formulation of LAMICTAL was discontinued.

5.4 Blood Dyscrasias

There have been reports of blood dyscrasias with the immediate-release formulation of LAMICTAL that may or may not be associated with the hypersensitivity syndrome. These have included neutropenia, leukopenia, anemia, thrombocytopenia, pancytopenia, and, rarely, aplastic anemia and pure red cell aplasia.

5.5 Suicidal Behavior and Ideation

Antiepileptic drugs (AEDs), including LAMICTAL XR, increase the risk of suicidal thoughts or behavior in patients taking these drugs for any indication. Patients treated with any AED for any indication should be monitored for the emergence or worsening of depression, suicidal thoughts or behavior, and/or any unusual changes in mood or behavior.
Pooled analyses of 199 placebo-controlled clinical trials (mono- and adjunctive therapy) of 11 different AEDs showed that patients randomized to one of the AEDs had approximately twice the risk (adjusted Relative Risk 1.8, 95% CI:1.2, 2.7) of suicidal thinking or behavior compared to patients randomized to placebo. In these trials, which had a median treatment duration of 12 weeks, the estimated incidence of suicidal behavior or ideation among 27,863 AED-treated patients was 0.43%, compared to 0.24% among 16,029 placebo-treated patients, representing an increase of approximately 1 case of suicidal thinking or behavior for every 530 patients treated. There were 4 suicides in drug-treated patients in the trials and none in placebo-treated patients, but the number of events is too small to allow any conclusion about drug effect on suicide.
The increased risk of suicidal thoughts or behavior with AEDs was observed as early as 1 week after starting treatment with AEDs and persisted for the duration of treatment assessed. Because most trials included in the analysis did not extend beyond 24 weeks, the risk of suicidal thoughts or behavior beyond 24 weeks could not be assessed.
The risk of suicidal thoughts or behavior was generally consistent among drugs in the data analyzed. The finding of increased risk with AEDs of varying mechanism of action and across a range of indications suggests that the risk applies to all AEDs used for any indication. The risk did not vary substantially by age (5 to 100 years) in the clinical trials analyzed.
Table 2 shows absolute and relative risk by indication for all evaluated AEDs.

Table 2. Risk by Indication for Antiepileptic Drugs in the Pooled Analysis

Indication	Placebo Patients With Events Per 1,000 Patients	Drug Patients With Events Per 1,000 Patients	Relative Risk: Incidence of Events in Drug Patients/Incidence in Placebo Patients	Risk Difference: Additional Drug Patients With Events Per 1,000 Patients
Epilepsy	1.0	3.4	3.5	2.4
Psychiatric	5.7	8.5	1.5	2.9
Other	1.0	1.8	1.9	0.9
Total	2.4	4.3	1.8	1.9

[See table 2 above]
The relative risk for suicidal thoughts or behavior was higher in clinical trials for epilepsy than in clinical trials for psychiatric or other conditions, but the absolute risk differences were similar for the epilepsy and psychiatric indications.
Anyone considering prescribing LAMICTAL XR or any other AED must balance the risk of suicidal thoughts or behavior with the risk of untreated illness. Epilepsy and many other illnesses for which AEDs are prescribed are themselves associated with morbidity and mortality and an increased risk of suicidal thoughts and behavior. Should suicidal thoughts and behavior emerge during treatment, the prescriber needs to consider whether the emergence of these symptoms in any given patient may be related to the illness being treated.
Patients, their caregivers, and families should be informed that AEDs increase the risk of suicidal thoughts and behavior and should be advised of the need to be alert for the emergence or worsening of the signs and symptoms of depression, any unusual changes in mood or behavior, or the emergence of suicidal thoughts, behavior, or thoughts about self-harm. Behaviors of concern should be reported immediately to healthcare providers.

5.6 Potential Medication Errors

Medication errors involving LAMICTAL have occurred. In particular, the names LAMICTAL or lamotrigine can be confused with the names of other commonly used medications. Medication errors may also occur between the different formulations of LAMICTAL. To reduce the potential of medication errors, write and say LAMICTAL XR clearly. Depictions of the LAMICTAL XR Extended-Release Tablets can be found in the Medication Guide *[see Patient Counseling Information (17.10)]*. Each LAMICTAL XR tablet has a distinct color and white center, and is printed with "LAMICTAL XR" and the tablet strength. These distinctive features serve to identify the different presentations of the drug and thus may help reduce the risk of medication errors. LAMICTAL XR is supplied in round, unit-of-use bottles with orange caps containing 30 tablets. The label on the bottle includes a depiction of the tablets which further communicates to patients and pharmacists that the medication is LAMICTAL XR and the specific tablet strength included in the bottle. The unit-of-use bottle with a distinctive orange cap and distinctive bottle label features serves to identify the different presentations of the drug and thus may help to reduce the risk of medication errors. To avoid the medication error of using the wrong drug or formulation, patients should be strongly advised to visually inspect their tablets to verify that they are LAMICTAL XR each time they fill their prescription.

5.7 Concomitant Use With Oral Contraceptives

Some estrogen-containing oral contraceptives have been shown to decrease serum concentrations of lamotrigine *[see Clinical Pharmacology (12.3)]*. **Dosage adjustments will be necessary in most patients who start or stop estrogen-containing oral contraceptives while taking LAMICTAL XR** *[see Dosage and Administration (2.1)]*. During the week of inactive hormone preparation ("pill-free" week) of oral contraceptive therapy, plasma lamotrigine levels are expected to rise, as much as doubling at the end of the week. Adverse reactions consistent with elevated levels of lamotrigine, such as dizziness, ataxia, and diplopia, could occur.

5.8 Withdrawal Seizures

As with other AEDs, LAMICTAL XR should not be abruptly discontinued. In patients with epilepsy there is a possibility of increasing seizure frequency. Unless safety concerns require a more rapid withdrawal, the dose of LAMICTAL XR should be tapered over a period of at least 2 weeks (approximately 50% reduction per week) *[see Dosage and Administration (2.1)]*.

5.9 Status Epilepticus

Valid estimates of the incidence of treatment-emergent status epilepticus among patients treated with immediate-release lamotrigine are difficult to obtain because reporters participating in clinical trials did not all employ identical rules for identifying cases. At a minimum, 7 of 2,343 adult patients had episodes that could unequivocally be described as status epilepticus. In addition, a number of reports of variably defined episodes of seizure exacerbation (e.g., seizure clusters, seizure flurries, etc.) were made.

5.10 Sudden Unexplained Death in Epilepsy (SUDEP)

During the premarketing development of the immediate-release formulation of LAMICTAL, 20 sudden and unexplained deaths were recorded among a cohort of 4,700 patients with epilepsy (5,747 patient-years of exposure).
Some of these could represent seizure-related deaths in which the seizure was not observed, e.g., at night. This represents an incidence of 0.0035 deaths per patient-year. Although this rate exceeds that expected in a healthy population matched for age and sex, it is within the range of estimates for the incidence of sudden unexplained deaths in patients with epilepsy not receiving lamotrigine (ranging from 0.0005 for the general population of patients with epilepsy, to 0.004 for a recently studied clinical trial population similar to that in the clinical development program for immediate-release lamotrigine, to 0.005 for patients with refractory epilepsy). Consequently, whether these figures are reassuring or suggest concern depends on the comparability of the populations reported upon to the cohort receiving immediate-release lamotrigine and the accuracy of the estimates provided. Probably most reassuring is the similarity of estimated SUDEP rates in patients receiving immediate-release lamotrigine and those receiving other AEDs, chemically unrelated to each other, that underwent clinical testing in similar populations. Importantly, that drug is chemically unrelated to lamotrigine. This evidence suggests, although it certainly does not prove, that the high SUDEP rates reflect population rates, not a drug effect.

5.11 Addition of LAMICTAL XR to a Multidrug Regimen That Includes Valproate

Because valproate reduces the clearance of lamotrigine, the dosage of lamotrigine in the presence of valproate is less than half of that required in its absence.

5.12 Binding in the Eye and Other Melanin-Containing Tissues

Because lamotrigine binds to melanin, it could accumulate in melanin-rich tissues over time. This raises the possibility that lamotrigine may cause toxicity in these tissues after extended use. Although ophthalmological testing was performed in one controlled clinical trial, the testing was inadequate to exclude subtle effects or injury occurring after long-term exposure. Moreover, the capacity of available tests to detect potentially adverse consequences, if any, of lamotrigine binding to melanin is unknown *[see Clinical Pharmacology (12.2)]*.
Accordingly, although there are no specific recommendations for periodic ophthalmological monitoring, prescribers should be aware of the possibility of long-term ophthalmologic effects.

5.13 Laboratory Tests

The value of monitoring plasma concentrations of lamotrigine in patients treated with LAMICTAL XR has not been established. Because of the possible pharmacokinetic interactions between lamotrigine and other drugs including AEDs (see Table 4), monitoring of the plasma levels of lamotrigine and concomitant drugs may be indicated, particularly during dosage adjustments. In general, clinical judgment should be exercised regarding monitoring of plasma levels of lamotrigine and other drugs and whether or not dosage adjustments are necessary.
Treatment with LAMICTAL XR caused an increased incidence of subnormal (below the reference range) values in some hematology analytes (e.g., total white blood cells, monocytes). The treatment effect (LAMICTAL XR % - Placebo %) incidence of subnormal counts was 3% for total white blood cells and 4% for monocytes.

6 ADVERSE REACTIONS

The following adverse reactions are described in more detail in the *Warnings and Precautions* section of the label:
- Serious skin rashes *[see Warnings and Precautions (5.1)]*
- Hypersensitivity reactions *[see Warnings and Precautions (5.2)]*
- Acute multiorgan failure *[see Warnings and Precautions (5.3)]*
- Blood dyscrasias *[see Warnings and Precautions (5.4)]*
- Suicidal behavior and ideation *[see Warnings and Precautions (5.5)]*
- Withdrawal seizures *[see Warnings and Precautions (5.8)]*
- Status epilepticus *[see Warnings and Precautions (5.9)]*
- Sudden unexplained death in epilepsy *[see Warnings and Precautions (5.10)]*

6.1 Clinical Trial Experience with LAMICTAL XR for Treatment of Partial Onset Seizures

Because clinical trials are conducted under widely varying conditions, adverse reaction rates observed in the clinical trials of a drug cannot be directly compared with rates in the clinical trials of another drug and may not reflect the rates observed in practice.
The most commonly observed adverse reactions (≥4% for LAMICTAL XR and more common on drug than placebo) in a double-blind, placebo-controlled trial of adjunctive therapy with LAMICTAL XR for the treatment of partial onset seizures were, in order of decreasing treatment difference (LAMICTAL XR % - Placebo %) incidence: dizziness, tremor/intention tremor, cerebellar coordination/balance disorder, nausea, asthenic conditions (asthenia, fatigue, malaise), vertigo/positional vertigo, and diplopia.

Continued on next page

Product information on these pages is effective as of June 2009. Further information is available at 1-888-825-5249 or www.gsk.com.

Lamictal XR—Cont.

Nine of 118 patients (8%) treated with LAMICTAL XR who participated in the double-blind, placebo-controlled trial in the treatment of partial onset seizures discontinued treatment due to adverse reactions compared to 2 of 121 patients (2%) who received placebo. Dizziness, nausea, and nystagmus were the most common adverse reactions (based upon treatment difference of ≥2%) that led to the withdrawal of subjects in the group treated with LAMICTAL XR.

Table 3 displays the incidence of adverse reactions in a 19-week, double-blind, placebo-controlled study of patients with partial onset seizures.

[See table 3 below]

Adverse reactions were also analyzed to assess the incidence of the onset of an event in the titration period, and in the maintenance period, and if adverse reactions occurring in the titration phase persisted in the maintenance phase.

The incidence for many adverse reactions caused by LAMICTAL XR treatment was increased relative to placebo (i.e., LAMICTAL XR % - Placebo % = treatment difference ≥3%) in either the titration or maintenance phases of the study. During the titration phase, an increased incidence (shown in descending order of % treatment difference) was observed for diarrhea, nausea, vertigo/positional vertigo, somnolence, myalgia, and hot flush. During the maintenance phase, an increased incidence was observed for dizziness, tremor/intention tremor, cerebellar coordination/balance disorder, vomiting, and diplopia. Some adverse reactions developing in the titration phase were notable for persisting (>7 days) into the maintenance phase. These "persistent" adverse reactions included somnolence, dizziness, and headache. In addition, some adverse reactions had an increased likelihood of recurring. Headache recurred predominantly in the titration period and vertigo and nausea recurred throughout the whole treatment period.

There were inadequate data to evaluate the effect of dose and/or concentration on the incidence of adverse reactions because although patients were randomized to different target doses based upon concomitant AED, the plasma exposure was expected to be generally similar among all patients receiving different doses. However, in a randomized, parallel study comparing placebo and 300 and 500 mg/day of immediate-release formulation of LAMICTAL, the incidence of the most common adverse reactions (≥5%) such as ataxia, blurred vision, diplopia, and dizziness were dose-related. Less common adverse reactions (<5%) were not assessed for dose-response relationships.

There were insufficient data to evaluate the effect of gender, age, and race on the adverse reaction profile for LAMICTAL XR.

6.2 Other Adverse Reactions Observed During the Clinical Development of the Immediate-Release Formulation of LAMICTAL

All reported reactions are included except those already listed in the previous tables or elsewhere in the labeling, those too general to be informative, and those not reasonably associated with the use of the drug.

Table 3. Treatment-Emergent Adverse Reaction Incidence in a Double-Blind, Placebo-Controlled Adjunctive Trial of Patients With Partial Onset Seizures (Adverse Reactions ≥ 2% of Patients Treated With LAMICTAL XR and More Common on Drug Than Placebo)

Body System/Adverse Reaction	LAMICTAL XR (n = 118) %	Placebo (n = 121) %
Ear and Labyrinth Disorders		
Vertigo/positional vertigo	4	0
Eye Disorders		
Vision blurred	4	2
Diplopia	4	0
Gastrointestinal Disorders		
Diarrhea	8	5
Nausea	7	2
Abdominal pain/discomfort	6	4
Vomiting	4	2
Constipation	3	1
Dry mouth	3	2
General Disorders and Administration Site Conditions		
Asthenic conditions (asthenia, fatigue, malaise)	9	5
Chest pain/discomfort	3	1
Gait disturbance	2	0
Pain	2	1
Infections and Infestations		
Influenza/influenza-like illness	3	2
Sinusitis	3	1
Metabolic and Nutrition Disorders		
Anorexia/decreased appetite	3	2
Weight increased	2	1
Musculoskeletal and Connective Tissue Disorders		
Myalgia	3	0
Nervous System		
Dizziness	19	5
Somnolence	7	5
Tremor/Intention tremor	7	2
Cerebellar coordination/balance disorder	5	0
Nystagmus	3	1
Psychiatric Disorders		
Depression	4	1
Anxiety	3	0
Respiratory, Thoracic, and Mediastinal Disorders		
Pharnygolaryngeal pain	3	2
Epistaxis	2	1
Sinus congestion	2	0
Skin and Subcutaneous Tissue Disorders		
Rash*	2	1
Alopecia	2	1
Vascular Disorder		
Hot flush	3	0

*All types of rash. In clinical trials evaluating the immediate-release formulation of LAMICTAL, the rate of serious rash was 0.3% in adults on adjunctive therapy for epilepsy *[see Boxed Warning]*.

Adjunctive Therapy in Adults With Epilepsy: In addition to the adverse reactions reported above from the development of LAMICTAL XR, the following adverse reactions with an uncertain relationship to lamotrigine were reported during the clinical development of the immediate-release formulation of LAMICTAL for treatment of epilepsy in adults. These reactions occurred in ≥2% of patients receiving the immediate-release formulation of LAMICTAL and more frequently than in the placebo group.

Body as a Whole: Fever, neck pain.

Musculoskeletal: Arthralgia.

Nervous: Insomnia, convulsion, irritability, speech disorder, concentration disturbance.

Respiratory: Rhinitis, pharyngitis, cough increased.

Skin and Appendages: Pruritus.

Urogenital: (female patients only) Vaginitis, amenorrhea, dysmenorrhea.

Other Clinical Trial Experience: The immediate-release formulation of LAMICTAL has been administered to 6,694 individuals for whom complete adverse reaction data was captured during all clinical trials, only some of which were placebo controlled. During these trials, all adverse reactions were recorded by the clinical investigators using terminology of their own choosing. To provide a meaningful estimate of the proportion of individuals having adverse reactions, similar types of reactions were grouped into a smaller number of standardized categories using modified COSTART dictionary terminology. The frequencies presented represent the proportion of the 6,694 individuals exposed to LAMICTAL who experienced an event of the type cited on at least one occasion while receiving LAMICTAL.

Adverse reactions are further classified within body system categories and enumerated in order of decreasing frequency using the following definitions: *frequent* adverse reactions are defined as those occurring in at least 1/100 patients; *infrequent* adverse reactions are those occurring in 1/100 to 1/1,000 patients; *rare* adverse reactions are those occurring in fewer than 1/1,000 patients.

Body as a Whole: *Infrequent:* Allergic reaction, chills, and malaise.

Cardiovascular System: *Infrequent:* Flushing, hypertension, palpitations, postural hypotension, syncope, tachycardia, and vasodilation.

Dermatological: *Infrequent:* Acne, hirsutism, maculopapular rash, skin discoloration, and urticaria. *Rare:* Angioedema, erythema, exfoliative dermatitis, fungal dermatitis, herpes zoster, leukoderma, multiforme erythema, petechial rash, pustular rash, Stevens-Johnson syndrome, and vesiculobullous rash.

Digestive System: *Infrequent:* Dysphagia, eructation, gastritis, gingivitis, increased appetite, increased salivation, liver function tests abnormal, and mouth ulceration. *Rare:* Gastrointestinal hemorrhage, glossitis, gum hemorrhage, gum hyperplasia, hematemesis, hemorrhagic colitis, hepatitis, melena, stomach ulcer, stomatitis, and tongue edema.

Endocrine System: *Rare:* Goiter and hypothyroidism.

Hematologic and Lymphatic System: *Infrequent:* Ecchymosis and leukopenia. *Rare:* Anemia, eosinophilia, fibrin decrease, fibrinogen decrease, iron deficiency anemia, leukocytosis, lymphocytosis, macrocytic anemia, petechia, and thrombocytopenia.

Metabolic and Nutritional Disorders: *Infrequent:* Aspartate transaminase increased. *Rare:* Alcohol intolerance, alkaline phosphatase increase, alanine transaminase increase, bilirubinemia, general edema, gamma glutamyl transpeptidase increase, and hyperglycemia.

Musculoskeletal System: Infrequent: Arthritis, leg cramps, myasthenia, and twitching. *Rare:* Bursitis, muscle atrophy, pathological fracture, and tendinous contracture.

Nervous System: *Frequent:* Confusion and paresthesia. *Infrequent:* Akathisia, apathy, aphasia, CNS depression, depersonalization, dysarthria, dyskinesia, euphoria, hallucinations, hostility, hyperkinesia, hypertonia, libido decreased, memory decrease, mind racing, movement disorder, myoclonus, panic attack, paranoid reaction, personality disorder, psychosis, stupor, and suicidal ideation. *Rare:* Choreoathetosis, delirium, delusions, dysphoria, dystonia, extrapyramidal syndrome, faintness, grand mal convulsions, hemiplegia, hyperalgesia, hyperesthesia, hypokinesia, hypotonia, manic depression reaction, muscle spasm, neuralgia, neurosis, paralysis, and peripheral neuritis.

Respiratory System: *Infrequent:* Yawn. *Rare:* Hiccup and hyperventilation.

Special Senses: *Frequent:* Amblyopia. *Infrequent:* Abnormality of accommodation, conjunctivitis, dry eyes, ear pain, photophobia, taste perversion, and tinnitus. *Rare:* Deafness, lacrimation disorder, oscillopsia, parosmia, ptosis, strabismus, taste loss, uveitis, and visual field defect.

Urogenital System: *Infrequent:* Abnormal ejaculation, hematuria, impotence, menorrhagia, polyuria, urinary incontinence. *Rare:* Acute kidney failure, anorgasmia, breast abscess, breast neoplasm, creatinine increase, cystitis, dysuria, epididymitis, female lactation, kidney failure, kidney pain, nocturia, urinary retention, urinary urgency.

6.3 Postmarketing Experience with the Immediate-Release Formulation of LAMICTAL

The following adverse events (not listed above in clinical trials or other sections of the prescribing information) have been identified during postapproval use of the immediate-release formulation of LAMICTAL. Because these events are reported voluntarily from a population of uncertain size, it is not always possible to reliably estimate their frequency or establish a causal relationship to drug exposure.

Blood and Lymphatic: Agranulocytosis, hemolytic anemia.
Gastrointestinal: Esophagitis.
Hepatobiliary Tract and Pancreas: Pancreatitis.
Immunologic: Lupus-like reaction, vasculitis.
Lower Respiratory: Apnea.
Musculoskeletal: Rhabdomyolysis has been observed in patients experiencing hypersensitivity reactions.
Neurology: Exacerbation of Parkinsonian symptoms in patients with pre-existing Parkinson's disease, tics.
Non-site Specific: Progressive immunosuppression.

7 DRUG INTERACTIONS

Significant drug interactions with lamotrigine are summarized in Table 4. Additional details of these drug interaction studies, which were conducted using the immediate-release formulation of LAMICTAL, are provided in the Clinical Pharmacology section *[see Clinical Pharmacology (12.3)]*.
[See table 4 above]

Table 4. Established and Other Potentially Significant Drug Interactions

Concomitant Drug	Effect on Concentration of Lamotrigine or Concomitant Drug	Clinical Comment
Estrogen-containing oral contraceptive preparations containing 30 mcg ethinylestradiol and 150 mcg levonorgestrel	↓ lamotrigine	Decreased lamotrigine levels approximately 50%.
	↓ levonorgestrel	Decrease in levonorgestrel component by 19%.
Carbamazepine (CBZ) and CBZ epoxide	↓ lamotrigine ? CBZ epoxide	Addition of carbamazepine decreases lamotrigine concentration approximately 40%. May increase CBZ epoxide levels.
Phenobarbital/Primidone	↓ lamotrigine	Decreased lamotrigine concentration approximately 40%.
Phenytoin (PHT)	↓ lamotrigine	Decreased lamotrigine concentration approximately 40%.
Rifampin	↓ lamotrigine	Decreased lamotrigine AUC approximately 40%.
Valproate	↑ lamotrigine ? valproate	Increased lamotrigine concentrations slightly more than 2-fold. Decreased valproate concentrations an average of 25% over a 3-week period then stabilized in healthy volunteers; no change in controlled clinical trials in epilepsy patients.

↓ = Decreased (induces lamotrigine glucuronidation).
↑ = Increased (inhibits lamotrigine glucuronidation).
? = Conflicting data.

8 USE IN SPECIFIC POPULATIONS

8.1 Pregnancy

Teratogenic Effects: Pregnancy Category C. No evidence of teratogenicity was found in mice, rats, or rabbits when lamotrigine was orally administered to pregnant animals during the period of organogenesis at doses up to 1.2, 0.5, and 1.1 times, respectively, on a mg/m^2 basis, the highest usual human maintenance dose (i.e., 500 mg/day). However, maternal toxicity and secondary fetal toxicity producing reduced fetal weight and/or delayed ossification were seen in mice and rats, but not in rabbits at these doses. Teratology studies were also conducted using bolus intravenous administration of the isethionate salt of lamotrigine in rats and rabbits. In rat dams administered an intravenous dose at 0.6 times the highest usual human maintenance dose, the incidence of intrauterine death without signs of teratogenicity was increased.

A behavioral teratology study was conducted in rats dosed during the period of organogenesis. At day 21 postpartum, offspring of dams receiving 5 mg/kg/day or higher displayed a significantly longer latent period for open field exploration and a lower frequency of rearing. In a swimming maze test performed on days 39 to 44 postpartum, time to completion was increased in offspring of dams receiving 25 mg/kg/day. These doses represent 0.1 and 0.5 times the clinical dose on a mg/m^2 basis, respectively.

Lamotrigine did not affect fertility, teratogenesis, or postnatal development when rats were dosed prior to and during mating, and throughout gestation and lactation at doses equivalent to 0.4 times the highest usual human maintenance dose on a mg/m^2 basis.

When pregnant rats were orally dosed at 0.1, 0.14, or 0.3 times the highest human maintenance dose (on a mg/m^2 basis) during the latter part of gestation (days 15 to 20), maternal toxicity and fetal death were seen. In dams, food consumption and weight gain were reduced, and the gestation period was slightly prolonged (22.6 vs. 22.0 days in the control group). Stillborn pups were found in all 3 drug-treated groups with the highest number in the high-dose group. Postnatal death was also seen, but only in the 2 highest doses, and occurred between day 1 and 20. Some of these deaths appear to be drug-related and not secondary to the maternal toxicity. A no-observed-effect level (NOEL) could not be determined for this study.

Although lamotrigine was not found to be teratogenic in the above studies, lamotrigine decreases fetal folate concentrations in rats, an effect known to be associated with teratogenesis in animals and humans. There are no adequate and well-controlled studies in pregnant women. Because animal reproduction studies are not always predictive of human response, this drug should be used during pregnancy only if the potential benefit justifies the potential risk to the fetus.

Non-Teratogenic Effects: As with other AEDs, physiological changes during pregnancy may affect lamotrigine concentrations and/or therapeutic effect. There have been reports of decreased lamotrigine concentrations during pregnancy and restoration of pre-partum concentrations after delivery. Dosage adjustments may be necessary to maintain clinical response.

Pregnancy Exposure Registry: To provide information regarding the effects of in utero exposure to LAMICTAL XR, physicians are advised to recommend that pregnant patients taking LAMICTAL XR enroll in the North American Antiepileptic Drug (NAAED) Pregnancy Registry. This can be done by calling the toll-free number 1-888-233-2334, and must be done by patients themselves. Information on the registry can also be found at the website http://www.aedpregnancyregistry.org/.

Physicians are also encouraged to register patients in the Lamotrigine Pregnancy Registry; enrollment in this registry must be done prior to any prenatal diagnostic tests and **before fetal outcome is known. Physicians** can obtain information by calling the Lamotrigine Pregnancy Registry at 1-800-336-2176 (toll-free).

8.2 Labor and Delivery

The effect of LAMICTAL XR on labor and delivery in humans is unknown.

8.3 Nursing Mothers

Preliminary data indicate that lamotrigine passes into human milk. Because the effects on the infant exposed to lamotrigine by this route are unknown, breastfeeding while taking LAMICTAL XR is not recommended.

8.4 Pediatric Use

LAMICTAL XR is indicated as adjunctive therapy for partial onset seizures with or without secondary generalization in patients ≥13 years of age. Safety and effectiveness of LAMICTAL XR for any use in patients below the age of 13 have not been established.

The immediate-release formulation of LAMICTAL is indicated for adjunctive therapy in patients ≥2 years of age for partial seizures, the generalized seizures of Lennox-Gastaut syndrome, and primary generalized tonic-clonic seizures.

Safety and efficacy of the immediate-release formulation of LAMICTAL, used as adjunctive treatment for partial seizures, were not demonstrated in a small randomized, double-blind, placebo-controlled, withdrawal study in very young pediatric patients (1 to 24 months). The immediate-release formulation of LAMICTAL was associated with an increased risk for infectious adverse reactions (LAMICTAL 37%, Placebo 5%), and respiratory adverse reactions (LAMICTAL 26%, Placebo 5%). Infectious adverse reactions included: bronchiolitis, bronchitis, ear infection, eye infection, otitis externa, pharyngitis, urinary tract infection, and viral infection. Respiratory adverse reactions included nasal congestion, cough, and apnea.

8.5 Geriatric Use

Clinical studies of LAMICTAL XR for epilepsy did not include sufficient numbers of subjects 65 years of age and over to determine whether they respond differently from younger subjects or exhibit a different safety profile than that of younger patients. In general, dose selection for an elderly patient should be cautious, usually starting at the low end of the dosing range, reflecting the greater frequency of decreased hepatic, renal, or cardiac function, and of concomitant disease or other drug therapy.

8.6 Patients With Hepatic Impairment

Experience in patients with hepatic impairment is limited. Based on a clinical pharmacology study with the immediate-release formulation of LAMICTAL in 24 patients with mild, moderate, and severe liver impairment *[see Clinical Pharmacology (12.4)]*, the following general recommendations can be made. No dosage adjustment is needed in patients with mild liver impairment. Initial, escalation, and maintenance doses should generally be reduced by approximately 25% in patients with moderate and severe liver impairment without ascites and 50% in patients with severe liver impairment with ascites. Escalation and maintenance doses may be adjusted according to clinical response *[see Dosage and Administration (2.1)]*.

8.7 Patients With Renal Impairment

Lamotrigine is metabolized mainly by glucuronic acid conjugation, with the majority of the metabolites being recovered in the urine. In a small study comparing a single dose of immediate-release lamotrigine in patients with varying degrees of renal impairment with healthy volunteers, the plasma half-life of lamotrigine was significantly longer in the patients with renal impairment *[see Clinical Pharmacology (12.3)]*.

Initial doses of LAMICTAL XR should be based on patients' AED regimens; reduced maintenance doses may be effective for patients with significant renal impairment. Few patients with severe renal impairment have been evaluated during chronic treatment with lamotrigine. Because there is inadequate experience in this population, LAMICTAL XR should be used with caution in these patients *[see Dosage and Administration (2.1)]*.

10 OVERDOSAGE

10.1 Human Overdose Experience

Overdoses involving quantities up to 15 g have been reported for the immediate-release formulation of LAMICTAL, some of which have been fatal. Overdose has resulted in ataxia, nystagmus, increased seizures, decreased level of consciousness, coma, and intraventricular conduction delay.

10.2 Management of Overdose

There are no specific antidotes for lamotrigine. Following a suspected overdose, hospitalization of the patient is advised. General supportive care is indicated, including frequent monitoring of vital signs and close observation of the patient. If indicated, emesis should be induced or gastric lavage should be performed; usual precautions should be taken to protect the airway. It is uncertain whether hemodialysis is an effective means of removing lamotrigine from the blood. In 6 renal failure patients, about 20% of the amount of lamotrigine in the body was removed by hemodialysis during a 4-hour session. A Poison Control Center should be contacted for information on the management of overdosage of LAMICTAL XR.

11 DESCRIPTION

LAMICTAL XR (lamotrigine), an AED of the phenyltriazine class, is chemically unrelated to existing AEDs. Its chemical name is 3,5-diamino-6-(2,3-dichlorophenyl)-*as*-triazine, its molecular formula is $C_9H_7N_5Cl_2$, and its molecular weight is 256.09. Lamotrigine is a white to pale cream-colored powder and has a pK_a of 5.7. Lamotrigine is very slightly soluble in water (0.17 mg/mL at 25°C) and slightly soluble in 0.1 M HCl (4.1 mg/mL at 25°C). The structural formula is:

Continued on next page

Product information on these pages is effective as of June 2009. Further information is available at 1-888-825-5249 or www.gsk.com.

Lamictal XR—Cont.

LAMICTAL XR Extended-Release Tablets are supplied for oral administration as 25-mg (yellow with white center), 50-mg (green with white center), 100-mg (orange with white center), and 200-mg (blue with white center) tablets. Each tablet contains the labeled amount of lamotrigine and the following inactive ingredients: glycerol monostearate, hypromellose, lactose monohydrate; magnesium stearate; methacrylic acid copolymer dispersion, polyethylene glycol 400, polysorbate 80, silicon dioxide (25-mg and 50-mg tablets only), titanium dioxide, triethyl citrate, iron oxide black (50-mg tablet only), iron oxide yellow (25-mg, 50-mg, 100-mg tablets only), iron oxide red (100-mg tablet only), FD&C Blue No. 2 Aluminum Lake (200-mg tablet only). Tablets are printed with edible black ink.

LAMICTAL XR Extended-Release Tablets contain a modified-release eroding formulation as the core. The tablets are coated with a clear enteric coat and have an aperture drilled through the coats on both faces of the tablet (DiffCORE™*) to enable a controlled release of drug in the acidic environment of the stomach. The combination of this and the modified-release core are designed to control the dissolution rate of lamotrigine over a period of approximately 12 to 15 hours, leading to a gradual increase in serum lamotrigine levels.

12 CLINICAL PHARMACOLOGY

12.1 Mechanism of Action

The precise mechanism(s) by which lamotrigine exerts its anticonvulsant action are unknown. In animal models designed to detect anticonvulsant activity, lamotrigine was effective in preventing seizure spread in the maximum electroshock (MES) and pentylenetetrazol (scMet) tests, and prevented seizures in the visually and electrically evoked after-discharge (EEAD) tests for antiepileptic activity. Lamotrigine also displayed inhibitory properties in the kindling model in rats both during kindling development and in the fully kindled state. The relevance of these models to human epilepsy, however, is not known.

One proposed mechanism of action of lamotrigine, the relevance of which remains to be established in humans, involves an effect on sodium channels. In vitro pharmacological studies suggest that lamotrigine inhibits voltage-sensitive sodium channels, thereby stabilizing neuronal membranes and consequently modulating presynaptic transmitter release of excitatory amino acids (e.g., glutamate and aspartate).

Although the relevance for human use is unknown, the following data characterize the performance of lamotrigine in receptor binding assays. Lamotrigine had a weak inhibitory effect on the serotonin 5-HT_3 receptor (IC_{50} = 18 μM). It does not exhibit high affinity binding (IC_{50}>100 μM) to the following neurotransmitter receptors: adenosine A_1 and A_2; adrenergic α_1, α_2, and β; dopamine D_1 and D_2; γ-aminobutyric acid (GABA) A and B; histamine H_1; kappa opioid; muscarinic acetylcholine; and serotonin 5-HT_2. Studies have failed to detect an effect of lamotrigine on dihydropyridine-sensitive calcium channels. It had weak effects at sigma opioid receptors (IC_{50} = 145 μM). Lamotrigine did not inhibit the uptake of norepinephrine, dopamine, or serotonin, (IC_{50}>200 μM) when tested in rat synaptosomes and/or human platelets in vitro.

Effect of Lamotrigine on N-Methyl d-Aspartate-Receptor Mediated Activity: Lamotrigine did not inhibit N-methyl d-aspartate (NMDA)-induced depolarizations in rat cortical slices or NMDA-induced cyclic GMP formation in immature rat cerebellum, nor did lamotrigine displace compounds that are either competitive or noncompetitive ligands at this glutamate receptor complex (CNQX, CGS, TCHP). The IC_{50} for lamotrigine effects on NMDA-induced currents (in the presence of 3 μM of glycine) in cultured hippocampal neurons exceeded 100 μM.

12.2 Pharmacodynamics

Folate Metabolism: In vitro, lamotrigine inhibited dihydrofolate reductase, the enzyme that catalyzes the reduction of dihydrofolate to tetrahydrofolate. Inhibition of this enzyme may interfere with the biosynthesis of nucleic acids and proteins. When oral daily doses of lamotrigine were given to pregnant rats during organogenesis, fetal, placental, and maternal folate concentrations were reduced. Significantly reduced concentrations of folate are associated with teratogenesis *[see Use in Specific Populations (8.1)]*. Folate concentrations were also reduced in male rats given repeated oral doses of lamotrigine. Reduced concentrations were partially returned to normal when supplemented with folinic acid.

Accumulation in Kidneys: Lamotrigine accumulated in the kidney of the male rat, causing chronic progressive nephrosis, necrosis, and mineralization. These findings are attributed to α-2 microglobulin, a species- and sex-specific protein that has not been detected in humans or other animal species.

Melanin Binding: Lamotrigine binds to melanin-containing tissues, e.g., in the eye and pigmented skin. It has been found in the uveal tract up to 52 weeks after a single dose in rodents.

Cardiovascular: In dogs, lamotrigine is extensively metabolized to a 2-N-methyl metabolite. This metabolite causes dose-dependent prolongations of the PR interval, widening of the QRS complex, and, at higher doses, complete AV conduction block. Similar cardiovascular effects are not anticipated in humans because only trace amounts of the 2-N-methyl metabolite (<0.6% of lamotrigine dose) have been found in human urine *[see Clinical Pharmacology (12.3)]*. However, it is conceivable that plasma concentrations of this metabolite could be increased in patients with a reduced capacity to glucuronidate lamotrigine (e.g., in patients with liver disease).

12.3 Pharmacokinetics

In comparison to immediate-release lamotrigine, the plasma lamotrigine levels following administration of LAMICTAL XR are not associated with any significant changes in trough plasma concentrations, and are characterized by lower peaks, longer time to peaks, and lower peak-to-trough fluctuation, as described in detail below.

Absorption: Lamotrigine is absorbed after oral administration with negligible first-pass metabolism. The bioavailability of lamotrigine is not affected by food.

In an open-label, crossover study of 44 subjects with epilepsy receiving concomitant AEDs, the steady-state pharmacokinetics of lamotrigine were compared following administration of equivalent total doses of LAMICTAL XR given once daily with those of lamotrigine immediate-release given twice daily. In this study, the median time to peak concentration (T_{max}) following administration of LAMICTAL XR was 4 to 6 hours in patients taking carbamazepine, phenytoin, phenobarbital, or primidone; 9 to 11 hours in patients taking VPA; and 6 to 10 hours in patients taking AEDs other than carbamazepine, phenytoin, phenobarbital, primidone, or VPA. In comparison, the median T_{max} following administration of immediate-release lamotrigine was between 1 and 1.5 hours.

The steady-state trough concentrations for extended-release lamotrigine were similar to or higher than those of immediate-release lamotrigine depending on concomitant AED (Table 5). A mean reduction in the lamotrigine C_{max} by 11% to 29% was observed for LAMICTAL XR compared to immediate-release lamotrigine resulting in a decrease in the peak-to-trough fluctuation in serum lamotrigine concentrations. However, in some subjects receiving enzyme-inducing AEDs, a reduction in C_{max} of 44% to 77% was observed. The degree of fluctuation was reduced by 17% in patients taking enzyme-inducing AEDs, 34% in patients taking VPA, and 37% in patients taking AEDs other than carbamazepine, phenytoin, phenobarbital, primidone, or VPA. LAMICTAL XR and immediate-release lamotrigine regimens were similar with respect to area under the curve (AUC, a measure of the extent of bioavailability) for patients receiving AEDs other than those known to induce the metabolism of lamotrigine. The relative bioavailability of extended-release lamotrigine was approximately 21% lower than immediate-release lamotrigine in subjects receiving enzyme-inducing AEDs. However, in some subjects in this group a reduction in exposure of up to 70% was observed when switched to LAMICTAL XR. Therefore, doses may need to be adjusted in some subjects based on therapeutic response.

Table 5. Steady-State Bioavailability of LAMICTAL XR Relative to Immediate-Release Lamotrigine at Equivalent Daily Doses (Ratio of XR to IR 90% CI)

Concomitant AED	$AUC_{(0-24ss)}$	C_{max}	C_{min}
EIAEDs*	0.79 (0.69, 0.90)	0.71 (0.61, 0.82)	0.99 (0.89, 1.09)
VPA	0.94 (0.81, 1.08)	0.88 (0.75, 1.03)	0.99 (0.88, 1.10)
AEDs other than EIAEDs* or VPA	1.00 (0.88, 1.14)	0.89 (0.78, 1.03)	1.14 (1.03, 1.25)

*EIAEDs include carbamazepine, phenytoin, phenobarbital, and primidone.

Dose Proportionality: In healthy volunteers not receiving any other medications and given LAMICTAL XR once daily, the systemic exposure to lamotrigine increased in direct proportion to the dose administered over the range of 50 to 200 mg. At doses between 25 and 50 mg, the increase was less than dose proportional, with a 2-fold increase in dose resulting in an approximately 1.6-fold increase in systemic exposure.

Distribution: Estimates of the mean apparent volume of distribution (Vd/F) of lamotrigine following oral administration ranged from 0.9 to 1.3 L/kg. Vd/F is independent of dose and is similar following single and multiple doses in both patients with epilepsy and in healthy volunteers.

Protein Binding: Data from in vitro studies indicate that lamotrigine is approximately 55% bound to human plasma proteins at plasma lamotrigine concentrations from 1 to 10 mcg/mL (10 mcg/mL is 4 to 6 times the trough plasma concentration observed in the controlled efficacy trials). Because lamotrigine is not highly bound to plasma proteins, clinically significant interactions with other drugs through competition for protein binding sites are unlikely. The binding of lamotrigine to plasma proteins did not change in the presence of therapeutic concentrations of phenytoin, phenobarbital, or valproate. Lamotrigine did not displace other AEDs (carbamazepine, phenytoin, phenobarbital) from protein binding sites.

Metabolism: Lamotrigine is metabolized predominantly by glucuronic acid conjugation; the major metabolite is an inactive 2-N-glucuronide conjugate. After oral administration of 240 mg of ^{14}C-lamotrigine (15 μCi) to 6 healthy volunteers, 94% was recovered in the urine and 2% was recovered in the feces. The radioactivity in the urine consisted of unchanged lamotrigine (10%), the 2-N-glucuronide (76%), a 5-N-glucuronide (10%), a 2-N-methyl metabolite (0.14%), and other unidentified minor metabolites (4%).

Enzyme Induction: The effects of lamotrigine on the induction of specific families of mixed-function oxidase isozymes have not been systematically evaluated.

Following multiple administrations (150 mg twice daily) to normal volunteers taking no other medications, lamotrigine induced its own metabolism, resulting in a 25% decrease in $t_{1/2}$ and a 37% increase in Cl/F at steady state compared with values obtained in the same volunteers following a single dose. Evidence gathered from other sources suggests that self-induction by lamotrigine may not occur when lamotrigine is given as adjunctive therapy in patients receiving enzyme-inducing drugs such as carbamazepine, phenytoin, phenobarbital, primidone, or other drugs such as rifampin that induce lamotrigine glucuronidation *[see Drug Interactions (7)]*.

Elimination: The elimination half-life and apparent clearance of lamotrigine following administration of immediate-release lamotrigine to adult patients with epilepsy and healthy volunteers is summarized in Table 6. Half-life and apparent oral clearance vary depending on concomitant AEDs.

Since the half-life of lamotrigine following administration of single doses of immediate-release lamotrigine is comparable to that observed following administration of LAMICTAL XR, similar changes in the half-life of lamotrigine would be expected for LAMICTAL XR.

[See table 6 at top of next page]

Drug Interactions: The apparent clearance of lamotrigine is affected by the coadministration of certain medications *[see Warnings and Precautions (5.7, 5.11), Drug Interactions (7)]*.

The net effects of drug interactions with lamotrigine are summarized in Table 7. Details of the drug interaction studies, which were done using immediate-release lamotrigine, are provided following Table 7.

[See table 7 at top of next page]

Estrogen-Containing Oral Contraceptives: In 16 female volunteers, an oral contraceptive preparation containing 30 mcg ethinylestradiol and 150 mcg levonorgestrel increased the apparent clearance of lamotrigine (300 mg/day) by approximately 2-fold with mean decreases in AUC of 52% and in C_{max} of 39%. In this study, trough serum lamotrigine concentrations gradually increased and were approximately 2-fold higher on average at the end of the week of the inactive hormone preparation compared with trough lamotrigine concentrations at the end of the active hormone cycle.

Gradual transient increases in lamotrigine plasma levels (approximate 2-fold increase) occurred during the week of inactive hormone preparation ("pill-free" week) for women not also taking a drug that increased the clearance of lamotrigine (carbamazepine, phenytoin, phenobarbital, primidone, or other drugs that induce lamotrigine glucuronidation *[see Drug Interactions (7)]*. The increase in lamotrigine plasma levels will be greater if the dose of LAMICTAL XR is increased in the few days before or during the "pill-free" week. Increases in lamotrigine plasma levels could result in dose-dependent adverse effects.

In the same study, coadministration of lamotrigine (300 mg/day) in 16 female volunteers did not affect the pharmacokinetics of the ethinylestradiol component of the oral contraceptive preparation. There were mean decreases in the AUC and C_{max} of the levonorgestrel component of 19% and 12%, respectively. Measurement of serum progesterone indicated that there was no hormonal evidence of ovulation in any of the 16 volunteers, although measurement of serum FSH, LH, and estradiol indicated that there was some loss of suppression of the hypothalamic-pituitary-ovarian axis.

The effects of doses of lamotrigine other than 300 mg/day have not been systematically evaluated in controlled clinical trials.

The clinical significance of the observed hormonal changes on ovulatory activity is unknown. However, the possibility of decreased contraceptive efficacy in some patients cannot be excluded. Therefore, patients should be instructed to promptly report changes in their menstrual pattern (e.g., break-through bleeding).

Dosage adjustments may be necessary for women receiving estrogen-containing oral contraceptive preparations *[see Dosage and Administration (2.1)]*.

Other Hormonal Contraceptives or Hormone Replacement Therapy: The effect of other hormonal contraceptive preparations or hormone replacement therapy on the pharmacokinetics of lamotrigine has not been systematically evaluated. It has been reported that ethinylestradiol, not progestogens, increased the clearance of lamotrigine up to 2-fold, and the progestin-only pills had no effect on lamotrigine plasma levels. Therefore, adjustments to the dosage of LAMICTAL XR in the presence of progestogens alone will likely not be needed.

Bupropion: The pharmacokinetics of a 100-mg single dose of lamotrigine in healthy volunteers (n = 12) were not changed by coadministration of bupropion sustained-release formulation (150 mg twice daily) starting 11 days before lamotrigine.

Carbamazepine: Lamotrigine has no appreciable effect on steady-state carbamazepine plasma concentration. Limited clinical data suggest there is a higher incidence of dizziness, diplopia, ataxia, and blurred vision in patients receiving carbamazepine with lamotrigine than in patients receiving other AEDs with lamotrigine *[see Adverse Reactions (6.1)]*. The mechanism of this interaction is unclear. The effect of lamotrigine on plasma concentrations of carbamazepine-epoxide is unclear. In a small subset of patients (n = 7) studied in a placebo-controlled trial, lamotrigine had no effect on carbamazepine-epoxide plasma concentrations, but in a small, uncontrolled study (n = 9), carbamazepine-epoxide levels increased.

The addition of carbamazepine decreases lamotrigine steady-state concentrations by approximately 40%.

Esomeprazole: In a study of 30 subjects, coadministration of LAMICTAL XR with esomeprazole resulted in no significant change in lamotrigine levels and a small decrease in T_{max}. The levels of gastric pH were not altered compared with pre-lamotrigine dosing.

Felbamate: In a study of 21 healthy volunteers, coadministration of felbamate (1,200 mg twice daily) with lamotrigine (100 mg twice daily for 10 days) appeared to have no clinically relevant effects on the pharmacokinetics of lamotrigine.

Folate Inhibitors: Lamotrigine is a weak inhibitor of dihydrofolate reductase. Prescribers should be aware of this action when prescribing other medications that inhibit folate metabolism.

Gabapentin: Based on a retrospective analysis of plasma levels in 34 patients who received lamotrigine both with and without gabapentin, gabapentin does not appear to change the apparent clearance of lamotrigine.

Levetiracetam: Potential drug interactions between levetiracetam and lamotrigine were assessed by evaluating serum concentrations of both agents during placebo-controlled clinical trials. These data indicate that lamotrigine does not influence the pharmacokinetics of levetiracetam and that levetiracetam does not influence the pharmacokinetics of lamotrigine.

Lithium: The pharmacokinetics of lithium were not altered in healthy subjects (n = 20) by coadministration of lamotrigine (100 mg/day) for 6 days.

Olanzapine: The AUC and C_{max} of olanzapine were similar following the addition of olanzapine (15 mg once daily) to lamotrigine (200 mg once daily) in healthy male volunteers (n = 16) compared with the AUC and C_{max} in healthy male volunteers receiving olanzapine alone (n = 16).

In the same study, the AUC and C_{max} of lamotrigine were reduced on average by 24% and 20%, respectively, following the addition of olanzapine to lamotrigine in healthy male volunteers compared with those receiving lamotrigine alone. This reduction in lamotrigine plasma concentrations is not expected to be clinically relevant.

Oxcarbazepine: The AUC and C_{max} of oxcarbazepine and its active 10-monohydroxy oxcarbazepine metabolite were not significantly different following the addition of oxcarbazepine (600 mg twice daily) to lamotrigine (200 mg once daily) in healthy male volunteers (n = 13) compared with healthy male volunteers receiving oxcarbazepine alone (n = 13).

In the same study, the AUC and C_{max} of lamotrigine were similar following the addition of oxcarbazepine (600 mg twice daily) to lamotrigine in healthy male volunteers compared with those receiving lamotrigine alone. Limited clinical data suggest a higher incidence of headache, dizziness, nausea, and somnolence with coadministration of lamotrigine and oxcarbazepine compared with lamotrigine alone or oxcarbazepine alone.

Phenobarbital, Primidone: The addition of phenobarbital or primidone decreases lamotrigine steady-state concentrations by approximately 40%.

Table 6. Mean* Pharmacokinetic Parameters of Immediate-Release Lamotrigine in Healthy Volunteers and Adult Patients With Epilepsy

Adult Study Population	Number of Subjects	$t_{1/2}$: Elimination Half-life (hr)	Cl/F: Apparent Plasma Clearance (mL/min/kg)
Healthy volunteers taking no other medications:			
Single-dose lamotrigine	179	32.8 (14.0-103.0)	0.44 (0.12-1.10)
Multiple-dose lamotrigine	36	25.4 (11.6-61.6)	0.58 (0.24-1.15)
Healthy volunteers taking valproate:			
Single-dose lamotrigine	6	48.3 (31.5-88.6)	0.30 (0.14-0.42)
Multiple-dose lamotrigine	18	70.3 (41.9-113.5)	0.18 (0.12-0.33)
Patients with epilepsy taking valproate only:			
Single-dose lamotrigine	4	58.8 (30.5-88.8)	0.28 (0.16-0.40)
Patients with epilepsy taking carbamazepine, phenytoin, phenobarbital, or primidone[†] plus valproate:			
Single-dose lamotrigine	25	27.2 (11.2-51.6)	0.53 (0.27-1.04)
Patients with epilepsy taking carbamazepine, phenytoin, phenobarbital, or primidone[†]:			
Single-dose lamotrigine	24	14.4 (6.4-30.4)	1.10 (0.51-2.22)
Multiple-dose lamotrigine	17	12.6 (7.5-23.1)	1.21 (0.66-1.82)

* The majority of parameter means determined in each study had coefficients of variation between 20% and 40% for half-life and Cl/F and between 30% and 70% for T_{max}. The overall mean values were calculated from individual study means that were weighted based on the number of volunteers/patients in each study. The numbers in parentheses below each parameter mean represent the range of individual volunteer/patient values across studies.

† Carbamazepine, phenobarbital, phenytoin, and primidone have been shown to increase the apparent clearance of lamotrigine. Estrogen-containing oral contraceptives and other drugs that induce lamotrigine glucuronidation have also been shown to increase the apparent clearance of lamotrigine *[see Drug Interactions (7)]*.

Table 7. Summary of Drug Interactions With Lamotrigine

Drug	Drug Plasma Concentration With Adjunctive Lamotrigine*	Lamotrigine Plasma Concentration With Adjunctive Drugs[†]
Oral contraceptives (e.g., ethinylestradiol/levonorgestrel[‡])	↔[§]	↓
Bupropion	Not assessed	↔
Carbamazepine (CBZ)	↔	↓
CBZ epoxide[‖]	?	
Felbamate	Not assessed	↔
Gabapentin	Not assessed	↔
Levetiracetam	↔	↔
Lithium	↔	Not assessed
Olanzapine	↔	↔[¶]
Oxcarbazepine	↔	↔
10-monohydroxy oxcarbazepine metabolite[#]	↔	
Phenobarbital/primidone	↔	↓
Phenytoin (PHT)	↔	↓
Pregabalin	↔	↔
Rifampin	Not assessed	↓
Topiramate	↔[**]	↔
Valproate	↓	↑
Valproate + PHT and/or CBZ	Not assessed	↔
Zonisamide	Not assessed	↔

* From adjunctive clinical trials and volunteer studies.

† Net effects were estimated by comparing the mean clearance values obtained in adjunctive clinical trials and volunteer studies.

‡ The effect of other hormonal contraceptive preparations or hormone replacement therapy on the pharmacokinetics of lamotrigine has not been systematically evaluated in clinical trials, although the effect may be similar to that seen with the ethinylestradiol/levonorgestrel combinations.

§ Modest decrease in levonorgestrel.

‖ Not administered, but an active metabolite of carbamazepine.

¶ Slight decrease, not expected to be clinically relevant.

\# Not administered, but an active metabolite of oxcarbazepine.

** Slight increase not expected to be clinically relevant.

↔= No significant effect.

? = Conflicting data.

Phenytoin: Lamotrigine has no appreciable effect on steady-state phenytoin plasma concentrations in patients with epilepsy. The addition of phenytoin decreases lamotrigine steady-state concentrations by approximately 40%.

Pregabalin: Steady-state trough plasma concentrations of lamotrigine were not affected by concomitant pregabalin (200 mg 3 times daily) administration. There are no pharmacokinetic interactions between lamotrigine and pregabalin.

Rifampin: In 10 male volunteers, rifampin (600 mg/day for 5 days) significantly increased the apparent clearance of a single 25-mg dose of lamotrigine by approximately 2-fold (AUC decreased by approximately 40%).

Topiramate: Topiramate resulted in no change in plasma concentrations of lamotrigine. Administration of lamotrigine resulted in a 15% increase in topiramate concentrations.

Valproate: When lamotrigine was administered to healthy volunteers (n = 18) receiving valproate, the trough steady-state valproate plasma concentrations decreased by an average of 25% over a 3-week period, and then stabilized. However, adding lamotrigine to the existing therapy did not cause a change in valproate plasma concentrations in either adult or pediatric patients in controlled clinical trials.

The addition of valproate increased lamotrigine steady-state concentrations in normal volunteers by slightly more than 2-fold. In one study, maximal inhibition of lamotrigine

Continued on next page

Product information on these pages is effective as of June 2009. Further information is available at 1-888-825-5249 or www.gsk.com.

Lamictal XR—Cont.

clearance was reached at valproate doses between 250 and 500 mg/day and did not increase as the valproate dose was further increased.

Zonisamide: In a study of 18 patients with epilepsy, coadministration of zonisamide (200 to 400 mg/day) with lamotrigine (150 to 500 mg/day for 35 days) had no significant effect on the pharmacokinetics of lamotrigine.

Known Inducers or Inhibitors of Glucuronidation: Drugs other than those listed above have not been systematically evaluated in combination with lamotrigine. Since lamotrigine is metabolized predominately by glucuronic acid conjugation, drugs that are known to induce or inhibit glucuronidation may affect the apparent clearance of lamotrigine, and doses of LAMICTAL XR may require adjustment based on clinical response.

Other: Results of in vitro experiments suggest that clearance of lamotrigine is unlikely to be reduced by concomitant administration of amitriptyline, clonazepam, clozapine, fluoxetine, haloperidol, lorazepam, phenelzine, risperidone, sertraline, or trazodone.

Results of in vitro experiments suggest that lamotrigine does not reduce the clearance of drugs eliminated predominantly by CYP2D6.

Special Populations: ***Patients With Renal Impairment:*** Twelve volunteers with chronic renal failure (mean creatinine clearance: 13 mL/min; range: 6 to 23) and another 6 individuals undergoing hemodialysis were each given a single 100 mg dose of immediate-release lamotrigine. The mean plasma half-lives determined in the study were 42.9 hours (chronic renal failure), 13.0 hours (during hemodialysis), and 57.4 hours (between hemodialysis) compared with 26.2 hours in healthy volunteers. On average, approximately 20% (range: 5.6 to 35.1) of the amount of lamotrigine present in the body was eliminated by hemodialysis during a 4-hour session *[see Dosage and Administration (2.1)]*.

Hepatic Disease: The pharmacokinetics of lamotrigine following a single 100-mg dose of immediate-release lamotrigine were evaluated in 24 subjects with mild, moderate, and severe hepatic impairment (Child-Pugh Classification system) and compared with 12 subjects without hepatic impairment. The patients with severe hepatic impairment were without ascites (n = 2) or with ascites (n = 5). The mean apparent clearances of lamotrigine in patients with mild (n = 12), moderate (n = 5), severe without ascites (n = 2), and severe with ascites (n = 5) liver impairment were 0.30 ± 0.09, 0.24 ± 0.1, 0.21 ± 0.04, and 0.15 ± 0.09 mL/min/kg, respectively, as compared with 0.37 ± 0.1 mL/min/kg in the healthy controls. Mean half-lives of lamotrigine in patients with mild, moderate, severe without ascites, and severe with ascites hepatic impairment were 46 ± 20, 72 ± 44, 67 ± 11, and 100 ± 48 hours, respectively, as compared with 33 ± 7 hours in healthy controls *[see Dosage and Administration (2.1)]*.

Elderly: The pharmacokinetics of lamotrigine following a single 150 mg dose of immediate-release lamotrigine were evaluated in 12 elderly volunteers between the ages of 65 and 76 years (mean creatinine clearance: 61 mL/min, range: 33 to 108 mL/min). The mean half-life of lamotrigine in these subjects was 31.2 hours (range: 24.5 to 43.4 hours), and the mean clearance was 0.40 mL/min/kg (range: 0.26 to 0.48 mL/min/kg).

Gender: The clearance of lamotrigine is not affected by gender. However, during dose escalation of immediate-release lamotrigine in one clinical trial in patients with epilepsy on a stable dose of valproate (n = 77), mean trough lamotrigine concentrations, unadjusted for weight, were 24% to 45% higher (0.3 to 1.7 mcg/mL) in females than in males.

Race: The apparent oral clearance of lamotrigine was 25% lower in non-Caucasians than Caucasians.

Pediatric Patients: Safety and effectiveness of LAMICTAL XR for use in patients below the age of 13 have not been established.

13 NONCLINICAL TOXICOLOGY

13.1 Carcinogenesis, Mutagenesis, Impairment of Fertility

No evidence of carcinogenicity was seen in 1 mouse study or 2 rat studies following oral administration of lamotrigine for up to 2 years at maximum tolerated doses (30 mg/kg/day for mice and 10 to 15 mg/kg/day for rats, doses that are equivalent to 90 mg/m^2 and 60 to 90 mg/m^2, respectively). Steady-state plasma concentrations ranged from 1 to 4 mcg/mL in the mouse study and 1 to 10 mcg/mL in the rat study. Plasma concentrations associated with the recommended human doses of 300 to 500 mg/day are generally in the range of 2 to 5 mcg/mL, but concentrations as high as 19 mcg/mL have been recorded.

Lamotrigine was not mutagenic in the presence or absence of metabolic activation when tested in 2 gene mutation assays (the Ames test and the in vitro mammalian mouse lymphoma assay). In 2 cytogenetic assays (the in vitro human lymphocyte assay and the in vivo rat bone marrow assay), lamotrigine did not increase the incidence of structural or numerical chromosomal abnormalities.

No evidence of impairment of fertility was detected in rats given oral doses of lamotrigine up to 2.4 times the highest usual human maintenance dose of 8.33 mg/kg/day or 0.4 times the human dose on a mg/m^2 basis. The effect of lamotrigine on human fertility is unknown.

14 CLINICAL STUDIES

14.1 Partial Onset Seizures

The effectiveness of immediate-release lamotrigine as adjunctive therapy (added to other AEDs) was initially established in 3 pivotal multicenter, placebo-controlled, double-blind clinical trials in 355 adults with refractory partial onset seizures.

The effectiveness of LAMICTAL XR as adjunctive therapy in partial onset seizures, with or without secondary generalization, was established in a 19-week, multicenter, double-blind, placebo-controlled trial in 236 patients, 13 years of age and older (approximately 93% of patients were 16 to 65 years old). Approximately 36% were from the U.S. and approximately 64% were from other countries including Argentina, Brazil, Chile, Germany, India, Korea, Russian Federation, and Ukraine. Patients with at least 8 partial onset seizures during an 8-week prospective baseline phase (or 4-week prospective baseline coupled with a 4-week historical baseline documented with seizure diary data) were randomized to treatment with LAMICTAL XR (n = 116) or placebo (n = 120) added to their current regimen of 1 or 2 AEDs. Approximately half of the patients were taking 2 concomitant AEDs at baseline. Target doses ranged from 200 to 500 mg/day of LAMICTAL XR based on concomitant AED (target dose = 200 mg for valproate, 300 mg for AEDs not altering plasma lamotrigine, and 500 mg for enzyme-inducing AEDs). The median partial seizure frequency per week at baseline was 2.3 for LAMICTAL XR and 2.1 for placebo.

The primary endpoint was the median percent change from baseline in partial onset seizure frequency during the entire double-blind treatment phase. The median percent reductions in weekly partial onset seizures were 47% in patients treated with LAMICTAL XR and 25% on placebo, a difference that was statistically significant.

Figure 1 presents the percentage of patients (X-axis) with a percent reduction in partial seizure frequency (responder rate) from baseline to the entire treatment period at least as great as that represented on the Y-axis. A positive value on the Y-axis indicates an improvement from baseline (i.e., a decrease in seizure frequency), while a negative value indicates a worsening from baseline (i.e., an increase in seizure frequency). Thus, in a display of this type, a curve for an effective treatment is shifted to the left of the curve for placebo. The proportion of patients achieving any particular level of reduction in seizure frequency was consistently higher for the group treated with LAMICTAL XR compared with the placebo group. For example, 44% of patients randomized to LAMICTAL XR experienced a 50% or greater reduction in seizure frequency, compared with 21% of patients randomized to placebo. Patients with an increase in seizure frequency >100% are represented on the Y-axis as equal to or greater than -100%.

Figure 1. Proportion of Patients by Responder Rate for LAMICTAL XR and Placebo Group

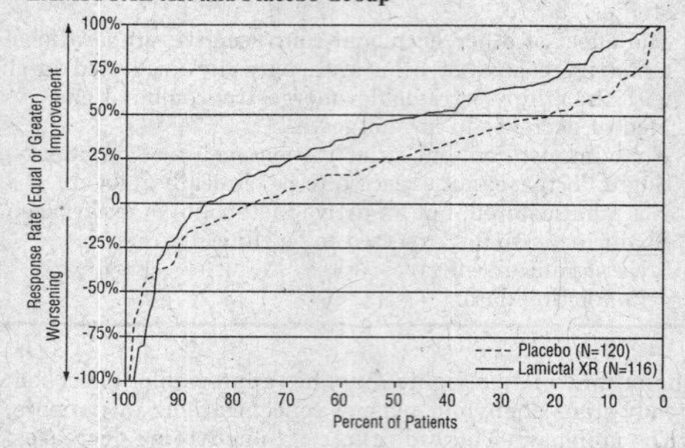

16 HOW SUPPLIED/STORAGE AND HANDLING

LAMICTAL XR (lamotrigine) Extended-Release Tablets

25 mg, yellow with a white center, round, biconvex, film-coated tablets printed on one face in black ink with "LAMICTAL" and "XR 25", unit-of-use bottles of 30 with orange caps (NDC 0173-0754-00).

50 mg, green with a white center, round, biconvex, film-coated tablets printed on one face in black ink with "LAMICTAL" and "XR 50", unit-of-use bottles of 30 with orange caps (NDC 0173-0755-00).

100 mg, orange with a white center, round, biconvex, film-coated tablets printed on one face in black ink with "LAMICTAL" and "XR 100", unit-of-use bottles of 30 with orange caps (NDC 0173-0756-00).

200 mg, blue with a white center, round, biconvex, film-coated tablets printed on one face in black ink with "LAMICTAL" and "XR 200", unit-of-use bottles of 30 with orange caps (NDC 0173-0757-00).

LAMICTAL XR (lamotrigine) Patient Titration Kit for Patients Taking Valproate (Blue XR Kit)

25 mg, yellow with a white center, round, biconvex, film-coated tablets printed on one face in black ink with "LAMICTAL" and "XR 25" and 50 mg, green with a white center, round, biconvex, film-coated tablets printed on one face in black ink with "LAMICTAL" and "XR 50"; blisterpack of 21/25-mg tablets and 7/50-mg tablets (NDC 0173-0758-00).

LAMICTAL XR (lamotrigine) Patient Titration Kit for Patients Taking Carbamazepine, Phenytoin, Phenobarbital, or Primidone, and Not Taking Valproate (Green XR Kit)

50 mg, green with a white center, round, biconvex, film-coated tablets printed on one face in black ink with "LAMICTAL" and "XR 50"; 100 mg, orange with a white center, round, biconvex, film-coated tablets printed on one face in black ink with "LAMICTAL" and "XR 100"; and 200 mg, blue with a white center, round, biconvex, film-coated tablets printed on one face in black ink with "LAMICTAL" and "XR 200"; blisterpack of 14/50-mg tablets, 14/100-mg tablets, and 7/200-mg tablets (NDC 0173-0759-00).

LAMICTAL XR (lamotrigine) Patient Titration Kit for Patients Not Taking Carbamazepine, Phenytoin, Phenobarbital, Primidone, or Valproate (Orange XR Kit)

25 mg, yellow with a white center, round, biconvex, film-coated tablets printed on one face in black ink with "LAMICTAL" and "XR 25"; 50 mg, green with a white center, round, biconvex, film-coated tablets printed on one face in black ink with "LAMICTAL" and "XR 50"; and 100 mg, orange with a white center, round, biconvex, film-coated tablets printed on one face in black ink with "LAMICTAL" and "XR 100"; blisterpack of 14/25-mg tablets, 14/50-mg tablets, and 7/100-mg tablets (NDC 0173-0760-00).

Storage: Store at 25°C (77°F); excursions permitted to 15-30°C (59-86°F) [see USP Controlled Room Temperature].

17 PATIENT COUNSELING INFORMATION

See Medication Guide (17.10).

17.1 Rash

Prior to initiation of treatment with LAMICTAL XR, the patient should be instructed that a rash or other signs or symptoms of hypersensitivity (e.g., fever, lymphadenopathy) may herald a serious medical event and that the patient should report any such occurrence to a physician immediately.

17.2 Suicidal Thinking and Behavior

Patients, their caregivers, and families should be counseled that AEDs, including LAMICTAL XR, may increase the risk of suicidal thoughts and behavior and should be advised of the need to be alert for the emergence or worsening of symptoms of depression, any unusual changes in mood or behavior, or the emergence of suicidal thoughts, behavior, or thoughts about self-harm. Behaviors of concern should be reported immediately to healthcare providers.

17.3 Worsening of Seizures

Patients should be advised to notify their physician if worsening of seizure control occurs.

17.4 CNS Adverse Effects

Patients should be advised that LAMICTAL XR may cause dizziness, somnolence, and other symptoms and signs of central nervous system (CNS) depression. Accordingly, they should be advised neither to drive a car nor to operate other complex machinery until they have gained sufficient experience on LAMICTAL XR to gauge whether or not it adversely affects their mental and/or motor performance.

17.5 Blood Dyscrasias and/or Acute Multiorgan Failure

Patients should be advised of the possibility of blood dyscrasias and/or acute multiorgan failure and to contact their physician immediately if they experience any signs or symptoms of these conditions *[see Warnings and Precautions (5.3, 5.4)]*.

17.6 Pregnancy

Patients should be advised to notify their physicians if they become pregnant or intend to become pregnant during therapy. Patients should be advised to notify their physicians if they intend to breastfeed or are breastfeeding an infant.

Patients should also be encouraged to enroll in the NAAED Pregnancy Registry if they become pregnant. This registry is collecting information about the safety of antiepileptic drugs during pregnancy. To enroll, patients can call the toll-free number 1-888-233-2334 *[see Use in Specific Populations (8.1)]*.

17.7 Oral Contraceptive Use

Women should be advised to notify their physician if they plan to start or stop use of oral contraceptives or other female hormonal preparations. Starting estrogen-containing oral contraceptives may significantly decrease lamotrigine plasma levels and stopping estrogen-containing oral contraceptives (including the "pill-free" week) may significantly increase lamotrigine plasma levels *[see Warnings and Pre-*

cautions (5.7), Clinical Pharmacology (12.3)]. Women should also be advised to promptly notify their physician if they experience adverse reactions or changes in menstrual pattern (e.g., break-through bleeding) while receiving LAMICTAL XR in combination with these medications.

17.8 Discontinuing LAMICTAL XR

Patients should be advised to notify their physician if they stop taking LAMICTAL XR for any reason and not to resume LAMICTAL XR without consulting their physician.

17.9 Potential Medication Errors

Medication errors involving LAMICTAL have occurred. In particular the names LAMICTAL or lamotrigine can be confused with the names of other commonly used medications. Medication errors may also occur between the different formulations of LAMICTAL. To reduce the potential of medication errors, write and say LAMICTAL XR clearly. Depictions of the LAMICTAL XR Extended-Release Tablets can be found in the Medication Guide in section 17.10. Each LAMICTAL XR tablet has a distinct color and white center, and is printed with "LAMICTAL XR" and the tablet strength. These distinctive features serve to identify the different presentations of the drug and thus may help reduce the risk of medication errors. LAMICTAL XR is supplied in round, unit-of-use bottles with orange caps containing 30 tablets. The label on the bottle includes a depiction of the tablets which further communicates to patients and pharmacists that the medication is LAMICTAL XR and the specific tablet strength included in the bottle. The unit-of-use bottle with a distinctive orange cap and distinctive bottle label features serves to identify the different presentations of the drug and thus may help to reduce the risk of medication errors. **To avoid a medication error of using the wrong drug or formulation, patients should be strongly advised to visually inspect their tablets to verify that they are LAMICTAL XR each time they fill their prescription and to immediately talk to their doctor/pharmacist if they receive a LAMICTAL XR tablet without a white center and without "LAMICTAL XR" and the strength printed on the tablet as they may have received the wrong medication** *[see Dosage Forms and Strengths (3), How Supplied/Storage and Handling (16)]*.

17.10 Medication Guide

A Medication Guide is provided as a separate leaflet accompanying the product. The full text of the Medication Guide is reprinted below.

MEDICATION GUIDE

LAMICTAL® (la-MIK-tal) XR™ (lamotrigine) Extended-Release Tablets

Read this Medication Guide before you start taking LAMICTAL XR and each time you get a refill. There may be new information. This information does not take the place of talking with your healthcare provider about your medical condition or treatment. If you have questions about LAMICTAL XR, ask your healthcare provider or pharmacist.

What is the most important information I should know about LAMICTAL XR?

1. **LAMICTAL XR may cause a serious skin rash that may cause you to be hospitalized or to stop LAMICTAL XR; it may rarely cause death.**
   There is no way to tell if a mild rash will develop into a more serious reaction. These serious skin reactions are more likely to happen when you begin taking LAMICTAL XR, within the first 2 to 8 weeks of treatment. But it can happen in people who have taken LAMICTAL XR for any period of time. Children between 2 to 16 years of age have a higher chance of getting this serious skin reaction while taking lamotrigine. LAMICTAL XR is not approved for use in children less than 13 years old.
   The risk of getting a rash is higher if you:
   - take LAMICTAL XR while taking valproate (DEPAKENE (valproic acid) or DEPAKOTE (divalproex sodium)).
   - take a higher starting dose of LAMICTAL XR than your healthcare provider prescribed.
   - increase your dose of LAMICTAL XR faster than prescribed.

   LAMICTAL XR can also cause other types of allergic reactions or serious problems which may affect organs and other parts of your body like the liver or blood cells. You may or may not have a rash with these types of reactions.
   Call your healthcare provider right away if you have any of the following:
   - **a skin rash**
   - **hives**
   - **fever**
   - **swollen lymph glands**
   - **painful sores in the mouth or around your eyes**
   - **swelling of your lips or tongue**
   - **yellowing of your skin or eyes**
   - **unusual bruising or bleeding**
   - **severe fatigue or weakness**
   - **severe muscle pain**
   - **frequent infections**

   These symptoms may be the first signs of a serious reaction. A healthcare provider should examine you to decide if you should continue taking LAMICTAL XR.
2. **Like other antiepileptic drugs, LAMICTAL XR may cause suicidal thoughts or actions in a very small number of people, about 1 in 500.**
   Call a healthcare provider right away if you have any of these symptoms, especially if they are new, worse, or worry you:
   - thoughts about suicide or dying
   - attempt to commit suicide
   - new or worse depression
   - new or worse anxiety
   - feeling agitated or restless
   - panic attacks
   - trouble sleeping (insomnia)
   - new or worse irritability
   - acting aggressive, being angry, or violent
   - acting on dangerous impulses
   - an extreme increase in activity and talking (mania)
   - other unusual changes in behavior or mood

   Do not stop LAMICTAL XR without first talking to a healthcare provider.
   - Stopping LAMICTAL XR suddenly can cause serious problems.
   - Suicidal thoughts or actions can be caused by things other than medicines. If you have suicidal thoughts or actions, your healthcare provider may check for other causes.

   How can I watch for early symptoms of suicidal thoughts and actions?
   - Pay attention to any changes, especially sudden changes, in mood, behaviors, thoughts, or feelings.
   - Keep all follow-up visits with your healthcare provider as scheduled.
   - Call your healthcare provider between visits as needed, especially if you are worried about symptoms.

   LAMICTAL XR can have other serious side effects. For more information ask your healthcare provider or pharmacist. Tell your healthcare provider if you have any side effect that bothers you. Be sure to read the section below entitled "What are the possible side effects of LAMICTAL XR?"
3. **Patients prescribed LAMICTAL have sometimes been given the wrong medicine because many medicines have names similar to LAMICTAL, so always check that you receive LAMICTAL XR.**
   Taking the wrong medication can cause serious health problems. When your healthcare provider gives you a prescription for LAMICTAL XR:
   - Make sure you can read it clearly.
   - Talk to your pharmacist to check that you are given the correct medicine.
   - Each time you fill your prescription, check the tablets you receive against the pictures of the tablets below.

   These pictures show the distinct wording, colors, and shapes of the tablets that help to identify the right strength of LAMICTAL XR. Immediately call your pharmacist if you receive a LAMICTAL XR tablet that does not look like one of the tablets shown below, as you may have received the wrong medication.

LAMICTAL XR (lamotrigine) Extended-Release Tablets

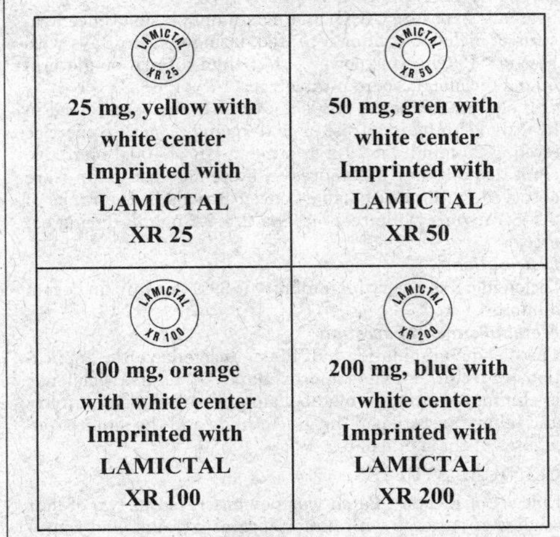

25 mg, yellow with white center Imprinted with LAMICTAL XR 25	50 mg, gren with white center Imprinted with LAMICTAL XR 50
100 mg, orange with white center Imprinted with LAMICTAL XR 100	200 mg, blue with white center Imprinted with LAMICTAL XR 200

What is LAMICTAL XR?

LAMICTAL XR is a prescription medicine used together with other medicines to treat partial seizures in people 13 years or older.

It is not known if LAMICTAL XR is safe or effective in children under the age of 13. Other forms of LAMICTAL can be used in children 2 to 12 years.

Who should not take LAMICTAL XR?

You should not take LAMICTAL XR if you have had an allergic reaction to lamotrigine or to any of the inactive ingredients in LAMICTAL XR. See the end of this leaflet for a complete list of ingredients in LAMICTAL XR.

What should I tell my healthcare provider before taking LAMICTAL XR?

Before taking LAMICTAL XR, tell your healthcare provider about all of your medical conditions, including if you:
- have had a rash or allergic reaction to another antiseizure medicine.
- have or have had depression, mood problems or suicidal thoughts or behavior.
- are taking oral contraceptives (birth control pills) or other female hormonal medicines. Do not start or stop taking birth control pills or other female hormonal medicine until you have talked with your healthcare provider. Tell your healthcare provider if you have any changes in your menstrual pattern such as breakthrough bleeding. Stopping these medicines may cause side effects (such as dizziness, lack of coordination, or double vision). Starting these medicines may lessen how well LAMICTAL XR works.
- are pregnant or plan to become pregnant. It is not known if LAMICTAL XR will harm your unborn baby. If you become pregnant while taking LAMICTAL XR, talk to your healthcare provider about registering with the North American Antiepileptic Drug Pregnancy Registry. You can enroll in this registry by calling 1-888-233-2334. The purpose of this registry is to collect information about the safety of antiepileptic drugs during pregnancy.
- are breastfeeding. LAMICTAL XR can pass into your breast milk. You and your healthcare provider should decide if you should take LAMICTAL XR or breastfeed. Breastfeeding while taking LAMICTAL XR is not recommended.

Tell your healthcare provider about all the medicines you take or if you are planning to take a new medicine, including prescription and non-prescription medicines, vitamins, and herbal supplements. Using LAMICTAL XR with certain other medicines can affect each other, causing side effects.

How should I take LAMICTAL XR?
- Take LAMICTAL XR exactly as prescribed.
- Your healthcare provider may change your dose. Do not change your dose without talking to your healthcare provider.
- Do not stop taking LAMICTAL XR without talking to your healthcare provider. Stopping LAMICTAL XR suddenly may cause serious problems. For example, if you have epilepsy and you stop taking LAMICTAL XR suddenly, you may get seizures that do not stop. Talk with your healthcare provider about how to stop LAMICTAL XR slowly.
- If you miss a dose of LAMICTAL XR, take it as soon as you remember. If it is almost time for your next dose, just skip the missed dose. Take the next dose at your regular time. **Do not take two doses at the same time.**
- You may not feel the full effect of LAMICTAL XR for several weeks.
- If you have epilepsy, tell your healthcare provider if your seizures get worse or if you have any new types of seizures.
- LAMICTAL XR can be taken with or without food.
- Do not chew, crush, or divide LAMICTAL XR.
- Swallow LAMICTAL XR tablets whole.
- If you have trouble swallowing LAMICTAL XR Tablets, tell your healthcare provider because there may be another form of LAMICTAL you can take.
- If you receive LAMICTAL XR in a blisterpack, examine the blisterpack before use. Do not use if blisters are torn, broken, or missing.

What should I avoid while taking LAMICTAL XR?
- Do not drive a car or operate complex, hazardous machinery until you know how LAMICTAL XR affects you.

What are possible side effects of LAMICTAL XR?
- See "What is the most important information I should know about LAMICTAL XR?"

Common side effects of LAMICTAL XR include:
- dizziness
- diarrhea
- weakness or fatigue
- difficulty with coordination or balance
- tremor
- hot flashes
- double vision

Continued on next page

Product information on these pages is effective as of June 2009. Further information is available at 1-888-825-5249 or www.gsk.com.

Lamictal XR—Cont.

- nausea
- depression
- muscle ache
- nervousness

Other common side effects that have been reported with another form of LAMICTAL include headache, sleepiness, and vomiting.

Tell your healthcare provider about any side effect that bothers you or that does not go away. These are not all the possible side effects of LAMICTAL XR. For more information, ask your healthcare provider or pharmacist.

Call your doctor for medical advice about side effects. You may report side effects to FDA at 1-800-FDA-1088.

How should I store LAMICTAL XR?

- Store LAMICTAL XR at room temperature between 59° F to 86° F (15° C to 30° C).
- **Keep LAMICTAL XR and all medicines out of the reach of children.**

General information about LAMICTAL XR

Medicines are sometimes prescribed for purposes other than those listed in a Medication Guide.

Do not use LAMICTAL XR for a condition for which it was not prescribed. Do not give LAMICTAL XR to other people, even if they have the same symptoms you have. It may harm them.

This Medication Guide summarizes the most important information about LAMICTAL XR. If you would like more information, talk with your healthcare provider. You can ask your healthcare provider or pharmacist for information about LAMICTAL XR that is written for healthcare professionals.

For more information, go to www.lamictalxr.com or call 1-888-825-5249.

What are the ingredients in LAMICTAL XR?

Active ingredient: Lamotrigine.

Inactive ingredients: glycerol monostearate, hypromellose, lactose monohydrate, magnesium stearate, methacrylic acid copolymer dispersion, polyethylene glycol 400, polysorbate 80, silicon dioxide (25 mg and 50 mg tablets only), titanium dioxide, triethyl citrate, iron oxide black (50 mg tablet only), iron oxide yellow (25 mg, 50 mg, 100 mg tablets only), iron oxide red (100 mg tablet only), FD&C Blue No. 2 Aluminum Lake (200 mg tablet only). Tablets are printed with edible black ink.

This Medication Guide has been approved by the U.S. Food and Drug Administration.

May 2009 LXR:1MG

GlaxoSmithKline
Research Triangle Park, NC 27709

LAMICTAL XR and DiffCORE are trademarks of GlaxoSmithKline.

DEPAKENE and DEPAKOTE are registered trademarks of Abbott Laboratories.

King Pharmaceuticals®, Inc.

501 FIFTH STREET
BRISTOL, TN 37620

Direct Inquiries to:
Customer Service:
Tel: 1 (888) 358-6436
Fax: 1 (866) 990-0545
To report an Adverse Drug Experience:
Tel: 1 (800) 546-4905
Fax: 1 (423) 990-8351
www.kingpharm.com

FLECTOR® PATCH ℞

(diclofenac epolamine topical patch)
1.3%
Rx Only

Cardiovascular Risk

- NSAIDs may cause an increased risk of serious cardiovascular thrombotic events, myocardial infarction, and stroke, which can be fatal. This risk may increase with duration of use. Patients with cardiovascular disease or risk factors for cardiovascular disease may be at greater risk (See **WARNINGS** and **CLINICAL TRIALS**).
- Flector® Patch is contraindicated for the treatment of peri-operative pain in the setting of coronary artery bypass graft (CABG) surgery (see **WARNINGS**).

Gastrointestinal Risk

- NSAIDs cause an increased risk of serious gastrointestinal adverse events including bleeding, ulceration, and perforation of the stomach or intestines, which can be fatal. These events can occur at any time during use and without warning symptoms. Elderly patients are at greater risk for serious gastrointestinal events (See **WARNINGS**).

DESCRIPTION

Flector® Patch (10 cm x 14 cm) is comprised of an adhesive material containing 1.3% diclofenac epolamine which is applied to a non-woven polyester felt backing and covered with a polypropylene film release liner. The release liner is removed prior to topical application to the skin.

Diclofenac epolamine is a non-opioid analgesic chemically designated as 2-[(2,6-dichlorophenyl) amino]benzeneacetic acid, (2-(pyrrolidin-1-yl) ethanol salt, with a molecular formula of $C_{20}H_{24}Cl_2N_2O_3$ (molecular weight 411.3), an n-octanol/water partition coefficient of 8 at pH 8.5, and the following structure:

DICLOFENAC

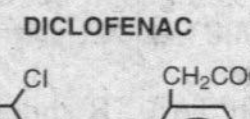
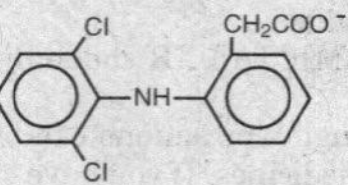

EPOLAMINE

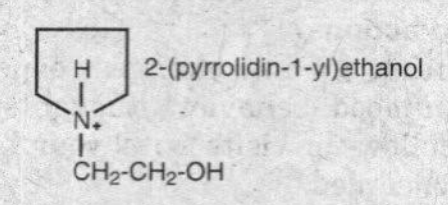

Each adhesive patch contains 180 mg of diclofenac epolamine (13 mg per gram adhesive) in an aqueous base. It also contains the following inactive ingredients: 1,3-butylene glycol, dihydroxyaluminum aminoacetate, disodium edetate, D-sorbitol, fragrance (Dalin PH), gelatin, kaolin, methylparaben, polysorbate 80, povidone, propylene glycol, propylparaben, sodium carboxymethylcellulose, sodium polyacrylate, tartaric acid, titanium dioxide, and purified water.

CLINICAL PHARMACOLOGY

Pharmacodynamics

Flector® Patch applied to intact skin provides local analgesia by releasing diclofenac epolamine from the patch into the skin. Diclofenac is a nonsteroidal anti-inflammatory drug (NSAID). In pharmacologic studies, diclofenac has shown anti-inflammatory, analgesic, and antipyretic activity. As with other NSAIDs, its mode of action is not known; its ability to inhibit prostaglandin synthesis, however, may be involved in its anti-inflammatory activity, as well as contribute to its efficacy in relieving pain associated with inflammation.

Pharmacokinetics

Absorption

Following a single application of the Flector® Patch on the upper inner arm, peak plasma concentrations of diclofenac (range 0.7–6 ng/mL) were noted between 10–20 hours of application. Plasma concentrations of diclofenac in the range of 1.3–8.8 ng/mL were noted after five days with twice-a-day Flector® Patch application.

Systemic exposure (AUC) and maximum plasma concentrations of diclofenac, after repeated dosing for four days with Flector® Patch, were lower (<1%) than after a single oral 50-mg diclofenac sodium tablet.

The pharmacokinetics of Flector® Patch has been tested in healthy volunteers at rest or undergoing moderate exercise (cycling 20 min/h for 12 h at a mean HR of 100.3 bpm). No clinically relevant differences in systemic absorption were observed, with peak plasma concentrations in the range of 2.2–8.1 ng/mL while resting, and 2.7–7.2 ng/mL during exercise.

Distribution

Diclofenac has a very high affinity (>99%) for human serum albumin.

Metabolism and Excretion

The plasma elimination half-life of diclofenac after application of Flector® Patch is approximately 12 hours. Diclofenac is eliminated through metabolism and subsequent urinary and biliary excretion of the glucuronide and the sulfate conjugates of the metabolites.

CLINICAL STUDIES

Efficacy of Flector® Patch was demonstrated in two of four studies of patients with minor sprains, strains, and contusions. Patients were randomly assigned to treatment with the Flector® Patch, or a placebo patch identical to the Flector® Patch minus the active ingredient. In the first of these two studies, patients with ankle sprains were treated once daily for a week. In the second study, patients with sprains, strains and contusions were treated twice daily for up to two weeks. Pain was assessed over the period of treatment. Patients treated with the Flector® Patch experienced a greater reduction in pain as compared to patients randomized to placebo patch as evidenced by the responder curves presented below.

Figure 1: Patients Achieving Various Levels of Pain Relief at Day 3; 14-Day Study

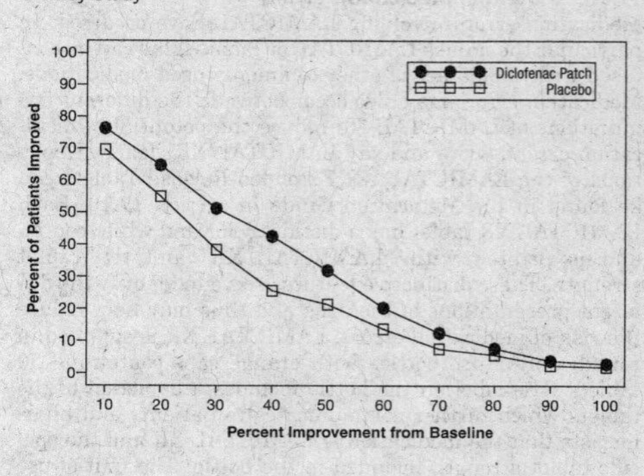

Figure 2: Patients Achieving Various Levels of Pain Relief at End of Study; 14-Day Study

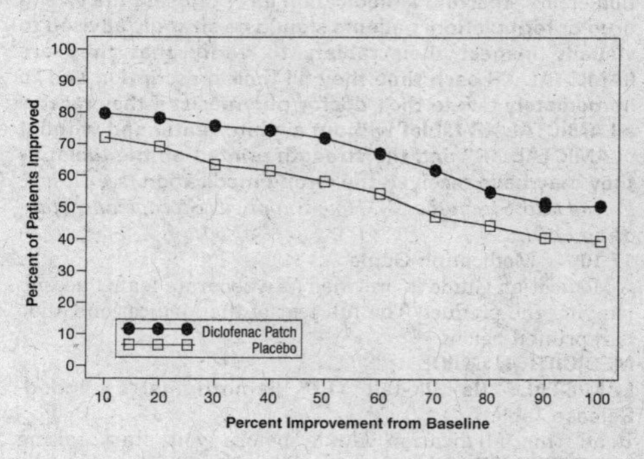

Figure 3: Patients Achieving Various Levels of Pain Relief at Day 3; 7-Day Study

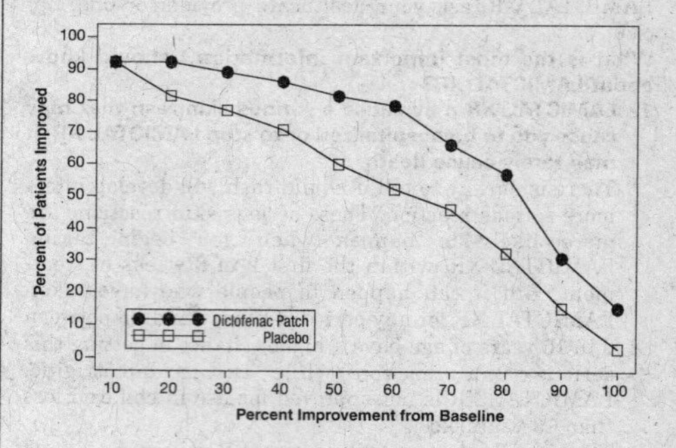

Figure 4: Patients Achieving Various Levels of Pain Relief at End of Study; 7-Day Study

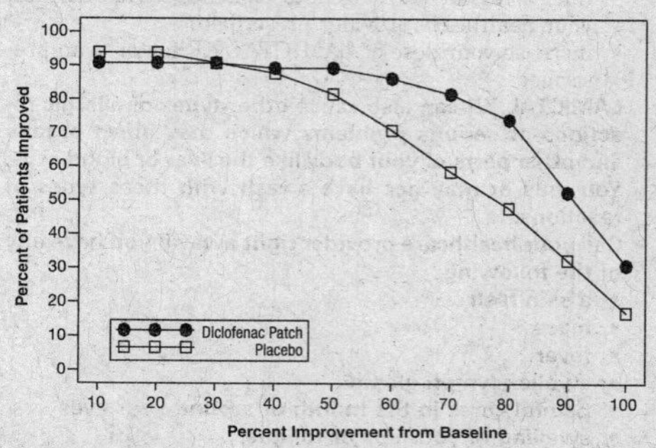

INDICATION AND USAGE

Carefully consider the potential benefits and risks of Flector® Patch and other treatment options before deciding

to use Flector® Patch. Use the lowest effective dose for the shortest duration consistent with individual patient treatment goals (see **WARNINGS**).
Flector® Patch is indicated for the topical treatment of acute pain due to minor strains, sprains, and contusions.

CONTRAINDICATIONS

Flector® Patch is contraindicated in patients with known hypersensitivity to diclofenac.
Flector® Patch should not be given to patients who have experienced asthma, urticaria, or allergic-type reactions after taking aspirin or other NSAIDs. Severe, rarely fatal, anaphylactic-like reactions to NSAIDs have been reported in such patients (see **WARNINGS - Anaphylactoid Reactions,** and **PRECAUTIONS - Preexisting Asthma**).
Flector® Patch is contraindicated for the treatment of perioperative pain in the setting of coronary artery bypass graft (CABG) surgery (see **WARNINGS**).
Flector® Patch should not be applied to non-intact or damaged skin resulting from any etiology e.g. exudative dermatitis, eczema, infected lesion, burns or wounds.

WARNINGS

CARDIOVASCULAR EFFECTS

Cardiovascular Thrombotic Events

Clinical trials of several COX-2 selective and nonselective NSAIDs of up to three years duration have shown an increased risk of serious cardiovascular (CV) thrombotic events, myocardial infarction, and stroke, which can be fatal. All NSAIDs, both COX-2 selective and nonselective, may have a similar risk. Patients with known CV disease or risk factors for CV disease may be at greater risk. To minimize the potential risk for an adverse CV event in patients treated with an NSAID, the lowest effective dose should be used for the shortest duration possible. Physicians and patients should remain alert for the development of such events, even in the absence of previous CV symptoms. Patients should be informed about the signs and/or symptoms of serious CV events and the steps to take if they occur.
There is no consistent evidence that concurrent use of aspirin mitigates the increased risk of serious CV thrombotic events associated with NSAID use. The concurrent use of aspirin and an NSAID does increase the risk of serious GI events (see **GI WARNINGS**).
Two large, controlled, clinical trials of a COX-2 selective NSAID for the treatment of pain in the first 10-14 days following CABG surgery found an increased incidence of myocardial infarction and stroke (see **CONTRAINDICATIONS**).

Hypertension

NSAIDs, including Flector® Patch, can lead to onset of new hypertension or worsening of preexisting hypertension, either of which may contribute to the increased incidence of CV events. Patients taking thiazides or loop diuretics may have impaired response to these therapies when taking NSAIDs. NSAIDs, including Flector® Patch, should be used with caution in patients with hypertension. Blood pressure (BP) should be monitored closely during the initiation of NSAID treatment and throughout the course of therapy.

Congestive Heart Failure and Edema

Fluid retention and edema have been observed in some patients taking NSAIDs. Flector® Patch should be used with caution in patients with fluid retention or heart failure.

Gastrointestinal Effects- Risk of Ulceration, Bleeding, and Perforation

NSAIDs, including Flector® Patch, can cause serious gastrointestinal (GI) adverse events including inflammation, bleeding, ulceration, and perforation of the stomach, small intestine, or large intestine, which can be fatal. These serious adverse events can occur at any time, with or without warning symptoms, in patients treated with NSAIDs. Only one in five patients, who develop a serious upper GI adverse event on NSAID therapy, is symptomatic. Upper GI ulcers, gross bleeding, or perforation caused by NSAIDs occur in approximately 1% of patients treated for 3-6 months, and in about 2-4% of patients treated for one year. These trends continue with longer duration of use, increasing the likelihood of developing a serious GI event at some time during the course of therapy. However, even short-term therapy is not without risk.
NSAIDs should be prescribed with extreme caution in those with a prior history of ulcer disease or gastrointestinal bleeding. Patients with a *prior history of peptic ulcer disease and/or gastrointestinal bleeding* who use NSAIDs have a greater than 10-fold increased risk for developing a GI bleed compared to patients with neither of these risk factors. Other factors that increase the risk for GI bleeding in patients treated with NSAIDs include concomitant use of oral corticosteroids or anticoagulants, longer duration of NSAID therapy, smoking, use of alcohol, older age, and poor general health status. Most spontaneous reports of fatal GI events are in elderly or debilitated patients and therefore, special care should be taken in treating this population.
To minimize the potential risk for an adverse GI event in patients treated with an NSAID, the lowest effective dose should be used for the shortest possible duration. Patients and physicians should remain alert for signs and symptoms of GI ulceration and bleeding during NSAID therapy and promptly initiate additional evaluation and treatment if a serious GI adverse event is suspected. This should include discontinuation of the NSAID until a serious GI adverse event is ruled out. For high risk patients, alternate therapies that do not involve NSAIDs should be considered.

Renal Effects

Long-term administration of NSAIDs has resulted in renal papillary necrosis and other renal injury. Renal toxicity has also been seen in patients in whom renal prostaglandins have a compensatory role in the maintenance of renal perfusion. In these patients, administration of a nonsteroidal anti-inflammatory drug may cause a dose-dependent reduction in prostaglandin formation and, secondarily, in renal blood flow, which may precipitate overt renal decompensation. Patients at greatest risk of this reaction are those with impaired renal function, heart failure, liver dysfunction, those taking diuretics and ACE inhibitors, and the elderly. Discontinuation of NSAID therapy is usually followed by recovery to the pretreatment state.

Advanced Renal Disease

No information is available from controlled clinical studies regarding the use of Flector® Patch in patients with advanced renal disease. Therefore, treatment with Flector® Patch is not recommended in these patients with advanced renal disease. If Flector® Patch therapy is initiated, close monitoring of the patient's renal function is advisable.

Anaphylactoid Reactions

As with other NSAIDs, anaphylactoid reactions may occur in patients without known prior exposure to Flector® Patch. Flector® Patch should not be given to patients with the aspirin triad. This symptom complex typically occurs in asthmatic patients who experience rhinitis with or without nasal polyps, or who exhibit severe, potentially fatal bronchospasm after taking aspirin or other NSAIDs (see **CONTRAINDICATIONS** and **PRECAUTIONS - Preexisting Asthma**). Emergency help should be sought in cases where an anaphylactoid reaction occurs.

Skin Reactions

NSAIDs, including Flector® Patch, can cause serious skin adverse events such as exfoliative dermatitis, Stevens-Johnson Syndrome (SJS), and toxic epidermal necrolysis (TEN), which can be fatal. These serious events may occur without warning. Patients should be informed about the signs and symptoms of serious skin manifestations and use of the drug should be discontinued at the first appearance of skin rash or any other sign of hypersensitivity.

Pregnancy

In late pregnancy, as with other NSAIDs, Flector® Patch should be avoided because it may cause premature closure of the ductus arteriosus.

PRECAUTIONS

General

Flector® Patch cannot be expected to substitute for corticosteroids or to treat corticosteroid insufficiency. Abrupt discontinuation of corticosteroids may lead to disease exacerbation. Patients on prolonged corticosteroid therapy should have their therapy tapered slowly if a decision is made to discontinue corticosteroids.
The pharmacological activity of Flector® Patch in reducing inflammation may diminish the utility of these diagnostic signs in detecting complications of presumed noninfectious, painful conditions.

Hepatic Effects

Borderline elevations of one or more liver tests may occur in up to 15% of patients taking NSAIDs including Flector® Patch. These laboratory abnormalities may progress, may remain unchanged, or may be transient with continuing therapy. Notable elevations of ALT or AST (approximately three or more times the upper limit of normal) have been reported in approximately 1% of patients in clinical trials with NSAIDs. In addition, rare cases of severe hepatic reactions, including jaundice and fatal fulminant hepatitis, liver necrosis and hepatic failure, some of them with fatal outcomes have been reported.
A patient with symptoms and/or signs suggesting liver dysfunction, or in whom an abnormal liver test has occurred, should be evaluated for evidence of the development of a more severe hepatic reaction while on therapy with Flector® Patch. If clinical signs and symptoms consistent with liver disease develop, or if systemic manifestations occur (e.g. eosinophilia, rash, etc.), Flector® Patch should be discontinued.

Hematological Effects

Anemia is sometimes seen in patients receiving NSAIDs. This may be due to fluid retention, occult or gross GI blood loss, or an incompletely described effect upon erythropoiesis. Patients on long-term treatment with NSAIDs, including Flector® Patch, should have their hemoglobin or hematocrit checked if they exhibit any signs or symptoms of anemia.
NSAIDs inhibit platelet aggregation and have been shown to prolong bleeding time in some patients. Unlike aspirin, their effect on platelet function is quantitatively less, of shorter duration, and reversible. Patients receiving Flector® Patch who may be adversely affected by alterations in platelet function, such as those with coagulation disorders or patients receiving anticoagulants, should be carefully monitored.

Preexisting Asthma

Patients with asthma may have aspirin-sensitive asthma. The use of aspirin in patients with aspirin-sensitive asthma has been associated with severe bronchospasm which can be fatal. Since cross reactivity, including bronchospasm, between aspirin and other nonsteroidal anti-inflammatory drugs has been reported in such aspirin-sensitive patients, Flector® Patch should not be administered to patients with this form of aspirin sensitivity and should be used with caution in patients with preexisting asthma.

Eye Exposure

Contact of Flector® Patch with eyes and mucosa, although not studied, should be avoided. If eye contact occurs, immediately wash out the eye with water or saline. Consult a physician if irritation persists for more than an hour.

Accidental Exposure in Children

Even a used Flector® Patch contains a large amount of diclofenac epolamine (as much as 170 mg). The potential therefore exists for a small child or pet to suffer serious adverse effects from chewing or ingesting a new or used Flector® Patch. It is important for patients to store and dispose of Flector® Patch out of the reach of children and pets.

Information for Patients

Patients should be informed of the following information before initiating therapy with an NSAID and periodically during the course of ongoing therapy. Patients should also be encouraged to read the NSAID Medication Guide that accompanies each prescription dispensed.

1. Flector® Patch, like other NSAIDs, may cause serious CV side effects, such as MI or stroke, which may result in hospitalization and even death. Although serious CV events can occur without warning symptoms, patients should be alert for the signs and symptoms of chest pain, shortness of breath, weakness, slurring of speech, and should ask for medical advice when observing any indicative sign or symptoms. Patients should be apprised of the importance of this follow-up (see **WARNINGS, Cardiovascular Effects**).
2. Flector® Patch, like other NSAIDs, may cause GI discomfort and, rarely, serious GI side effects, such as ulcers and bleeding, which may result in hospitalization and even death. Although serious GI tract ulcerations and bleeding can occur without warning symptoms, patients should be alert for the signs and symptoms of ulcerations and bleeding, and should ask for medical advice when observing any indicative sign or symptoms including epigastric pain, dyspepsia, melena, and hematemesis. Patients should be apprised of the importance of this follow-up (see **WARNINGS, Gastrointestinal Effects: Risk of Ulceration, Bleeding, and Perforation**).
3. Flector® Patch, like other NSAIDs, may cause serious skin side effects such as exfoliative dermatitis, SJS, and TEN, which may result in hospitalizations and even death. Although serious skin reactions may occur without warning, patients should be alert for the signs and symptoms of skin rash and blisters, fever, or other signs of hypersensitivity such as itching, and should ask for medical advice when observing any indicative signs or symptoms. Patients should be advised to stop the drug immediately if they develop any type of rash and contact their physicians as soon as possible.
4. Patients should be instructed to promptly report signs or symptoms of unexplained weight gain or edema to their physicians (see **WARNINGS, Cardiovascular Effects**).
5. Patients should be informed of the warning signs and symptoms of hepatotoxicity (e.g., nausea, fatigue, lethargy, pruritus, jaundice, right upper quadrant tenderness, and "flu-like" symptoms). If these occur, patients should be instructed to stop therapy and seek immediate medical therapy.
6. Patients should be informed of the signs of an anaphylactoid reaction (e.g. difficulty breathing, swelling of the face or throat). If these occur, patients should be instructed to seek immediate emergency help (see **WARNINGS**).
7. In late pregnancy, as with other NSAIDs, Flector® Patch should be avoided because it may cause premature closure of the ductus arteriosus.
8. Patients should be advised not to use Flector® Patch if they have an aspirin-sensitive asthma. Flector® Patch, like other NSAIDs, could cause severe and even fatal bronchospasm in these patients (see **PRECAUTIONS, Preexisting asthma**). Patients should discontinue use of

Continued on next page

Flector Patch—Cont.

Flector® Patch and should immediately seek emergency help if they experience wheezing or shortness of breath.
9. Patients should be informed that Flector® Patch should be used only on intact skin.
10. Patients should be advised to avoid contact of Flector® Patch with eyes and mucosa. Patients should be instructed that if eye contact occurs, they should immediately wash out the eye with water or saline, and consult a physician if irritation persists for more than an hour.
11. Patients and caregivers should be instructed to wash their hands after applying, handling or removing the patch.
12. Patients should be informed that, if Flector® Patch begins to peel off, the edges of the patch may be taped down.
13. Patients should be instructed not to wear Flector® Patch during bathing or showering. Bathing should take place in between scheduled patch removal and application (see **DOSAGE AND ADMINISTRATION**).
14. Patients should be advised to store Flector® Patch and to discard used patches out of the reach of children and pets. If a child or pet accidentally ingests Flector® Patch, medical help should be sought immediately (see **PRECAUTIONS, Accidental Exposure in Children**).

Laboratory Tests
Because serious GI tract ulcerations and bleeding can occur without warning symptoms, physicians should monitor for signs or symptoms of GI bleeding. Patients on long-term treatment with NSAIDs, should have their CBC and a chemistry profile checked periodically. If clinical signs and symptoms consistent with liver or renal disease develop, systemic manifestations occur (e.g., eosinophilia, rash, etc.) or if abnormal liver tests persist or worsen, Flector® Patch should be discontinued.

Drug Interactions
ACE-inhibitors
Reports suggest that NSAIDs may diminish the antihypertensive effect of ACE-inhibitors. This interaction should be given consideration in patients taking NSAIDs concomitantly with ACE-inhibitors.

Aspirin
When Flector® Patch is administered with aspirin, the binding of diclofenac to protein is reduced, although the clearance of free diclofenac is not altered. The clinical significance of this interaction is not known; however, as with other NSAIDs, concomitant administration of diclofenac and aspirin is not generally recommended because of the potential of increased adverse effects.

Diuretics
Clinical studies, as well as post marketing observations, have shown that Flector® Patch may reduce the natriuretic effect-of furosemide and thiazides in some patients. This response has been attributed to inhibition of renal prostaglandin synthesis. During concomitant therapy with NSAIDs, the patient should be observed closely for signs of renal failure (see **WARNINGS, Renal Effects**), as well as to assure diuretic efficacy.

Lithium
NSAIDs have produced an elevation of plasma lithium levels and a reduction in renal lithium clearance. The mean minimum lithium concentration increased 15% and the renal clearance was decreased by approximately 20%. These effects have been attributed to inhibition of renal prostaglandin synthesis by the NSAID. Thus, when NSAIDs and lithium are administered concurrently, subjects should be observed carefully for signs of lithium toxicity.

Methotrexate
NSAIDs have been reported to competitively inhibit methotrexate accumulation in rabbit kidney slices. This may indicate that they could enhance the toxicity of methotrexate. Caution should be used when NSAIDs are administered concomitantly with methotrexate.

Warfarin
The effects of warfarin and NSAIDs on GI bleeding are synergistic, such that users of both drugs together have a risk of serious GI bleeding higher than users of either drug alone.

Carcinogenesis, Mutagenesis, Impairment of Fertility
Carcinogenesis
Long-term studies in animals have not been performed to evaluate the carcinogenic potential of either diclofenac epolamine or Flector® Patch.

Mutagenesis
Diclofenac epolamine is not mutagenic in *Salmonella Typhimurium* strains, nor does it induce an increase in metabolic aberrations in cultured human lymphocytes, or the frequency of micronucleated cells in the bone marrow micronucleus test performed in rats.

Impairment of Fertility
Male and female Sprague Dawley rats were administered 1, 3, or 6 mg/kg/day diclofenac epolamine via oral gavage (males treated for 60 days prior to conception and during mating period, females treated for 14 days prior to mating through day 19 of gestation). Diclofenac epolamine treatment with 6 mg/kg/day resulted in increased early resorptions and postimplantation losses; however, no effects on the mating and fertility indices were found. The 6 mg/kg/day dose corresponds to 3-times the maximum recommended daily exposure in humans based on a body surface area comparison.

Pregnancy
Teratogenic Effects. Pregnancy Category C.
Pregnant Sprague Dawley rats were administered 1, 3, or 6 mg/kg diclofenac epolamine via oral gavage daily from gestation days 6-15. Maternal toxicity, embryotoxicity, and increased incidence of skeletal anomalies were noted with 6 mg/kg/day diclofenac epolamine, which corresponds to 3-times the maximum recommended daily exposure in humans based on a body surface area comparison. Pregnant New Zealand White rabbits were administered 1, 3, or 6 mg/kg diclofenac epolamine via oral gavage daily from gestation days 6-18. No maternal toxicity was noted; however, embryotoxicity was evident at 6 mg/kg/day group which corresponds to 6.5-times the maximum recommended daily exposure in humans based on a body surface area comparison.
There are no adequate and well-controlled studies in pregnant women. Flector® Patch should be used during pregnancy only if the potential benefit justifies the potential risk to the fetus.

Nonteratogenic Effects
Because of the known effects of nonsteroidal anti-inflammatory drugs on the fetal cardiovascular system (closure of ductus arteriosus), use during pregnancy (particularly late pregnancy) should be avoided.
Male rats were orally administered diclofenac epolamine (1, 3, 6 mg/kg) for 60 days prior to mating and throughout the mating period, and females were given the same doses 14 days prior to mating and through mating, gestation, and lactation. Embryotoxicity was observed at 6 mg/kg diclofenac epolamine (3-times the maximum recommended daily exposure in humans based on a body surface area comparison), and was manifested as an increase in early resorptions, post-implantation losses, and a decrease in live fetuses. The number of live born and total born were also reduced as was F1 postnatal survival, but the physical and behavioral development of surviving F1 pups in all groups was the same as the deionized water control, nor was reproductive performance adversely affected despite a slight treatment-related reduction in body weight.

Labor and Delivery
In rat studies with NSAIDs, as with other drugs known to inhibit prostaglandin synthesis, an increased incidence of dystocia, delayed parturition, and decreased pup survival occurred. The effects of Flector® Patch on labor and delivery in pregnant women are unknown.

Nursing Mothers
It is not known whether this drug is excreted in human milk. Because many drugs are excreted in human-milk and because of the potential for serious adverse reactions in nursing infants from Flector® Patch, a decision should be made whether to discontinue nursing or to discontinue the drug, taking into account the importance of the drug to the mother.

Pediatric Use
Safety and effectiveness in pediatric patients have not been established.

Geriatric Use
Clinical studies of Flector® Patch did not include sufficient numbers of subjects aged 65 and over to determine whether they respond differently from younger subjects. Other reported clinical experience has not identified differences in responses between the elderly and younger patients.
Diclofenac, as with any NSAID, is known to be substantially excreted by the kidney, and the risk of toxic reactions to Flector® Patch may be greater in patients with impaired renal function. Because elderly patients are more likely to have decreased renal function, care should be taken when using Flector® Patch in the elderly, and it may be useful to monitor renal function.

ADVERSE REACTIONS

In controlled trials during the premarketing development of Flector® Patch, approximately 600 patients with minor sprains, strains, and contusions have been treated with Flector® Patch for up to two weeks.

Adverse Events Leading to Discontinuation of Treatment
In the controlled trials, 3% of patients in both the Flector® Patch and placebo patch groups discontinued treatment due to an adverse event. The most common adverse events leading to discontinuation were application site reactions, occurring in 2% of both the Flector® Patch and placebo patch groups. Application site reactions leading to dropout included pruritus, dermatitis, and burning.

Common Adverse Events
Localized Reactions
Overall, the most common adverse events associated with Flector® Patch treatment were skin reactions at the site of treatment.
Table 1 lists all adverse events, regardless of causality, occurring in ≥ 1% of patients in controlled trials of Flector®

Table 1. Common Adverse Events (by body system and preferred term) in ≥ 1% of Patients treated with Flector® Patch or Placebo Patch[1]

	Diclofenac (N=572)		Placebo (N=564)	
	N	%	N	%
Application Site Conditions	64	11	70	12
Pruritus	31	5	44	8
Dermatitis	9	2	3	<1
Burning	2	<1	8	1
Other[2]	22	4	15	3
Gastrointestinal Disorders	49	9	33	6
Nausea	17	3	11	2
Dysgeusia	10	2	3	<1
Dyspepsia	7	1	8	1
Other[3]	15	3	11	2
Nervous System Disorders	13	2	18	3
Headache	7	1	10	2
Paresthesia	6	1	8	1
Somnolence	4	1	6	1
Other[4]	4	1	3	<1

[1] The table lists adverse events occurring in placebo-treated patients because the placebo-patch was comprised of the same ingredients as Flector® Patch except for diclofenac. Adverse events in the placebo group may therefore reflect effects of the non-active ingredients.
[2] Includes: application site dryness, irritation, erythema, atrophy, discoloration, hyperhidriosis, and vesicles.
[3] Includes: gastritis, vomiting, diarrhea, constipation, upper abdominal pain, and dry mouth.
[4] Includes: hypoaesthesia, dizziness, and hyperkinesias.

Patch. A majority of patients treated with Flector® Patch had adverse events with a maximum intensity of "mild" or "moderate."

[See table 1 at bottom of previous page]

Foreign labeling describes that dermal allergic reactions may occur with Flector® Patch treatment. Additionally, the treated area may become irritated or develop itching, erythema, edema, vesicles, or abnormal sensation.

DRUG ABUSE AND DEPENDENCE

Controlled Substance Class

Flector® Patch is not a controlled substance.

Physical and Psychological Dependence

Diclofenac, the active ingredient in Flector® Patch, is an NSAID that does not lead to physical or psychological dependence.

OVERDOSAGE

There is limited experience with overdose of Flector® Patch. In clinical studies, the maximum single dose administered was one Flector® Patch containing 180 mg of diclofenac epolamine. There were no serious adverse events.

Should systemic side effects occur due to incorrect use or accidental overdose of this product, the general measures recommended for intoxication with non-steroidal anti-inflammatory drugs should be taken.

DOSAGE AND ADMINISTRATION

Carefully consider the potential benefits and risks of Flector® Patch and other treatment options before deciding to use Flector® Patch. Use the lowest effective dose for the shortest duration consistent with individual patient treatment goals (see **WARNINGS**).

The recommended dose of Flector® Patch is one (1) patch to the most painful area twice a day.

Flector® Patch should not be applied to damaged or non-intact skin.

Flector® Patch should not be worn when bathing or showering.

HANDLING AND DISPOSAL

Patients and caregivers should wash their hands after applying, handling or removing the patch. Eye contact should be avoided.

HOW SUPPLIED

The Flector® Patch is supplied in resealable envelopes, each containing 5 patches (10 cm × 14 cm), with 6 envelopes per box (NDC 63857-111-33). Each individual patch is embossed with "FLECTOR PATCH <DICLOFENAC EPOLAMINE TOPICAL PATCH> 1.3%".

- Each patch contains 180 mg of diclofenac epolamine in an aqueous base (13 mg of active per gram of adhesive or 1.3%).
- The product is intended for topical use only.
- Keep out of reach of children and pets.
- The ENVELOPES SHOULD BE SEALED AT ALL TIMES WHEN NOT IN USE.
- Store at 25°C (77°F); excursions permitted to 15°-30°C (59°-86°F). [See USP Controlled Room Temperature].

Distributed by: Alpharma Pharmaceuticals LLC
One New England Avenue, Piscataway, NJ 08854 USA
(Telephone: 1-888-840-8884) • www.FlectorPatch.com
Manufactured for: IBSA Institut Biochimique SA, CH-6903 Lugano, Switzerland
Manufactured by: Teikoku Seiyaku Co., Ltd., Sanbonmatsu, Kagawa 769-2695 Japan
Version June 2008
FI/161 1086
Ed III/01.09

Medication Guide for Non-Steroidal Anti-Inflammatory Drugs (NSAIDs)

(See the end of this Medication Guide for a list of prescription NSAID medicines.)

What is the most important information I should know about medicines called Non-Steroidal Anti-Inflammatory Drugs (NSAIDs)?

NSAID medicines may increase the chance of a heart attack or stroke that can lead to death. This chance increases:
- with longer use of NSAID medicines
- in people who have heart disease

NSAID medicines should never be used right before or after a heart surgery called a "coronary artery bypass graft" (CABG).

NSAID medicines can cause ulcers and bleeding in the stomach and intestines at any time during treatment. Ulcers and bleeding:
- can happen without warning symptoms
- may cause death

The chance of a person getting an ulcer or bleeding increases with:
- taking medicines called "corticosteroids" and "anticoagulants"
- longer use
- smoking
- drinking alcohol
- older age
- having poor health

NSAID medicines should only be used:
- exactly as prescribed
- at the lowest dose possible for your treatment
- for the shortest time needed

What are Non-Steroidal Anti-Inflammatory Drugs (NSAIDs)?

NSAID medicines are used to treat pain and redness, swelling and heat (inflammation) from medical conditions such as:
- different types of arthritis
- menstrual cramps and other types of short-term pain

Who should not take a Non-Steroidal Anti-Inflammatory Drug (NSAID)?

Do not take an NSAID medicine:
- if you had an asthma attack, hives, or other allergic reaction with aspirin or any other NSAID medicine
- for pain right before or after heart bypass surgery

Tell your healthcare provider:
- about all of your medical conditions.
- about all of the medicines you take. NSAIDs and some other medicines can interact with each other and cause serious side effects. **Keep a list of your medicines to show to your healthcare provider and pharmacist.**
- if you are pregnant. **NSAID medicines should not be used by pregnant women late in their pregnancy.**
- if you are breastfeeding. **Talk to your doctor.**

What are the possible side effects of Non-Steroidal Anti-Inflammatory Drugs (NSAIDs)?

Serious side effects include:
- heart attack
- stroke
- high blood pressure
- heart failure from body swelling (fluid retention)
- kidney problems including kidney failure
- bleeding and ulcers in the stomach and intestine
- low red blood cells (anemia)
- life-threatening skin reactions
- life-threatening allergic reactions
- liver problems including liver failure
- asthma attacks in people who have asthma

Other side effects include:
- stomach pain
- constipation
- diarrhea
- gas
- heartburn
- nausea
- vomiting
- dizziness

Get emergency help right away if you have any of the following symptoms:
- shortness of breath or trouble breathing
- chest pain
- weakness in one part or side of your body
- slurred speech
- swelling of the face or throat

Stop your NSAID medicine and call your healthcare provider right away if you have any of the following symptoms:
- nausea
- more tired or weaker than usual
- itching
- your skin or eyes look yellow
- stomach pain
- flu-like symptoms
- vomit blood
- there is blood in your bowel movement or it is black and sticky like tar
- unusual weight gain
- skin rash or blisters with fever
- swelling of the arms and legs, hands and feet

These are not all the side effects with NSAID medicines. Talk to your healthcare provider or pharmacist for more information about NSAID medicines.

Call your doctor for medical advice about side effects. You may report side effects to FDA at 1-800-FDA-1088.

Other information about Non-Steroidal Anti-Inflammatory Drugs (NSAIDs)
- Aspirin is an NSAID medicine but it does not increase the chance of a heart attack. Aspirin can cause bleeding in the brain, stomach, and intestines. Aspirin can also cause ulcers in the stomach and intestines.
- Some of these NSAID medicines are sold in lower doses without a prescription (over–the–counter). Talk to your healthcare provider before using over–the–counter NSAIDs for more than 10 days.

NSAID medicines that need a prescription

Generic Name	Tradename
Celecoxib	Celebrex
Diclofenac	Flector, Cataflam, Voltaren, Arthrotec (combined with misoprostol)
Diflunisal	Dolobid
Etodolac	Lodine, Lodine XL
Fenoprofen	Nalfon, Nalfon 200
Flurbirofen	Ansaid
Ibuprofen	Motrin, Tab-Profen, Vicoprofen (combined with hydrocodone), Combunox (combined with oxycodone)
Indomethacin	Indocin, Indocin SR, Indo-Lemmon, Indomethagan
Ketoprofen	Oruvail
Ketorolac	Toradol
Mefenamic Acid	Ponstel
Meloxicam	Mobic
Nabumetone	Relafen
Naproxen	Naprosyn, Anaprox, Anaprox DS, EC-Naproxyn, Naprelan, Naprapac (copackaged with lansoprazole)
Oxaprozin	Daypro
Piroxicam	Feldene
Sulindac	Clinoril
Tolmetin	Tolectin, Tolectin DS, Tolectin 600

This Medication Guide has been approved by the U.S. Food and Drug Administration.

Distributed by: Alpharma Pharmaceuticals LLC
One New England Avenue, Piscataway, NJ 08854 USA
(Telephone: 1-888-840-8884) • www.FlectorPatch.com
Manufactured for: IBSA Institut Biochimique SA, CH-6903 Lugano, Switzerland
Manufactured by: Teikoku Seiyaku Co., Ltd., Sanbonmatsu, Kagawa 769-2695 Japan
Version January 2009
FI/161 1086
Ed III/01.09

PriCara®

Division of Ortho-McNeil-Janssen Pharmaceuticals, Inc.
1000 ROUTE 202
RARITAN, NJ 08869

For Medical Information Contact:
(800) 682-6532
In Emergencies:
(908) 218-7325
For Patient Education Materials Contact:
(877) 323-2200
For Customer Service (Sales and Ordering):
(800) 631-5273

NUCYNTA™ Ⓒ ℞

[new-sinn'-tah]
(tapentadol)
Immediate-release oral tablets C-II

HIGHLIGHTS OF PRESCRIBING INFORMATION

These highlights do not include all the information needed to use NUCYNTA™ safely and effectively. See full prescribing information for NUCYNTA™.

NUCYNTA™ (tapentadol) immediate-release oral tablets C-II

Initial U.S. Approval: 2008

------------RECENT MAJOR CHANGES------------

Warnings and Precautions (5.4) March 2009

------------INDICATIONS AND USAGE------------

NUCYNTA™ is an opioid analgesic indicated for the relief of moderate to severe acute pain in patients 18 years of age or older. (1)

Continued on next page

Nucynta—Cont.

----------DOSAGE AND ADMINISTRATION----------
- As with many centrally-acting analgesic medications, the dosing regimen of NUCYNTA™ should be individualized according to the severity of pain being treated, the previous experience with similar drugs and the ability to monitor the patient. (2)
- Initiate NUCYNTA™ with or without food at a dose of 50 mg, 75 mg, or 100 mg every 4 to 6 hours depending upon pain intensity. On the first day of dosing, the second dose may be administered as soon as one hour after the first dose, if adequate pain relief is not attained with the first dose. Subsequent dosing is 50 mg, 75 mg, or 100 mg every 4 to 6 hours and should be adjusted to maintain adequate analgesia with acceptable tolerability. Daily doses greater than 700 mg on the first day of therapy and 600 mg on subsequent days have not been studied and are, therefore, not recommended. (2)

---------DOSAGE FORMS AND STRENGTHS---------
Tablets: 50 mg, 75 mg, 100 mg (3)

----------------CONTRAINDICATIONS----------------
- Impaired pulmonary function (significant respiratory depression, acute or severe bronchial asthma or hypercapnia in unmonitored settings or the absence of resuscitative equipment) (4.1)
- Paralytic ileus (4.2)
- Concomitant use with monoamine oxidase inhibitors (MAOI) or use within 14 days (4.3)

-----------WARNINGS AND PRECAUTIONS-----------
- Respiratory depression: Increased risk in elderly, debilitated patients, those suffering from conditions accompanied by hypoxia, hypercapnia, or upper airway obstruction. (5.1)
- CNS effects: Additive CNS depressive effects when used in conjunction with alcohol, other opioids, or illicit drugs. (5.2)
- Elevation of intracranial pressure: May be markedly exaggerated in the presence of head injury, other intracranial lesions. (5.3)
- Abuse potential may occur. Monitor patients closely for signs of abuse and addiction. (5.4)
- Impaired mental/physical abilities: Caution must be used with potentially hazardous activities. (5.5)
- Seizures: Use with caution in patients with a history of seizures. (5.7)
- Serotonin Syndrome: Potentially life-threatening condition could result from concomitant serotonergic administration. (5.8)

---------------ADVERSE REACTIONS---------------
The most common adverse events were nausea, dizziness, vomiting and somnolence. (6)

To report SUSPECTED ADVERSE REACTIONS, contact PriCara, Division of Ortho-McNeil-Janssen Pharmaceuticals, Inc. at 1-800-526-7736 or FDA at 1-800-FDA-1088 or *www.fda.gov/medwatch*.

---------------DRUG INTERACTIONS---------------
- Use NUCYNTA™ with caution in patients currently using specified centrally-acting drugs or alcohol. (7.3)
- Do not use NUCYNTA™ in patients currently using or within 14 days of using a monoamine oxidase inhibitor (MAOI). (7.4)

---------USE IN SPECIFIC POPULATIONS---------
- Labor and delivery: should not use during and immediately prior to labor and delivery. Monitor neonates, whose mothers have been taking NUCYNTA™, for respiratory depression. (8.2)
- Nursing mothers: should not breast-feed. (8.3)
- Pediatric use: safety and effectiveness not established in patients less than 18 years of age. (8.4)
- Renal or hepatic impairment: not recommended in patients with severe renal or hepatic impairment. Use with caution in patients with moderate hepatic impairment. (8.6, 8.7)
- Elderly: care should be taken when selecting an initial dose. (2.3)

See 17 for PATIENT COUNSELING INFORMATION AND MEDICATION GUIDE.

Revised: 03/2009

FULL PRESCRIBING INFORMATION: CONTENTS*

[* Sections or subsections omitted from the full prescribing information are not listed]

FULL PRESCRIBING INFORMATION

1 INDICATIONS AND USAGE

NUCYNTA™ (tapentadol) is indicated for the relief of moderate to severe acute pain in patients 18 years of age or older.

2 DOSAGE AND ADMINISTRATION

As with many centrally-acting analgesic medications, the dosing regimen should be individualized according to the severity of pain being treated, the previous experience with similar drugs and the ability to monitor the patient.

The dose is 50 mg, 75 mg, or 100 mg every 4 to 6 hours depending upon pain intensity.

On the first day of dosing, the second dose may be administered as soon as one hour after the first dose, if adequate pain relief is not attained with the first dose. Subsequent dosing is 50 mg, 75 mg, or 100 mg every 4 to 6 hours and should be adjusted to maintain adequate analgesia with acceptable tolerability.

Daily doses greater than 700 mg on the first day of therapy and 600 mg on subsequent days have not been studied and are not recommended.

NUCYNTA™ may be given with or without food *[see Clinical Pharmacology (12.3)]*.

2.1 Renal Impairment

No dosage adjustment is recommended in patients with mild or moderate renal impairment *[see Clinical Pharmacology (12.3)]*.

NUCYNTA™ has not been studied in patients with severe renal impairment. The use in this population is not recommended.

2.2 Hepatic Impairment

No dosage adjustment is recommended in patients with mild hepatic impairment *[see Clinical Pharmacology (12.3)]*.

NUCYNTA™ should be used with caution in patients with moderate hepatic impairment. Treatment in these patients should be initiated at 50 mg with the interval between doses no less than every 8 hours (maximum of three doses in 24 hours). Further treatment should reflect maintenance of analgesia with acceptable tolerability, to be achieved by either shortening or lengthening the dosing interval *[see Clinical Pharmacology (12.3)]*.

NUCYNTA™ has not been studied in patients with severe hepatic impairment and use in this population is not recommended *[see Warnings and Precautions (5.10)]*.

2.3 Elderly Patients

In general, recommended dosing for elderly patients with normal renal and hepatic function is the same as for younger adult patients with normal renal and hepatic function. Because elderly patients are more likely to have decreased renal and hepatic function, consideration should be given to starting elderly patients with the lower range of recommended doses.

3 DOSAGE FORMS AND STRENGTHS

NUCYNTA™ Tablets are round, biconvex and film-coated and are available in the following strengths, colors, and debossings: 50 mg of tapentadol (yellow with "O-M" on one side and "50" on the other side), 75 mg of tapentadol (yellow-orange with "O-M" on one side and "75" on the other side), and 100 mg of tapentadol (orange with "O-M" on one side and "100" on the other side).

4 CONTRAINDICATIONS

4.1 Impaired Pulmonary Function

Like other drugs with mu-opioid agonist activity, NUCYNTA™ is contraindicated in patients with significant respiratory depression in unmonitored settings or the absence of resuscitative equipment. NUCYNTA™ is also contraindicated in patients with acute or severe bronchial asthma or hypercapnia in unmonitored settings or the absence of resuscitative equipment *[see Warnings and Precautions (5.1)]*.

4.2 Paralytic Ileus

Like drugs with mu-opioid agonist activity, NUCYNTA™ is contraindicated in any patient who has or is suspected of having paralytic ileus.

4.3 Monoamine Oxidase Inhibitors

NUCYNTA™ is contraindicated in patients who are receiving monoamine oxidase (MAO) inhibitors or who have taken them within the last 14 days due to potential additive effects on norepinephrine levels which may result in adverse cardiovascular events *[see Drug Interactions (7.4)]*.

5 WARNINGS AND PRECAUTIONS

5.1 Respiratory Depression

Respiratory depression is the primary risk of mu-opioid agonists. Respiratory depression occurs more frequently in elderly or debilitated patients and in those suffering from conditions accompanied by hypoxia, hypercapnia, or upper airway obstruction, in whom even moderate therapeutic doses may significantly decrease pulmonary ventilation.

NUCYNTA™ should be administered with caution to patients with conditions accompanied by hypoxia, hypercapnia or decreased respiratory reserve such as: asthma, chronic obstructive pulmonary disease or cor pulmonale, severe obesity, sleep apnea syndrome, myxedema, kyphoscoliosis, central nervous system (CNS) depression, or coma. In such patients, even usual therapeutic doses of NUCYNTA™ may increase airway resistance and decrease respiratory drive to the point of apnea. Alternative non-mu-opioid agonist analgesics should be considered and NUCYNTA™ should be employed only under careful medical supervision at the lowest effective dose in such patients. If respiratory depression occurs, it should be treated as any mu-opioid agonist-induced respiratory depression *[see Overdosage (10.2)]*.

5.2 CNS Depression

Patients receiving other mu-opioid agonist analgesics, general anesthetics, phenothiazines, other tranquilizers, sedatives, hypnotics, or other CNS depressants (including alcohol) concomitantly with NUCYNTA™ may exhibit additive CNS depression. Interactive effects resulting in respiratory depression, hypotension, profound sedation, coma or death may result if these drugs are taken in combination with NUCYNTA™. When such combined therapy is contemplated, a dose reduction of one or both agents should be considered.

5.3 Head Injury and Increased Intracranial Pressure

Opioid analgesics can raise cerebrospinal fluid pressure as a result of respiratory depression with carbon dioxide retention. Therefore, NUCYNTA™ should not be used in patients who may be susceptible to the effects of raised cerebrospinal fluid pressure such as those with evidence of head injury and increased intracranial pressure. Opioid analgesics may obscure the clinical course of patients with head injury due to effects on pupillary response and consciousness. NUCYNTA™ should be used with caution in patients with head injury, intracranial lesions, or other sources of pre-existing increased intracranial pressure.

5.4 Misuse and Abuse

Tapentadol is a mu-opioid agonist and is a Schedule II controlled substance. Such drugs are sought by drug abusers and people with addiction disorders. Diversion of Schedule II products is an act subject to criminal penalty.

NUCYNTA™ can be abused in a manner similar to other opioid agonists, legal or illicit. This should be considered when prescribing or dispensing NUCYNTA™ in situations where the physician or pharmacist is concerned about an increased risk of misuse and abuse. Concerns about abuse and addiction should not prevent the proper management of pain. However, all patients treated with mu-opioid agonists require careful monitoring for signs of abuse and addiction, since use of mu-opioid agonist analgesic products carry the risk of addiction even under appropriate medical use *[see Drug Abuse and Dependence (9.2)]*.

NUCYNTA™ may be abused by crushing, chewing, snorting or injecting the product. These practices pose a significant risk to the abuser that could result in overdose and death *[see Drug Abuse and Dependence (9)]*.

5.5 Driving and Operating Machinery

Patients should be cautioned that NUCYNTA™ may impair the mental and/or physical abilities required for the performance of potentially hazardous tasks such as driving a car or operating machinery. This is to be expected especially at the beginning of treatment, at any change of dosage as well as in combination with alcohol or tranquilizers *[see Drug Interactions (7.3)]*.

5.6 Interactions with Alcohol and Drugs of Abuse

Due to its mu-opioid agonist activity, NUCYNTA™ may be expected to have additive effects when used in conjunction with alcohol, opioids, or illicit drugs that cause central nervous system depression, respiratory depression, hypotension, and profound sedation, coma or death *[see Drug Interactions (7.3)]*.

5.7 Seizures

NUCYNTA™ has not been systematically evaluated in patients with a seizure disorder, and such patients were excluded from clinical studies. NUCYNTA™ should be prescribed with care in patients with a history of a seizure disorder or any condition that would put the patient at risk of seizures.

5.8 Serotonin Syndrome Risk

The development of a potentially life-threatening serotonin syndrome may occur with use of Serotonin and Norepinephrine Reuptake Inhibitor (SNRI) products, including NUCYNTA™, particularly with concomitant use of serotonergic drugs such as Selective Serotonin Reuptake Inhibitors (SSRIs), SNRIs, tricyclic antidepressants (TCAs), MAOIs and triptans, and with drugs that impair metabolism of serotonin (including MAOIs). This may occur within the recommended dose. Serotonin syndrome may include mental-status changes (e.g., agitation, hallucinations, coma), autonomic instability (e.g., tachycardia, labile blood pressure, hyperthermia), neuromuscular aberrations (e.g., hyperreflexia, incoordination) and/or gastrointestinal symptoms (e.g., nausea, vomiting, diarrhea).

5.9 Withdrawal

Withdrawal symptoms may occur if NUCYNTA™ is discontinued abruptly. These symptoms may include: anxiety, sweating, insomnia, rigors, pain, nausea, tremors, diarrhea, upper respiratory symptoms, piloerection, and rarely, hallucinations. Withdrawal symptoms may be reduced by tapering NUCYNTA™ *[see Drug Abuse and Dependence (9.3)]*.

5.10 Hepatic Impairment

A study of NUCYNTA™ in subjects with hepatic impairment showed higher serum concentrations than in those with normal hepatic function. NUCYNTA™ should be used with caution in patients with moderate hepatic impairment *[see Dosage and Administration (2.2) and Clinical Pharmacology (12.3)]*.

NUCYNTA™ has not been studied in patients with severe hepatic impairment and, therefore, use in this population is not recommended.

5.11 Use in Pancreatic/Biliary Tract Disease

Like other drugs with mu-opioid agonist activity, NUCYNTA™ may cause spasm of the sphincter of Oddi and should be used with caution in patients with biliary tract disease, including acute pancreatitis.

6 ADVERSE REACTIONS

The following treatment-emergent adverse events are discussed in more detail in other sections of the labeling:

- Respiratory Depression *[see Contraindications (4.1) and Warnings and Precautions (5.1)]*
- CNS Depression *[see Warnings and Precautions (5.2)]*

Because clinical studies are conducted under widely varying conditions, adverse event rates observed in the clinical studies of a drug cannot be directly compared to rates in the clinical studies of another drug and may not reflect the rates observed in clinical practice. A treatment-emergent adverse event refers to any untoward medical event associated with the use of the drug in humans, whether or not considered drug-related.

Based on data from nine Phase 2/3 studies that administered multiple doses (seven placebo- and/or active-controlled, one noncontrolled and one Phase 3 active-controlled safety study) the most common adverse events (reported by ≥ 10% in any NUCYNTA™ dose group) were: nausea, dizziness, vomiting and somnolence.

The most common reasons for discontinuation due to adverse events in the studies described above (reported by ≥ 1% in any NUCYNTA™ dose group) were dizziness (2.6% vs. 0.5%), nausea (2.3% vs. 0.6%), vomiting (1.4% vs. 0.2%), somnolence (1.3% vs. 0.2%) and headache (0.9% vs. 0.2%) for NUCYNTA™- and placebo-treated patients, respectively.

Seventy-six percent of NUCYNTA™-treated patients from the nine studies experienced adverse events.

NUCYNTA™ was studied in multiple-dose, active- or placebo-controlled studies, or noncontrolled studies (n = 2178), in single-dose studies (n = 870), in open-label study extension (n = 483) and in Phase 1 studies (n = 597). Of these, 2034 patients were treated with doses of 50 mg to 100 mg of NUCYNTA™ dosed every 4 to 6 hours.

The data described below reflect exposure to NUCYNTA™ in 3161 patients, including 449 exposed for 45 days. NUCYNTA™ was studied primarily in placebo- and active-controlled studies (n = 2266, and n = 2944, respectively). The population was 18 to 85 years old (mean age 46 years), 68% were female, 75% white and 67% were postoperative. Most patients received NUCYNTA™ doses of 50 mg, 75 mg, or 100 mg every 4 to 6 hours.

6.1 Commonly-Observed Treatment-Emergent Adverse Events in Double-Blind Controlled Clinical Trials

Table 1 lists the adverse events reported in ≥ 1% or more of NUCYNTA™-treated patients with acute moderate to severe pain in the pooled safety data from nine Phase 2/3 studies that administered multiple doses (seven placebo- and/or active-controlled, one noncontrolled, and one Phase 3 active-controlled safety study).

Table 1 Treatment-Emergent Adverse Events* Reported by ≥ 1% of NUCYNTA™-Treated Patients In Seven Phase 2/3 Placebo- and/or Oxycodone-Controlled, One Noncontrolled, and One Phase 3 Oxycodone-Controlled Safety, Multiple-Dose Clinical Studies

System/Organ Class MedDRA Preferred Term	NUCYNTA™ 21 mg–120 mg (n=2178) %	Placebo (n=619) %
Gastrointestinal disorders		
Nausea	30	13
Vomiting	18	4
Constipation	8	3
Dry mouth	4	<1
Dyspepsia	2	<1
General disorders and administration site conditions		
Fatigue	3	<1
Feeling hot	1	<1
Infections and infestations		
Nasopharyngitis	1	<1
Upper respiratory tract infection	1	<1
Urinary tract infection	1	<1
Metabolism and nutrition disorders		
Decreased appetite	2	0
Musculoskeletal and connective tissue disorders		
Arthralgia	1	<1
Nervous system disorders		
Dizziness	24	8
Somnolence	15	3
Tremor	1	<1
Lethargy	1	<1
Psychiatric disorders		
Insomnia	2	<1
Confusional state	1	0
Abnormal dreams	1	<1
Anxiety	1	<1
Skin and subcutaneous tissue disorders		
Pruritus	5	1
Hyperhidrosis	3	<1
Pruritus generalized	3	<1
Rash	1	<1
Vascular disorders		
Hot flush	1	<1

*A treatment-emergent adverse event refers to any untoward medical event associated with the use of the drug in humans, whether or not considered drug-related.

6.2 Other Adverse Reactions Observed During the Premarketing Evaluation of NUCYNTA™

The following adverse drug reactions occurred in <1% of NUCYNTA™-treated patients in the pooled safety data from nine Phase 2/3 studies that administered multiple doses (seven were placebo- and/or active-controlled, one noncontrolled, and one Phase 3 active-controlled safety study):

Cardiac disorders: heart rate increased, heart rate decreased

Eye disorders: visual disturbance

Gastrointestinal disorders: abdominal discomfort, impaired gastric emptying

General disorders and administration site conditions: irritability, edema, drug withdrawal syndrome, feeling drunk

Immune system disorders: hypersensitivity

Investigations: gamma-glutamyltransferase increased, alanine aminotransferase increased, aspartate aminotransferase increased

Musculoskeletal and connective tissue disorders: involuntary muscle contractions, sensation of heaviness

Nervous system disorders: hypoesthesia, paresthesia, disturbance in attention, sedation, dysarthria, depressed level of consciousness, memory impairment, ataxia, presyncope, syncope, coordination abnormal, seizure

Psychiatric disorders: euphoric mood, disorientation, restlessness, agitation, nervousness, thinking abnormal

Renal and urinary disorders: urinary hesitation, pollakiuria

Respiratory, thoracic and mediastinal disorders: oxygen saturation decreased, cough, dyspnea, respiratory depression

Skin and subcutaneous tissue disorders: urticaria

Vascular disorders: blood pressure decreased

In the pooled safety data, the overall incidence of adverse reactions increased with increased dose of NUCYNTA™, as did the percentage of patients with adverse reactions of nausea, dizziness, vomiting, somnolence, and pruritus.

7 DRUG INTERACTIONS

NUCYNTA™ is mainly metabolized by glucuronidation. The following substances have been included in a set of interaction studies without any clinically significant finding: acetaminophen, acetylsalicylic acid, naproxen and probenecid *[see Clinical Pharmacology (12.3)]*.

The pharmacokinetics of tapentadol were not affected when gastric pH or gastrointestinal motility were increased by omeprazole and metoclopramide, respectively *[see Clinical Pharmacology (12.3)]*.

7.1 Drugs Metabolized by Cytochrome P450 Enzymes

In vitro investigations indicate that NUCYNTA™ does not inhibit or induce P450 enzymes. Thus, clinically relevant interactions mediated by the cytochrome P450 system are unlikely to occur *[see Clinical Pharmacology (12.3)]*.

Continued on next page

Nucynta—Cont.

7.2 Drugs That Inhibit or Induce Cytochrome P450 Enzymes

The major pathway of tapentadol metabolism is conjugation with glucuronic acid to produce glucuronides. To a lesser extent, tapentadol is additionally metabolized to N-desmethyl tapentadol (13%) by CYP2C9 and CYP2C19 to hydroxy tapentadol (2%) by CYP2D6, which are further metabolized by conjugation. Since only a minor amount of NUCYNTA™ is metabolized via the oxidative pathway clinically relevant interactions mediated by the cytochrome P450 system are unlikely to occur *[see Clinical Pharmacology (12.3)]*.

7.3 Centrally-Acting Drugs and Alcohol

Patients receiving other opioid agonist analgesics, general anesthetics, phenothiazines, antiemetics, other tranquilizers, sedatives, hypnotics, or other CNS depressants (including alcohol) concomitantly with NUCYNTA™ may exhibit an additive CNS depression. Interactive effects resulting in respiratory depression, hypotension, profound sedation, or coma may result if these drugs are taken in combination with NUCYNTA™. When such combined therapy is contemplated, a dose reduction of one or both agents should be considered *[see Warnings and Precautions (5.2) and (5.6)]*.

7.4 Monoamine Oxidase Inhibitors

NUCYNTA™ is contraindicated in patients who are receiving monoamine oxidase (MAO) inhibitors or who have taken them within the last 14 days due to potential additive effects on norepinephrine levels which may result in adverse cardiovascular events *[see Contraindications (4.3)]*.

8 USE IN SPECIFIC POPULATIONS

8.1 Pregnancy

Pregnancy Category C.

Tapentadol HCl was evaluated for teratogenic effects in pregnant rats and rabbits following intravenous and subcutaneous exposure during the period of embryofetal organogenesis. When tapentadol was administered twice daily by the subcutaneous route in rats at dose levels of 10, 20, or 40 mg/kg/day [producing up to 1 times the plasma exposure at the maximum recommended human dose (MRHD) of 700 mg/day based on an area under the time-curve (AUC) comparison], no teratogenic effects were observed. Evidence of embryofetal toxicity included transient delays in skeletal maturation (i.e. reduced ossification) at the 40 mg/kg/day dose which was associated with significant maternal toxicity. Administration of tapentadol HCl in rabbits at doses of 4, 10, or 24 mg/kg/day by subcutaneous injection [producing 0.2, 0.6, and 1.85 times the plasma exposure at the MRHD based on an AUC comparison] revealed embryofetal toxicity at doses ≥10 mg/kg/day. Findings included reduced fetal viability, skeletal delays and other variations. In addition, there were multiple malformations including gastroschisis/thoracogastroschisis, amelia/phocomelia, and cleft palate at doses ≥ 10 mg/kg/day and above, and ablepharia, encephalopathy, and spina bifida at the high dose of 24 mg/kg/day. Embryofetal toxicity, including malformations, may be secondary to the significant maternal toxicity observed in the study.

In a study of pre- and postnatal development in rats, oral administration of tapentadol at doses of 20, 50, 150, or 300 mg/kg/day to pregnant and lactating rats during the late gestation and early postnatal period [resulting in up to 1.7 times the plasma exposure at the MRHD on an AUC basis] did not influence physical or reflex development, the outcome of neurobehavioral tests or reproductive parameters. Treatment-related developmental delay was observed, including incomplete ossification, and significant reductions in pup body weights and body weight gains at doses associated with maternal toxicity (150 mg/kg/day and above). At maternal tapentadol doses ≥ 150 mg/kg/day, a dose-related increase in pup mortality was observed through postnatal Day 4.

There are no adequate and well controlled studies of NUCYNTA™ in pregnant women. NUCYNTA™ should be used during pregnancy only if the potential benefit justifies the potential risk to the fetus.

8.2 Labor and Delivery

The effect of tapentadol on labor and delivery in humans is unknown. NUCYNTA™ is not recommended for use in women during and immediately prior to labor and delivery. Due to the mu-opioid receptor agonist activity of NUCYNTA™, neonates whose mothers have been taking NUCYNTA™ should be monitored for respiratory depression. A specific opioid antagonist, such as naloxone, should be available for reversal of opioid induced respiratory depression in the neonate.

8.3 Nursing Mothers

There is insufficient/limited information on the excretion of tapentadol in human or animal breast milk. Physicochemical and available pharmacodynamic/toxicological data on tapentadol point to excretion in breast milk and risk to the suckling child cannot be excluded. NUCYNTA™ should not be used during breast-feeding.

8.4 Pediatric Use

The safety and effectiveness of NUCYNTA™ in pediatric patients less than 18 years of age have not been established. NUCYNTA™ is not recommended in this population.

8.5 Geriatric Use

Of the total number of patients in Phase 2/3 double-blind, multiple-dose clinical studies of NUCYNTA™, 19% were 65 and over, while 5% were 75 and over. No overall differences in effectiveness were observed between these patients and younger patients. The rate of constipation was higher in subjects greater than or equal to 65 years than those less than 65 years (12% vs. 7%).

In general, recommended dosing for elderly patients with normal renal and hepatic function is the same as for younger adult patients with normal renal and hepatic function. Because elderly patients are more likely to have decreased renal and hepatic function, consideration should be given to starting elderly patients with the lower range of recommended doses *[see Clinical Pharmacology (12.3)]*.

8.6 Renal Impairment

In patients with severe renal impairment, the safety and effectiveness of NUCYNTA™ has not been established. NUCYNTA™ is not recommended in this population *[see Dosage and Administration (2.1)]*.

8.7 Hepatic Impairment

Administration of NUCYNTA™ resulted in higher exposures and serum levels to tapentadol in subjects with impaired hepatic function compared to subjects with normal hepatic function *[see Clinical Pharmacology (12.3)]*. NUCYNTA™ should be used with caution in patients with moderate hepatic impairment *[see Dosage and Administration (2.2)]*.

NUCYNTA™ has not been studied in patients with severe hepatic impairment, therefore, use of NUCYNTA™ is not recommended in this population *[see Warnings and Precautions (5.10)]*.

9 DRUG ABUSE AND DEPENDENCE

9.1 Controlled Substance

NUCYNTA™ contains tapentadol, a mu-opioid agonist and is a Schedule II controlled substance. NUCYNTA™ has an abuse potential similar to hydromorphone, can be abused and is subject to criminal diversion.

9.2 Abuse

Addiction is a primary, chronic, neurobiologic disease, with genetic, psychosocial, and environmental factors influencing its development and manifestations. It is characterized by behaviors that include one or more of the following: impaired control over drug use, compulsive use, continued use despite harm, and craving. Drug addiction is a treatable disease, utilizing a multidisciplinary approach, but relapse is common.

Concerns about abuse and addiction should not prevent the proper management of pain. However, all patients treated with opioids require careful monitoring for signs of abuse and addiction, because use of opioid analgesic products carries the risk of addiction even under appropriate medical use.

"Drug seeking" behavior is very common in addicts, and drug abusers. Drug-seeking tactics include emergency calls or visits near the end of office hours, refusal to undergo appropriate examination, testing or referral, repeated claims of loss of prescriptions, tampering with prescriptions and reluctance to provide prior medical records or contact information for other treating physician(s). "Doctor shopping" (visiting multiple prescribers) to obtain additional prescriptions is common among drug abusers and people suffering from untreated addiction. Preoccupation with achieving adequate pain relief can be appropriate behavior in a patient with poor pain control.

Abuse and addiction are separate and distinct from physical dependence and tolerance. Physicians should be aware that addiction may not be accompanied by concurrent tolerance and symptoms of physical dependence in all addicts. In addition, abuse of mu-opioid agonists can occur in the absence of true addiction and is characterized by misuse for non-medical purposes, often in combination with other psychoactive substances. Careful recordkeeping of prescribing information, including quantity, frequency, and renewal requests is strongly advised.

Abuse of NUCYNTA™ poses a risk of overdose and death. This risk is increased with concurrent abuse of NUCYNTA™ with alcohol and other substances. In addition, parenteral drug abuse is commonly associated with transmission of infectious diseases such as hepatitis and HIV.

Proper assessment of the patient, proper prescribing practices, periodic re-evaluation of therapy, and proper dispensing and storage are appropriate measures that help to limit abuse of drugs with mu-opioid agonist properties.

Infants born to mothers physically dependent on opioids will also be physically dependent and may exhibit respiratory difficulties and withdrawal symptoms *[see Warnings and Precautions (5.1)]*. Use of NUCYNTA™ in this population has not been characterized. As NUCYNTA™ has mu-opioid agonist activity, infants whose mothers have taken NUCYNTA™, should be carefully monitored.

9.3 Dependence

Tolerance is the need for increasing doses of opioids to maintain a defined effect such as analgesia (in the absence of disease progression or other external factors). Physical dependence is manifested by withdrawal symptoms after abrupt discontinuation of a drug or upon administration of an antagonist.

The opioid abstinence or withdrawal syndrome is characterized by some or all of the following: restlessness, lacrimation, rhinorrhea, yawning, perspiration, chills, myalgia, and mydriasis. Other symptoms also may develop, including irritability, anxiety, backache, joint pain, weakness, abdominal cramps, insomnia, nausea, anorexia, vomiting, diarrhea, increased blood pressure, respiratory rate, or heart rate.

Generally, tolerance and/or withdrawal are more likely to occur the longer a patient is on continuous opioid therapy. In a safety study where drug was administered up to 90 days, 82.7% of patients taking NUCYNTA™ who stopped abruptly without initiating alternative therapy and were assessed 2 to 4 days after discontinuation, did not have objective signs of opioid withdrawal using the Clinical Opiate Withdrawal Scale. Moderate withdrawal symptoms were seen in 0.3% of patients with the rest (17%) experiencing mild symptoms. Withdrawal symptoms may be reduced by tapering NUCYNTA™.

10 OVERDOSAGE

10.1 Human Experience

Experience with NUCYNTA™ overdose is very limited. Preclinical data suggest that symptoms similar to those of other centrally acting analgesics with mu-opioid agonist activity are to be expected upon intoxication with tapentadol. In principle, these symptoms may particularly appear in the clinical setting: miosis, vomiting, cardiovascular collapse, consciousness disorders up to coma, convulsions and respiratory depression up to respiratory arrest.

10.2 Management of Overdose

Management of overdose should be focused on treating symptoms of mu-opioid agonism. Primary attention should be given to re-establishment of a patent airway and institution of assisted or controlled ventilation when overdose of NUCYNTA™ is suspected. Supportive measures (including oxygen and vasopressors) should be employed in the management of circulatory shock and pulmonary edema accompanying overdose as indicated. Cardiac arrest or arrhythmias may require cardiac massage or defibrillation.

Pure opioid antagonists, such as naloxone, are specific antidotes to respiratory depression resulting from opioid overdose. Respiratory depression following an overdose may outlast the duration of action of the opioid antagonist. Administration of an opioid antagonist is not a substitute for continuous monitoring of airway, breathing, and circulation following an opioid overdose. If the response to opioid antagonists is suboptimal or only brief in nature, an additional antagonist should be administered as directed by the manufacturer of the product.

Gastrointestinal decontamination may be considered in order to eliminate unabsorbed drug. Gastrointestinal decontamination with activated charcoal or by gastric lavage is only recommended within 2 hours after intake. Gastrointestinal decontamination at a later time point may be useful in case of intoxication with exceptionally large quantities. Before attempting gastrointestinal decontamination, care should be taken to secure the airway.

11 DESCRIPTION

NUCYNTA™ (tapentadol) Tablets are immediate-release film-coated tablets for oral administration. The chemical name is 3-[(1*R*,2*R*)-3-(dimethylamino)-1-ethyl-2-methylpropyl]phenol monohydrochloride. The structural formula is:

The molecular weight of tapentadol HCl is 257.80, and the molecular formula is $C_{14}H_{23}NO \bullet HCl$. The n-octanol:water partition coefficient log P value is 2.87. The pKa values are 9.34 and 10.45. In addition to the active ingredient tapentadol HCl, tablets also contain the following inactive ingredients: microcrystalline cellulose, lactose monohydrate, croscarmellose sodium, povidone, magnesium stearate, and Opadry® II, a proprietary film-coating mixture containing polyvinyl alcohol, titanium dioxide, polyethylene glycol, talc, and aluminum lake coloring.

12 CLINICAL PHARMACOLOGY

12.1 Mechanism of Action

Tapentadol is a centrally-acting synthetic analgesic. Although its exact mechanism is unknown, analgesic efficacy is thought to be due to mu-opioid agonist activity and the inhibition of norepinephrine reuptake.

12.2 Pharmacodynamics

Tapentadol is a centrally-acting synthetic analgesic. It is 18 times less potent than morphine in binding to the human mu-opioid receptor and is 2-3 times less potent in producing analgesia in animal models. Tapentadol has been shown to inhibit norepinephrine reuptake in the brains of rats resulting in increased norepinephrine concentrations. In preclinical models, the analgesic activity due to the mu-opioid receptor agonist activity of tapentadol can be antagonized by selective mu-opioid antagonists (e.g., naloxone), whereas the norepinephrine reuptake inhibition is sensitive to norepinephrine modulators. Tapentadol exerts its analgesic effects without a pharmacologically active metabolite.

Effects on the cardiovascular system: There was no effect of therapeutic and supratherapeutic doses of tapentadol on the QT interval. In a randomized, double-blind, placebo- and positive-controlled crossover study, healthy subjects were administered five consecutive doses of NUCYNTA™ 100 mg every 6 hours, NUCYNTA™ 150 mg every 6 hours, placebo and a single oral dose of moxifloxacin. Similarly, NUCYNTA™ had no relevant effect on other ECG parameters (heart rate, PR interval, QRS duration, T-wave or U-wave morphology).

12.3 Pharmacokinetics

Absorption

Mean absolute bioavailability after single-dose administration (fasting) is approximately 32% due to extensive first-pass metabolism. Maximum serum concentrations of tapentadol are typically observed at around 1.25 hours after dosing.

Dose-proportional increases in the C_{max} and AUC values of tapentadol have been observed over the 50 to 150 mg dose range.

A multiple (every 6 hour) dose study with doses ranging from 75 to 175 mg tapentadol showed a mean accumulation factor of 1.6 for the parent drug and 1.8 for the major metabolite tapentadol-O-glucuronide, which are primarily determined by the dosing interval and apparent half-life of tapentadol and its metabolite.

Food Effect

The AUC and C_{max} increased by 25% and 16%, respectively, when NUCYNTA™ was administered after a high-fat, high-calorie breakfast. NUCYNTA™ may be given with or without food.

Distribution

Tapentadol is widely distributed throughout the body. Following intravenous administration, the volume of distribution (Vz) for tapentadol is 540 +/- 98 L. The plasma protein binding is low and amounts to approximately 20%.

Metabolism and Elimination

In humans, the metabolism of tapentadol is extensive. About 97% of the parent compound is metabolized. Tapentadol is mainly metabolized via Phase 2 pathways, and only a small amount is metabolized by Phase 1 oxidative pathways. The major pathway of tapentadol metabolism is conjugation with glucuronic acid to produce glucuronides. After oral administration approximately 70% (55% O-glucuronide and 15% sulfate of tapentadol) of the dose is excreted in urine in the conjugated form. A total of 3% of drug was excreted in urine as unchanged drug. Tapentadol is additionally metabolized to N-desmethyl tapentadol (13%) by CYP2C9 and CYP2C19 and to hydroxy tapentadol (2%) by CYP2D6, which are further metabolized by conjugation. Therefore, drug metabolism mediated by cytochrome P450 system is of less importance than phase 2 conjugation.

None of the metabolites contributes to the analgesic activity.

Tapentadol and its metabolites are excreted almost exclusively (99%) via the kidneys. The terminal half-life is on average 4 hours after oral administration. The total clearance is 1530 +/- 177 ml/min.

Special Populations

Elderly

The mean exposure (AUC) to tapentadol was similar in elderly subjects compared to young adults, with a 16% lower mean C_{max} observed in the elderly subject group compared to young adult subjects.

Renal Impairment

AUC and C_{max} of tapentadol were comparable in subjects with varying degrees of renal function (from normal to severely impaired). In contrast, increasing exposure (AUC) to tapentadol-O-glucuronide was observed with increasing degree of renal impairment. In subjects with mild, moderate, and severe renal impairment, the AUC of tapentadol-O-glucuronide are 1.5-, 2.5-, and 5.5-fold higher compared with normal renal function, respectively.

Hepatic Impairment

Administration of NUCYNTA™ resulted in higher exposures and serum levels to tapentadol in subjects with impaired hepatic function compared to subjects with normal hepatic function. The ratio of tapentadol pharmacokinetic parameters for the mild and moderate hepatic impairment groups in comparison to the normal hepatic function group were 1.7 and 4.2, respectively, for AUC; 1.4 and 2.5, respectively, for C_{max}; and 1.2 and 1.4, respectively, for $t_{1/2}$. The rate of formation of tapentadol-O-glucuronide was lower in subjects with increased liver impairment.

Pharmacokinetic Drug Interactions

Tapentadol is mainly metabolized by Phase 2 glucuronidation, a high capacity/low affinity system, therefore, clinically relevant interactions caused by Phase 2 metabolism are unlikely to occur. Naproxen and probenecid increased the AUC of tapentadol by 17% and 57%, respectively. These changes are not considered clinically relevant and no change in dose is required.

No changes in the pharmacokinetic parameters of tapentadol were observed when acetaminophen and acetylsalicylic acid were given concomitantly.

In vitro studies did not reveal any potential of tapentadol to either inhibit or induce cytochrome P450 enzymes. Thus, clinically relevant interactions mediated by the cytochrome P450 system are unlikely to occur.

The pharmacokinetics of tapentadol were not affected when gastric pH or gastrointestinal motility were increased by omeprazole and metoclopramide, respectively.

Plasma protein binding of tapentadol is low (approximately 20%). Therefore, the likelihood of pharmacokinetic drug-drug interactions by displacement from the protein binding site is low.

13 NON-CLINICAL TOXICOLOGY

13.1 Carcinogenesis, Mutagenesis, Impairment of Fertility

Carcinogenesis

Tapentadol was administered to rats (diet) and mice (oral gavage) for two years.

In mice, tapentadol HCl was administered by oral gavage at dosages of 50, 100 and 200 mg/kg/day for 2 years (up to 0.2 times the plasma exposure at the maximum recommended human dose [MRHD] on an area under the time-curve [AUC] basis). No increase in tumor incidence was observed at any dose level.

In rats, tapentadol HCl was administered in diet at dosages of 10, 50, 125 and 250 mg/kg/day for two years (up to 0.2 times in the male rats and 0.6 times in the female rats the MRHD on an AUC basis). No increase in tumor incidence was observed at any dose level.

Mutagenesis

Tapentadol did not induce gene mutations in bacteria, but was clastogenic with metabolic activation in a chromosomal aberration test in V79 cells. The test was repeated and was negative in the presence and absence of metabolic activation. The one positive result for tapentadol was not confirmed *in vivo* in rats, using the two endpoints of chromosomal aberration and unscheduled DNA synthesis, when tested up to the maximum tolerated dose.

Impairment of Fertility

Tapentadol HCl was administered intravenously to male or female rats at dosages of 3, 6, or 12 mg/kg/day (representing exposures of up to approximately 0.4 times the exposure at the MRHD on an AUC basis, based on extrapolation from toxicokinetic analyses in a separate 4-week intravenous study in rats). Tapentadol did not alter fertility at any dose level. Maternal toxicity and adverse effects on embryonic development, including decreased number of implantations, decreased numbers of live conceptuses, and increased pre- and post-implantation losses occurred at dosages ≥ 6 mg/kg/day.

13.2 Animal Toxicology and/or Pharmacology

In toxicological studies with tapentadol, the most common systemic effects of tapentadol were related to the mu-opioid receptor agonist and norepinephrine reuptake inhibition pharmacodynamic properties of the compound. Transient, dose-dependent and predominantly CNS-related findings were observed, including impaired respiratory function and convulsions, the latter occurring in the dog at plasma levels (C_{max}) which are in the range associated with the maximum recommended human dose (MRHD).

14 CLINICAL STUDIES

The efficacy and safety of NUCYNTA™ in the treatment of moderate to severe acute pain has been established in two randomized, double-blind, placebo- and active-controlled studies of moderate to severe pain from first metatarsal bunionectomy and end-stage degenerative joint disease.

14.1 Orthopedic Surgery – Bunionectomy

A randomized, double-blind, parallel-group, active- and placebo-controlled, multiple-dose study demonstrated the efficacy of 50 mg, 75 mg, and 100 mg NUCYNTA™ given every 4 to 6 hours for 72 hours in patients aged 18 to 80 years experiencing moderate to severe pain following unilateral, first metatarsal bunionectomy surgery. Patients who qualified for the study with a baseline pain score of ≥ 4 on an 11-point rating scale ranging from 0 to 10 were randomized to 1 of 5 treatments. Patients were allowed to take a second dose of study medication as soon as 1 hour after the first dose on study Day 1, with subsequent dosing every 4 to 6 hours. If rescue analgesics were required, the patients were discontinued for lack of efficacy. Efficacy was evaluated by comparing the sum of pain intensity difference over the first 48 hours (SPID48) versus placebo. NUCYNTA™ at each dose provided a greater reduction in pain compared to placebo based on SPID48 values.

For various degrees of improvement from baseline to the 48-hour endpoint, Figure 1 shows the fraction of patients achieving that level of improvement. The figures are cumulative, such that every patient that achieves a 50% reduction in pain from baseline is included in every level of improvement below 50%. Patients who did not complete the 48-hour observation period in the study were assigned 0% improvement.

Figure 1: Percentage of Patients Achieving Various Levels of Pain Relief as Measured by Pain Severity at 48 Hours Compared to Baseline- Post Operative Bunionectomy

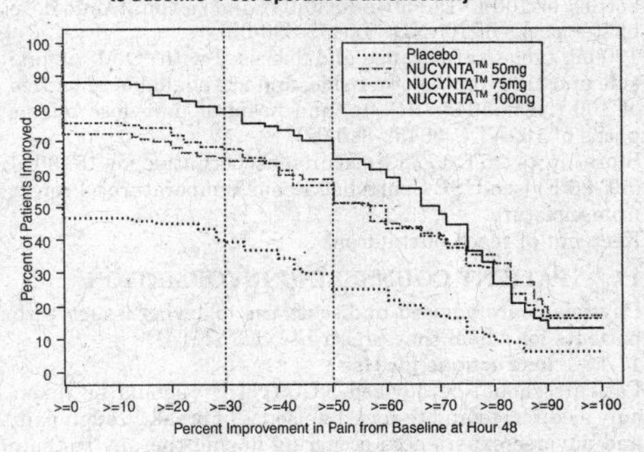

The proportions of patients who showed reduction in pain intensity at 48 hours of 30% or greater, or 50% or greater were significantly higher in patients treated with NUCYNTA™ at each dose versus placebo.

14.2 End-Stage Degenerative Joint Disease

A randomized, double-blind, parallel-group, active- and placebo-controlled, multiple-dose study evaluated the efficacy and safety of 50 mg and 75 mg NUCYNTA™ given every 4 to 6 hours during waking hours for 10 days in patients aged 18 to 80 years, experiencing moderate to severe pain from end stage degenerative joint disease of the hip or knee, defined as a 3-day mean pain score of ≥ 5 on an 11-point pain intensity scale, ranging from 0 to 10. Pain scores were assessed twice daily and assessed the pain the patient had experienced over the previous 12 hours. Patients were allowed to continue non-opioid analgesic therapy for which they had been on a stable regimen before screening throughout the study. Eighty-three percent (83%) of patients in the tapentadol treatment groups and the placebo group took such analgesia during the study. The 75 mg treatment group was dosed at 50 mg for the first day of the study, followed by 75 mg for the remaining nine days. Patients requiring rescue analgesics other than study medication were discontinued for lack of efficacy. Efficacy was evaluated by comparing the sum of pain intensity difference (SPID) versus placebo over the first five days of treatment. NUCYNTA™ 50 mg and 75 mg provided improvement in pain compared with placebo based on the 5-Day SPID.

For various degrees of improvement from baseline to the Day 5 endpoint, Figure 2 shows the fraction of patients achieving that level of improvement. The figures are cumulative, such that every patient that achieves a 50% reduction in pain from baseline is included in every level of improvement below 50%. Patients who did not complete the 5-day observation period in the study were assigned 0% improvement.

Figure 2: Percentage of Patients Achieving Various Levels of Pain Relief as Measured by Average Pain Severity for the Previous 12 hours, Measured on Study Day 5 Compared to Baseline — End Stage Degenerative Joint Disease

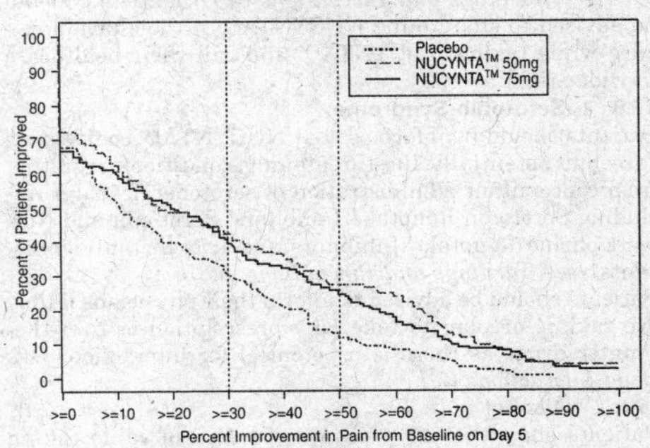

The proportions of patients who showed reduction in pain intensity at 5 days of 30% or greater, or 50% or greater were significantly higher in patients treated with NUCYNTA™ at each dose versus placebo.

Continued on next page

Nucynta—Cont.

16 HOW SUPPLIED/STORAGE AND HANDLING

NUCYNTA™ Tablets are available in the following strengths and packages. All tablets are round and biconvex-shaped.

50 mg tablets are yellow and debossed with "O-M" on one side and "50" on the other side, and are available in bottles of 100 (NDC 50458-820-04) and hospital unit dose blister packs of 10 (NDC 50458-820-02).

75 mg tablets are yellow-orange and debossed with "O-M" on one side and "75" on the other side, and are available in bottles of 100 (NDC 50458-830-04) and hospital unit dose blister packs of 10 (NDC 50458-830-02).

100 mg tablets are orange and debossed with "O-M" on one side and "100" on the other side, and are available in bottles of 100 (NDC 50458-840-04) and hospital unit dose blister packs of 10 (NDC 50458-840-02).

Store up to 25°C (77°F); excursions permitted to 15°-30°C (59°-86°F) [see USP Controlled Room Temperature]. Protect from moisture.

Keep out of reach of children.

17 PATIENT COUNSELING INFORMATION

Physicians are advised to discuss the following issues with patients for whom they prescribe NUCYNTA™:

17.1 Instructions for Use

Patients should be advised NUCYNTA™ should be taken only as directed and to report episodes of breakthrough pain and adverse experiences occurring during therapy to their physician. Individualization of dosage is essential to make optimal use of this medication. Patients should be advised not to adjust the dose of NUCYNTA™ without consulting their physician *[see Dosage and Administration (2)]*. Patients should be advised that it may be appropriate to taper dosing when discontinuing treatment with NUCYNTA™ as withdrawal symptoms may occur *[see Drug Abuse and Dependence (9.3)]*. The physician can provide a dose schedule to accomplish a gradual discontinuation of the medication.

17.2 Misuse and Abuse

Patients should be advised that NUCYNTA™ is a potential drug of abuse. Patients should protect NUCYNTA™ from theft, and NUCYNTA™ should never be given to anyone other than the individual for whom NUCYNTA™ was prescribed *[see Warnings and Precautions (5.4)]*.

17.3 Interference with Cognitive and Motor Performance

As NUCYNTA™ has the potential to impair judgment, thinking, or motor skills, patients should be cautioned about operating hazardous machinery, including automobiles *[see Warnings and Precautions (5.5)]*.

17.4 Pregnancy

Patients should be advised to notify their physician if they become pregnant or intend to become pregnant during treatment with NUCYNTA™ *[see Use in Specific Populations (8.1)]*.

17.5 Nursing

Patients should be advised not to breast-feed an infant during treatment with NUCYNTA™ *[see Use in Specific Populations (8.3)]*.

17.6 Monoamine Oxidase Inhibitors

Patients should be informed not to take NUCYNTA™ while using any drugs that inhibit monoamine oxidase. Patients should not start any new medications while taking NUCYNTA™ until they are assured by their healthcare provider that the new medication is not a monoamine oxidase inhibitor.

17.7 Seizures

Patients should be informed that NUCYNTA™ could cause seizures if they are at risk for seizures or have epilepsy. Such patients should be advised to use NUCYNTA™ with care *[see Warnings and Precautions (5.7)]*. Patients should be advised to stop taking NUCYNTA™ if they have a seizure while taking NUCYNTA™ and call their healthcare provider right away.

17.8 Serotonin Syndrome

Patients should be informed that NUCYNTA™ could cause rare but potentially life-threatening conditions resulting from concomitant administration of serotonergic drugs (including Serotonin Reuptake Inhibitors, Serotonin and Norepinephrine Reuptake Inhibitors and tricyclic antidepressants) *[see Warnings and Precautions (5.8)]*.

Patients should be advised to inform their physicians if they are taking, or plan to take, any prescription or over-the-counter drugs as there is a potential for interactions *[see Drug Interactions (7)]*.

17.9 Alcohol

Patients should be advised to avoid alcohol while taking NUCYNTA™ *[see Drug Interactions (7.3)]*.

17.10 Medication Guide

See Medication Guide.

Revised: March 2009

Manufactured by:
Janssen Ortho, LLC
Gurabo, PR 00778

Manufactured for:
PriCara®, Division of Ortho-McNeil-Janssen Pharmaceuticals, Inc.
Raritan, NJ 08869

10169800

MEDICATION GUIDE
NUCYNTA™ (new-SINN-tah)
(tapentadol)
Immediate-release oral tablets C-II

- **NUCYNTA™ is a federally controlled substance (C-II) because it can be abused. Keep NUCYNTA™ in a safe place to prevent theft. Selling or giving away NUCYNTA™ may harm others, and is against the law.**
- **Tell your doctor if you (or a family member) have ever abused or been dependent on alcohol, prescription medicines, or street drugs.**

Read the Medication Guide that comes with NUCYNTA™ before you start taking it and each time you get a new prescription. There may be new information. This Medication Guide does not take the place of talking to your doctor about your medical condition or your treatment. Talk to your doctor if you have any questions.

What is the most important information I should know about NUCYNTA™?

NUCYNTA™ is a tablet that contains tapentadol, a strong medicine that is a pain medicine.

Use NUCYNTA™ exactly how your doctor tells you to. Do not use NUCYNTA™ if it has not been prescribed for you. You should not take NUCYNTA™ if your pain is mild and can be controlled with other pain medicines such as non-steroidal anti-inflammatory medicines (NSAIDS) or acetaminophen.

What is NUCYNTA™?

- NUCYNTA™ is a prescription medicine that is used in adults 18 years of age or older to treat moderate to severe pain that is expected to last a short time.
  NUCYNTA™ is for short-term use only because the risks for withdrawal symptoms, abuse and addiction are higher when NUCYNTA™ is used longer.

Who should not take NUCYNTA™?

Do not take NUCYNTA™ if you:

- have severe lung problems
- have a gastrointestinal problem called paralytic ileus in which the intestines are not working normally.
- take a monoamine oxidase inhibitor (MAOI) medicine or have taken an MAOI within the last 14 days. Ask your doctor or pharmacist if any of your medicines is an MAOI.

What should I tell my doctor before taking NUCYNTA™?

NUCYNTA™ may not be right for you. Tell your doctor about all your medical conditions, including if you have:

- trouble breathing or lung problems
- or had a head injury
- liver or kidney problems
- convulsions or seizures
- dependency problems with alcohol
- pancreas or gall bladder problems
- past or present substance abuse or drug addiction. There is a risk of abuse or addiction with narcotic pain medicines. If you have abused drugs in the past, you may have a higher chance of developing abuse or addiction again while using NUCYNTA™.
- are pregnant or plan to become pregnant
- are breast-feeding. You should not breast-feed while taking NUCYNTA™.

Tell your doctor about all the medicines you take, including prescription and nonprescription medicines, vitamins, and herbal supplements. Using NUCYNTA™ with other medicines can cause serious side effects. The doses of some other medicines may need to be changed. Your doctor can tell you what medicines can be safely taken with NUCYNTA™. Especially tell your doctor if you take:

- **Monoamine Oxidase Inhibitors (MAOIs).** See "Who should not take NUCYNTA™."
- **any medicine that makes you sleepy.** NUCYNTA™ can make you sleepy and affect your breathing. Taking these medicines together can be dangerous.

How should I take NUCYNTA™?

- Do not take NUCYNTA™ unless it has been prescribed for you by your doctor.
- Take NUCYNTA™ exactly as prescribed by your doctor.
- **Do not change the dose of NUCYNTA™ unless your doctor tells you to.** Your doctor may change your dose after seeing how the medicine affects you. Do not use NUCYNTA™ more often than prescribed. Call your doctor if your pain is not well controlled while taking NUCYNTA™.
- Follow your doctor's instructions about how to slowly stop taking NUCYNTA™ to help lessen withdrawal symptoms.
- NUCYNTA™ can be taken with or without food.

What should I avoid while taking NUCYNTA™?

- Do not drive, operate machinery, or participate in any other possibly dangerous activities until you know how you react to this medicine. NUCYNTA™ can make you sleepy.
- You should not drink alcohol while using NUCYNTA™. Alcohol increases your chance of having dangerous side effects.

What are the possible side effects of NUCYNTA™?

NUCYNTA™ can cause serious side effects including:

- **Life-threatening breathing problems. Call your doctor right away or get emergency medical help if you:**
  - have trouble breathing, or have slow or shallow breathing
  - have a slow heartbeat
  - have severe sleepiness
  - have cold, clammy skin
  - feel faint, dizzy, confused, or can not think, walk or talk normally
  - have a seizure
  - have hallucinations
- **Physical Dependence.** NUCYNTA™ can cause physical dependence. Talk to your doctor about slowly stopping NUCYNTA™ to avoid getting sick with withdrawal symptoms. You could become sick with uncomfortable symptoms because your body has become used to the medicine. Tell your doctor if you have any of these symptoms of withdrawal: feeling anxious, sweating, sleep problems, shivering, pain, nausea, tremors, diarrhea, upper respiratory symptoms, hallucinations, hair "standing on end." Physical dependence is not the same as drug addiction. Your doctor can tell you more about the differences between physical dependence and drug addiction.
- **Serotonin syndrome.** Serotonin syndrome is a rare, life-threatening problem that could happen if you take NUCYNTA™ with Selective Serotonin Reuptake Inhibitors (SSRIs), Serotonin and Norepinephrine Reuptake Inhibitors (SNRIs), Monoamine Oxidase Inhibitors (MAOIs), triptans or certain other medicines. Call your doctor or get medical help right away if you have any one or more of the these symptoms: you feel agitated, have hallucinations, coma, rapid heart beat, feel overheated, loss of coordination, over active reflexes, nausea, vomiting, or diarrhea.
- **Seizures.** NUCYNTA™ can cause seizures in people who are at risk for seizures or who have epilepsy. Tell your doctor right away if you have a seizure and stop taking NUCYNTA™.
- **Low blood pressure.** This can make you feel dizzy if you get up too fast from sitting or lying down.

The common side effects with NUCYNTA™ are nausea, dizziness, vomiting, sleepiness, and itching.

Constipation is a common side effect of all opioid medicines. Talk to your doctor about the use of laxatives and stool softeners to prevent or treat constipation while taking NUCYNTA™.

Tell your doctor about any side effect that bothers you or that does not go away. These are not all the possible side effects of NUCYNTA™. For a complete list, ask your doctor or pharmacist.

Call your doctor for medical advice about side effects. You may report side effects to FDA at 1-800-FDA-1088.

How should I store NUCYNTA™?

- Store NUCYNTA™ at 59°F to 86°F (15°C to 30°C). Keep NUCYNTA™ tablets dry.
- Dispose of NUCYNTA™ tablets you no longer need.

Keep NUCYNTA™ in a safe place out of the reach of children.

General information about NUCYNTA™

Medicines are sometimes prescribed for purposes other than those listed in a Medication Guide. Do not use NUCYNTA™ for a condition for which it was not prescribed. **Do not give NUCYNTA™ to other people, even if they have the same symptoms you have. Sharing NUCYNTA™ could be harmful and is against the law.**

This Medication Guide summarizes the most important information about NUCYNTA™. If you would like more information, talk with your doctor. You can ask your doctor or pharmacist for information about NUCYNTA™ that is written for doctors. For more information about NUCYNTA™ call 1-800-526-7736.

What are the ingredients in NUCYNTA™?

Active Ingredient: tapentadol

Inactive ingredients: microcrystalline cellulose, lactose monohydrate, croscarmellose sodium, povidone, magnesium stearate, and Opadry® II, a proprietary film-coating mixture containing polyvinyl alcohol, titanium dioxide, polyethylene glycol, talc, and aluminum lake coloring.

This Medication Guide has been approved by the U.S. Food and Drug Administration.

Revised: March 2009

Manufactured by:
Janssen Ortho, LLC
Gurabo, PR 00778

Manufactured for:
PriCara®, Division of Ortho-McNeil-Janssen Pharmaceuticals, Inc.
Raritan, NJ 08869

10169800

REVISED INFORMATION

As new research data and clinical findings become available, the product information in *PDR* is revised accordingly. Revisions submitted since the 2009 edition went to press can be found below. To remind yourself of a revision, write "See Supplement B" next to the product's heading in the 2009 *PDR*.

GlaxoSmithKline

FIVE MOORE DRIVE
RESEARCH TRIANGLE PARK, NC 27709

For all inquiries, including adverse event and quality assurance reporting, contact the GSK Response Center at 1-888-825-5249.
For updates to the product information listed below, also consult www.gsk.com.

ADVAIR® HFA 45/21 ℞

[*ad' vair*]
(fluticasone propionate 45 mcg and salmeterol 21 mcg*)
Inhalation Aerosol

ADVAIR® HFA 115/21 ℞

(fluticasone propionate 115 mcg and salmeterol 21 mcg*)
Inhalation Aerosol

ADVAIR® HFA 230/21 ℞

(fluticasone propionate 230 mcg and salmeterol 21 mcg*)
Inhalation Aerosol

*As salmeterol xinafoate salt 30.45 mcg, equivalent to salmeterol base 21 mcg
For Oral Inhalation Only

Prescribing information for this product, which appears on pages 1288-1297 of the 2009 PDR, has been revised as follows. Please write "See Supplement B" next to the product heading.

The following sentences were revised in the **DESCRIPTION** *section:*

ADVAIR HFA 45/21 Inhalation Aerosol, ADVAIR HFA 115/21 Inhalation Aerosol, and ADVAIR HFA 230/21 Inhalation Aerosol are pressurized metered-dose aerosol units fitted with a counter.

In cases where the inhaler has not been used for more than 4 weeks or when it has been dropped, prime the inhaler again by releasing 2 test sprays into the air away from the face, shaking well for 5 seconds before each spray.

The following paragraphs were added to the **WARNINGS** *section:*

13. Pneumonia. Lower respiratory tract infections, including pneumonia, have been reported in patients with COPD following the inhaled administration of corticosteroids, including fluticasone propionate and ADVAIR DISKUS. In 2 replicate 12-month studies of 1,579 patients with COPD, there was a higher incidence of pneumonia reported in patients receiving ADVAIR DISKUS 250/50 (7%) than in those receiving salmeterol 50 mcg (3%). The incidence of pneumonia in the patients treated with ADVAIR DISKUS was higher in patients over 65 years of age (9%) compared with the incidence in patients less than 65 years of age (4%).

In a 3-year study of 6,184 patients with COPD, there was a higher incidence of pneumonia reported in patients receiving ADVAIR DISKUS 500/50 compared with placebo (16% with ADVAIR DISKUS 500/50, 14% with fluticasone propionate 500 mcg, 11% with salmeterol 50 mcg, and 9% with placebo). Similar to what was seen in the 1-year studies with ADVAIR DISKUS 250/50, the incidence of pneumonia was higher in patients over 65 years of age (18% with ADVAIR DISKUS 500/50 versus 10% with placebo) compared with patients less than 65 years of age (14% with ADVAIR DISKUS 500/50 versus 8% with placebo.)

The following sentences were revised in the **PRECAUTIONS: Information for Patients** *section:*

12. In cases where the inhaler has not been used for more than 4 weeks or when it has been dropped, prime the inhaler again by releasing 2 test sprays into the air away from the face, shaking well for 5 seconds before each spray.
15. When the counter reads 020, contact the pharmacist for a refill of medication or consult the physician to determine whether a prescription refill is needed. Discard the inhaler when the counter reads 000. Never try to alter the numbers or remove the counter from the metal canister.

The following paragraph was deleted from the **PRECAUTIONS: Information for Patients** *section:*

16. Patients should never immerse the canister into water to determine the amount remaining in the canister ("float test").

The following sentence was revised in the **DOSAGE AND ADMINISTRATION** *section:*

In cases where the inhaler has not been used for more than 4 weeks or when it has been dropped, prime the inhaler again by releasing 2 test sprays into the air away from the face, shaking well for 5 seconds before each spray.

The following paragraphs were revised in the **HOW SUPPLIED** *section, including new NDC numbers assigned to the inhaler with counter:*

Each strength of ADVAIR HFA Inhalation Aerosol is supplied in a 12-g pressurized aluminum canister containing 120 metered actuations in a box of 1.* Each canister is fitted with a counter, supplied with a purple actuator with a light purple strapcap, and sealed in a plastic-coated, moisture-protective foil pouch with a desiccant that should be discarded when the pouch is opened. Each canister is packaged with a Medication Guide leaflet.

*NDC 0173-0715-20 ADVAIR HFA 45/21 Inhalation Aerosol
*NDC 0173-0716-20 ADVAIR HFA 115/21 Inhalation Aerosol
*NDC 0173-0717-20 ADVAIR HFA 230/21 Inhalation Aerosol

The correct amount of medication in each actuation cannot be assured after the counter reads 000, even though the canister is not completely empty and will continue to operate. The inhaler should be discarded when the counter reads 000.

The following text was added to the MEDICATION GUIDE *in the* **PATIENT INFORMATION** *section. In addition, the instructions for use were extensively revised to describe the appearance and use of the counter attached to the inhaler canister; consult the Prescribing Information accompanying the product or www.advair.com/asthma for the full text of the revised Medication Guide.*

Who should not use ADVAIR HFA?

Do not use ADVAIR HFA:
- to treat sudden severe symptoms of asthma
- if you are allergic to any of the ingredients in ADVAIR HFA. See the end of this Medication Guide for a list of ingredients in ADVAIR HFA.

What are the possible side effects with ADVAIR HFA?

- **pneumonia.** ADVAIR HFA contains the same medicine found in ADVAIR DISKUS. ADVAIR DISKUS is used to treat people with asthma and people with chronic obstructive pulmonary disease (COPD). People with COPD have a higher chance of getting pneumonia. ADVAIR DISKUS may increase the chance of getting pneumonia. ADVAIR HFA has not been studied in people with COPD.

Common side effects of ADVAIR HFA include upper respiratory tract infection and headache.

GlaxoSmithKline
Research Triangle Park, NC 27709

November 2008 ADH:3PI-3MG

ALBENZA® ℞

[*ăl-ben'zə*]
(albendazole)
Tablets

Prescribing information for this product, which appears on pages 1297–1299 of the 2009 PDR, has been revised as follows. Please write "See Supplement B" next to the product heading.

The following paragraph was added to the **PRECAUTIONS: General** *section:*

Pre-existing neurocysticercosis may also be uncovered in patients treated with albendazole for other conditions. Patients may experience neurological symptoms (e.g. seizures, increased intracranial pressure and focal signs) as a result of an inflammatory reaction caused by death of the parasite within the brain. Symptoms may occur soon after treatment; appropriate steroid and anticonvulsant therapy should be started immediately.

GlaxoSmithKline
Research Triangle Park, NC 27709

June 2009 ALB:8PI

COREG® ℞

[*kor' eg*]
(carvedilol)
Tablets

Prescribing information for this product, which appears on pages 1387–1393 of the 2009 PDR, has been revised as follows. Please write "See Supplement B" next to the product heading.

In the **HIGHLIGHTS OF PRESCRIBING INFORMATION**, *the following bullet point under* **CONTRAINDICATIONS** *was revised to:*

- History of serious hypersensitivity reaction (e.g., Stevens-Johnson syndrome, anaphylactic reaction, angioedema) to any component of this medication or other medications containing carvedilol. (4)

Continued on next page

Product information on these pages is effective as of June 2009. Further information is available at 1-888-825-5249 or www.gsk.com.

Coreg—Cont.

The following bullet point under **DRUG INTERACTIONS** *was revised to:*

- Amiodarone may increase carvedilol levels resulting in further slowing of the heart rate or cardiac conduction. (7.6)

In the **FULL PRESCRIBING INFORMATION**, *the last bullet point under* **4 CONTRAINDICATIONS** *was revised to:*

- Patients with a history of a serious hypersensitivity reaction (e.g., Stevens-Johnson syndrome, anaphylactic reaction, angioedema) to any component of this medication or other medications containing carvedilol.

The following was added under **7 DRUG INTERACTIONS**:

7.6 Amiodarone

Amiodarone, and its metabolite desethyl amiodarone, inhibitors of CYP2C9 and P-glycoprotein, increased concentrations of the S(-)-enantiomer of carvedilol by at least 2-fold *[see Clinical Pharmacology (12.5)]*. The concomitant administration of amiodarone or other CYP2C9 inhibitors such as fluconazole with COREG may enhance the β-blocking properties of carvedilol resulting in further slowing of the heart rate or cardiac conduction. Patients should be observed for signs of bradycardia or heart block, particularly when one agent is added to pre-existing treatment with the other.

The following paragraph was added under **12 CLINICAL PHARMACOLOGY: 12.5 Drug-Drug Interactions:**

Amiodarone: In a pharmacokinetic study conducted in 106 Japanese patients with heart failure, coadministration of small loading and maintenance doses of amiodarone with carvedilol resulted in at least a 2-fold increase in the steady-state trough concentrations of S(-)-carvedilol *[see Drug Interactions (7.6)]*.

In the **PATIENT INFORMATION** *leaflet under* **17 PATIENT COUNSELING INFORMATION: 17.2 FDA-Approved Patient Labeling**, *the following information under* **What are possible side effects of COREG?** *was revised and made a bullet point:*

- Rare but serious allergic reactions (including hives or swelling of the face, lips, tongue, and/or throat that may cause difficulty in breathing or swallowing) have happened in patients who were on COREG. These reactions can be life-threatening.

Manufactured for
GlaxoSmithKline
Research Triangle Park, NC 27709
Manufactured by
Patheon Puerto Rico, Inc.
Manati, PR 00674 USA

June 2009 CRG:18PI-3PIL

COREG CR® ℞

[kor' eg]
(carvedilol phosphate)
Extended-release Capsules

Prescribing information for this product, which appears on pages 1393–1400 of the 2009 PDR, has been revised as follows. Please write "See Supplement B" next to the product heading.

In the **HIGHLIGHTS OF PRESCRIBING INFORMATION**, *the following sentence was added to* **DOSAGE AND ADMINISTRATION**:

Do not crush or chew capsules.

The following bullet point was added to **DOSAGE AND ADMINISTRATION**:

- Elderly patients (> 65 years of age): When switching from higher doses of immediate-release carvedilol to COREG CR, a lower starting dose should be considered to reduce the risk of hypotension and syncope. (2.5)

The first paragraph of **ADVERSE REACTIONS** *was revised to:*

The safety profile of COREG CR was similar to that observed for immediate-release carvedilol. Most common adverse events seen with immediate-release carvedilol. (6.1):

The following bullet point was added to **DRUG INTERACTIONS**:

- Both digitalis glycosides and β-blockers slow atrioventricular conduction and decrease heart rate. Concomitant use can increase the risk of bradycardia. (7.4)

The following bullet point was revised in **DRUG INTERACTIONS**:

- Amiodarone may increase carvedilol levels resulting in further slowing of the heart rate or cardiac conduction. (7.6)

In the **FULL PRESCRIBING INFORMATION**, *the following table was revised in* **2 DOSAGE AND ADMINISTRATION**:

Table 1. Dosing Conversion

Daily Dose of Immediate-Release Carvedilol Tablets	Daily Dose of COREG CR Capsules*
6.25 mg (3.125 mg twice daily)	10 mg once daily
12.5 mg (6.25 mg twice daily)	20 mg once daily
25 mg (12.5 mg twice daily)	40 mg once daily
50 mg (25 mg twice daily)	80 mg once daily

* When switching from carvedilol 12.5 mg or 25 mg twice daily, a starting dose of COREG CR 20 mg or 40 mg once daily, respectively, may be warranted for elderly patients or those at increased risk of hypotension, dizziness, or syncope. Subsequent titration to higher doses should, as appropriate, be made after an interval of at least 2 weeks.

The following was added to **2 DOSAGE AND ADMINISTRATION**:

2.5 Geriatric Use

When switching elderly patients (65 years of age or older) who are taking the higher doses of immediate-release carvedilol tablets (25 mg twice daily) to COREG CR, a lower starting dose (40 mg) of COREG CR is recommended to minimize the potential for dizziness, syncope, or hypotension *[see Dosage and Administration (2)]*. Patients who have switched and who tolerate COREG CR should, as appropriate, have their dose increased after an interval of at least 2 weeks *[see Use in Specific Populations (8.5)]*.

The following sentence was added to the end of the first paragraph under **5 WARNINGS AND PRECAUTIONS: 5.3 Hypotension**:

In a trial comparing heart failure patients switched to COREG CR or maintained on immediate-release carvedilol, there was a 2-fold increase in the combined incidence of hypotension, syncope or dizziness in elderly patients switched from the highest dose of carvedilol (25 mg twice daily) to COREG CR 80 mg once daily *[see Use in Specific Populations (8.5)]*.

The last paragraph under **5 WARNINGS AND PRECAUTIONS: 5.3 Hypotension** *was revised to:*

Starting with a low dose, administration with food, and gradual up-titration should decrease the likelihood of syncope or excessive hypotension *[see Dosage and Administration (2.1, 2.2, 2.3)]*. During initiation of therapy, the patient should be cautioned to avoid situations such as driving or hazardous tasks, where injury could result should syncope occur.

The following was added to **7 DRUG INTERACTIONS**:

7.6 Amiodarone

Amiodarone, and its metabolite desethyl amiodarone, inhibitors of CYP2C9 and P-glycoprotein, increased concentrations of the S(-) enantiomer of carvedilol by at least 2-fold *[see Clinical Pharmacology (12.5)]*. The concomitant administration of amiodarone or other CYP2C9 inhibitors such as fluconazole with COREG CR may enhance β-blocking properties of carvedilol resulting in further slowing of the heart rate or cardiac conduction. Patients should be observed for signs of bradycardia or heart block, particularly when one agent is added to pre-existing treatment with the other.

The following paragraph was added to **8 USE IN SPECIFIC POPULATIONS: 8.5 Geriatric Use**:

A randomized study (n = 405) comparing mild to severe heart failure patients switched to COREG CR or maintained on immediate-release carvedilol included 220 patients who were 65 years of age or older. In this elderly subgroup, the combined incidence of dizziness, hypotension, or syncope was 24% (18/75) in patients switched from the highest dose of immediate-release carvedilol (25 mg twice daily) to the highest dose of COREG CR (80 mg once daily) compared to 11% (4/36) in patients maintained on immediate-release carvedilol (25 mg twice daily). When switching from the highest dose of immediate-release carvedilol to COREG CR, a lower starting dose is recommended for elderly patients *[see Dosage and Administration (2.5)]*.

The following paragraph was added to **12 CLINICAL PHARMACOLOGY: 12.5 Drug-Drug Interactions**:

Amiodarone: In a pharmacokinetic study conducted in 106 Japanese patients with heart failure, coadministration of small loading and maintenance doses of amiodarone with carvedilol resulted in at least a 2-fold increase in the steady-state trough concentrations of S(-)-carvedilol *[see Drug Interactions (7.6)]*.

In the **PATIENT INFORMATION** *leaflet, the following information under* **What are possible side effects of COREG CR?** *was revised and made a bullet point:*

- **rare but serious allergic reactions** (including hives or swelling of the face, lips, tongue, and/or throat that may cause difficulty in breathing or swallowing) have happened in patients who were on COREG or COREG CR. These reactions can be life-threatening. In some cases, these reactions happened in patients who had been on COREG before taking COREG CR.

GlaxoSmithKline
Research Triangle Park, NC 27709

June 2009 CCR:11PI-4PIL

ENGERIX-B® ℞

[in' jə-rix]
Hepatitis B Vaccine (Recombinant)

Prescribing information for this product, which appears on pages 1409–1412 of the 2009 PDR, has been revised as follows. Please write "See Supplement B" next to the product heading.

The **DOSAGE AND ADMINISTRATION: Preparation for Administration** *section was revised to:*

Shake well before withdrawal and use. Inspect ENGERIX-B visually for particulate matter, discoloration, and cracks in the vial or syringe prior to administration, whenever solution and container permit. If any of these conditions exist, the vaccine should not be administered. With thorough agitation, ENGERIX-B is a slightly turbid white suspension. Discard if it appears otherwise. The vaccine should be used as supplied; no dilution is necessary. The full recommended dose of the vaccine should be used. Any vaccine remaining in a single-dose vial should be discarded.

Manufactured by **GlaxoSmithKline Biologicals**
Rixensart, Belgium, US License No. 1617
Distributed by **GlaxoSmithKline**
Research Triangle Park, NC 27709

June 2009 ENG:39PI

HAVRIX® ℞

[hav' rix]
(Hepatitis A Vaccine)
Suspension for Intramuscular Injection

Prescribing information for this product, which appears on pages 1454–1456 of the 2009 PDR, has been revised as follows. Please write "See Supplement B" next to the product heading.

The following **6 ADVERSE REACTIONS: 6.2 Postmarketing Experience** *subsections were added or revised:*

Infections and Infestations: Rhinitis.
Immune System Disorders: Anaphylactic reaction, anaphylactoid reaction, serum sickness–like syndrome.
Nervous System Disorders: Convulsion, dizziness, encephalopathy, Guillain-Barré syndrome, hypoesthesia, multiple sclerosis, myelitis, neuropathy, paresthesia, somnolence, syncope.
Vascular Disorders: Vasculitis.
Musculoskeletal and Connective Tissue Disorders: Musculoskeletal stiffness.
General Disorders and Administration Site Conditions: Chills, influenza-like symptoms, injection site reaction, local swelling.

Manufactured by GlaxoSmithKline Biologicals
Rixensart, Belgium, US License No. 1617
Distributed by GlaxoSmithKline
Research Triangle Park, NC 27709

May 2009 HVX:25PI

KINRIX™ ℞

[kin' rix]
(Diphtheria and Tetanus Toxoids and Acellular Pertussis Adsorbed and Inactivated Poliovirus Vaccine)
Suspension for Intramuscular Injection

Prescribing information for this product, which appears on pages 1486–1488 of the 2009 PDR, has been revised as follows. Please write "See Supplement B" next to the product heading.

The **2 DOSAGE AND ADMINISTRATION: 2.1 Preparation for Administration** *section was revised to:*

Shake vigorously to obtain a homogeneous, turbid, white suspension. Do not use if resuspension does not occur with vigorous shaking. Parenteral drug products should be inspected visually for particulate matter and discoloration prior to administration, whenever solution and container permit. KINRIX also should be inspected visually for cracks in the vial or syringe prior to administration. If any of these conditions exist, the vaccine should not be administered. After removal of the dose, any vaccine remaining in the vial should be discarded.

Manufactured by GlaxoSmithKline Biologicals
Rixensart, Belgium, US License No. 1617

LAMICTAL® ℞
[la-mĭk'tal]
(lamotrigine)
Tablets
LAMICTAL® ℞
(lamotrigine)
Chewable Dispersible Tablets
LAMICTAL® ODT™ ℞
(lamotrigine)
Orally Disintegrating Tablets

Prescribing information for these products, which appears on pages 1488–1498 of the 2009 PDR, has been completely revised as follows, including the addition of the new product LAMICTAL® ODT™ Orally Disintegrating Tablets. *Please write "See Supplement B" next to the product heading.*

HIGHLIGHTS OF PRESCRIBING INFORMATION
These highlights do not include all the information needed to use LAMICTAL safely and effectively. See full prescribing information for LAMICTAL.
LAMICTAL (lamotrigine) Tablets
LAMICTAL (lamotrigine) Chewable Dispersible Tablets
LAMICTAL ODT (lamotrigine) Orally Disintegrating Tablets
Initial U.S. Approval: 1994

WARNING: SERIOUS SKIN RASHES
See full prescribing information for complete boxed warning.
Cases of life-threatening serious rashes, including Stevens-Johnson syndrome, toxic epidermal necrolysis, and/or rash-related death, have been caused by LAMICTAL. The rate of serious rash is greater in pediatric patients than in adults. Additional factors that may increase the risk of rash include (5.1):
- **coadministration with valproate**
- **exceeding recommended initial dose of LAMICTAL**
- **exceeding recommended dose escalation of LAMICTAL**

Benign rashes are also caused by LAMICTAL; however, it is not possible to predict which rashes will prove to be serious or life-threatening. LAMICTAL should be discontinued at the first sign of rash, unless the rash is clearly not drug-related. (5.1)

RECENT MAJOR CHANGES

Dosage and Administration, LAMICTAL ODT (2.6) May/2009
Warnings and Precautions, Suicidal Behavior and Ideation (5.5) April/2009

INDICATIONS AND USAGE

LAMICTAL is an antiepileptic drug (AED) indicated for:
Epilepsy—adjunctive therapy in patients ≥2 years of age: (1.1)
- partial seizures.
- primary generalized tonic-clonic seizures.
- generalized seizures of Lennox-Gastaut syndrome.

Epilepsy—monotherapy in patients ≥16 years of age: conversion to monotherapy in patients with partial seizures who are receiving treatment with carbamazepine, phenobarbital, phenytoin, primidone, or valproate as the single AED. (1.1)
Bipolar Disorder in patients ≥18 years of age: maintenance treatment of Bipolar I Disorder to delay the time to occurrence of mood episodes in patients treated for acute mood episodes with standard therapy. (1.2)

DOSAGE AND ADMINISTRATION

- Dosing is based on concomitant medications, indication, and patient age. (2.2, 2.4)
- To avoid an increased risk of rash, the recommended initial dose and subsequent dose escalations should not be exceeded. LAMICTAL Starter Kits and LAMICTAL ODT Patient Titration Kits are available for the first 5 weeks of treatment. (2.1, 16)
- Do not restart LAMICTAL in patients who discontinued due to rash unless the potential benefits clearly outweigh the risks. (2.1)
- Adjustments to maintenance doses will in most cases be required in patients starting or stopping estrogen-containing oral contraceptives. (2.1, 5.8)
- LAMICTAL should be discontinued over a period of at least 2 weeks (approximately 50% reduction per week). (2.1, 5.9)

Epilepsy
Adjunctive therapy—See Table 1 for patients >12 years of age and Tables 2 and 3 for patients 2 to 12 years. (2.2)
Conversion to monotherapy—See Table 4. (2.3)
Bipolar Disorder: See Tables 5 and 6. (2.4)

DOSAGE FORMS AND STRENGTHS

Tablets: 25 mg, 100 mg, 150 mg, and 200 mg scored. (3.1, 16)
Chewable Dispersible Tablets: 2 mg, 5 mg, and 25 mg. (3.2, 16)
Orally Disintegrating Tablets: 25 mg, 50 mg, 100 mg, and 200 mg. (3.3, 16)

CONTRAINDICATIONS

Hypersensitivity to the drug or its ingredients. (Boxed Warning, 4)

WARNINGS AND PRECAUTIONS

- Life-threatening serious rash and/or rash-related death may result. (Boxed Warning, 5.1)
- Hypersensitivity reaction may be fatal or life-threatening. Early signs of hypersensitivity (e.g., fever, lymphadenopathy) may present without rash; if signs present, patient should be evaluated immediately. LAMICTAL should be discontinued if alternate etiology for hypersensitivity signs is not found. (5.2)
- Acute multiorgan failure has resulted (some cases fatal). (5.3)
- Blood dyscrasias (e.g., neutropenia, thrombocytopenia, pancytopenia), may result either with or without an associated hypersensitivity syndrome. (5.4)
- Suicidal behavior and ideation. (5.5)
- Clinical worsening, emergence of new symptoms, and suicidal ideation/behaviors may be associated with treatment of bipolar disorder. Patients should be closely monitored, particularly early in treatment or during dosage changes. (5.6)
- Medication errors involving LAMICTAL have occurred. In particular the names LAMICTAL or lamotrigine can be confused with names of other commonly used medications. Medication errors may also occur between the different formulations of LAMICTAL. (3.4, 5.7, 16, 17.9)

ADVERSE REACTIONS

- Most common adverse reactions (incidence ≥10%) in adult epilepsy clinical studies were dizziness, headache, diplopia, ataxia, nausea, blurred vision, somnolence, rhinitis, and rash. Additional adverse reactions (incidence ≥10%) reported in children in epilepsy clinical studies included vomiting, infection, fever, accidental injury, pharyngitis, abdominal pain, and tremor. (6.1)
- Most common adverse reactions (incidence >5%) in adult bipolar clinical studies were nausea, insomnia, somnolence, back pain, fatigue, rash, rhinitis, abdominal pain, and xerostomia. (6.1)

To report SUSPECTED ADVERSE REACTIONS, contact GlaxoSmithKline at 1-888-825-5249 or FDA at 1-800-FDA-1088 or www.fda.gov/medwatch.

DRUG INTERACTIONS

- Valproate increases lamotrigine concentrations more than 2-fold. (7, 12.3)
- Carbamazepine, phenytoin, phenobarbital, and primidone decrease lamotrigine concentrations by approximately 40%. (7, 12.3)
- Oral estrogen-containing contraceptives and rifampin also decrease lamotrigine concentrations by approximately 50%. (7, 12.3)

USE IN SPECIFIC POPULATIONS

- Hepatic impairment: Dosage adjustments required. (2.1)
- Healthcare professionals can enroll patients in the Lamotrigine Pregnancy Registry (1-800-336-2176). Patients can enroll themselves in the North American Antiepileptic Drug Pregnancy Registry (1-888-233-2334). (8.1)
- Efficacy of LAMICTAL, used as adjunctive treatment for partial seizures, was not demonstrated in a small randomized, double-blind, placebo-controlled study in very young pediatric patients (1 to 24 months). (8.4)

See 17 for PATIENT COUNSELING INFORMATION and Medication Guide.

Revised: May 2009
LMT:2PI

FULL PRESCRIBING INFORMATION CONTENTS*:

*Sections or subsections omitted from the full prescribing information are not listed.

FULL PRESCRIBING INFORMATION

WARNING: SERIOUS SKIN RASHES
LAMICTAL® can cause serious rashes requiring hospitalization and discontinuation of treatment. The incidence of these rashes, which have included Stevens-Johnson syndrome, is approximately 0.8% (8 per 1,000) in pediatric patients (2 to 16 years of age) receiving LAMICTAL as adjunctive therapy for epilepsy and 0.3% (3 per 1,000) in adults on adjunctive therapy for epilepsy. In clinical trials of bipolar and other mood disorders, the rate of serious rash was 0.08% (0.8 per 1,000) in adult patients receiving LAMICTAL as initial monotherapy and 0.13% (1.3 per 1,000) in adult patients receiving LAMICTAL as adjunctive therapy. In a prospectively followed cohort of 1,983 pediatric patients (2 to 16 years of age) with epilepsy taking adjunctive LAMICTAL, there was 1 rash-related death. In worldwide postmarketing experience, rare cases of toxic epidermal necrolysis and/or rash-related death have been reported in adult and pediatric patients, but their numbers are too few to permit a precise estimate of the rate.
Other than age, there are as yet no factors identified that are known to predict the risk of occurrence or the

Continued on next page

Product information on these pages is effective as of June 2009. Further information is available at 1-888-825-5249 or www.gsk.com.

Lamictal—Cont.

severity of rash caused by LAMICTAL. There are suggestions, yet to be proven, that the risk of rash may also be increased by (1) coadministration of LAMICTAL with valproate (includes valproic acid and divalproex sodium), (2) exceeding the recommended initial dose of LAMICTAL, or (3) exceeding the recommended dose escalation for LAMICTAL. However, cases have occurred in the absence of these factors.
Nearly all cases of life-threatening rashes caused by LAMICTAL have occurred within 2 to 8 weeks of treatment initiation. However, isolated cases have occurred after prolonged treatment (e.g., 6 months). Accordingly, duration of therapy cannot be relied upon as means to predict the potential risk heralded by the first appearance of a rash.
Although benign rashes are also caused by LAMICTAL, it is not possible to predict reliably which rashes will prove to be serious or life-threatening. Accordingly, LAMICTAL should ordinarily be discontinued at the first sign of rash, unless the rash is clearly not drug-related. Discontinuation of treatment may not prevent a rash from becoming life-threatening or permanently disabling or disfiguring *[see Warnings and Precautions (5.1)]*.

1 INDICATIONS AND USAGE

1.1 Epilepsy

Adjunctive Therapy: LAMICTAL is indicated as adjunctive therapy for the following seizure types in patients ≥2 years of age:
- partial seizures
- primary generalized tonic-clonic seizures
- generalized seizures of Lennox-Gastaut syndrome

Monotherapy: LAMICTAL is indicated for conversion to monotherapy in adults (≥16 years of age) with partial seizures who are receiving treatment with carbamazepine, phenytoin, phenobarbital, primidone, or valproate as the single antiepileptic drug (AED).
Safety and effectiveness of LAMICTAL have not been established (1) as initial monotherapy; (2) for conversion to monotherapy from AEDs other than carbamazepine, phenytoin, phenobarbital, primidone, or valproate; or (3) for simultaneous conversion to monotherapy from 2 or more concomitant AEDs.

1.2 Bipolar Disorder

LAMICTAL is indicated for the maintenance treatment of Bipolar I Disorder to delay the time to occurrence of mood episodes (depression, mania, hypomania, mixed episodes) in adults (≥18 years of age) treated for acute mood episodes with standard therapy. The effectiveness of LAMICTAL in the acute treatment of mood episodes has not been established.
The effectiveness of LAMICTAL as maintenance treatment was established in 2 placebo-controlled trials in patients with Bipolar I Disorder as defined by DSM-IV *[see Clinical Studies (14.2)]*. The physician who elects to prescribe LAMICTAL for periods extending beyond 16 weeks should periodically re-evaluate the long-term usefulness of the drug for the individual patient.

2 DOSAGE AND ADMINISTRATION

2.1 General Dosing Considerations

Rash: There are suggestions, yet to be proven, that the risk of severe, potentially life-threatening rash may be increased by (1) coadministration of LAMICTAL with valproate, (2) exceeding the recommended initial dose of LAMICTAL, or (3) exceeding the recommended dose escalation for LAMICTAL. However, cases have occurred in the absence of these factors *[see Boxed Warning]*. Therefore, it is important that the dosing recommendations be followed closely.
The risk of nonserious rash may be increased when the recommended initial dose and/or the rate of dose escalation of LAMICTAL is exceeded and in patients with a history of allergy or rash to other AEDs.
LAMICTAL Starter Kits and LAMICTAL® ODT™ Patient Titration Kits provide LAMICTAL at doses consistent with the recommended titration schedule for the first 5 weeks of treatment, based upon concomitant medications for patients with epilepsy (>12 years of age) and Bipolar I Disorder (≥18 years of age) and are intended to help reduce the potential for rash. The use of LAMICTAL Starter Kits and LAMICTAL ODT Patient Titration Kits is recommended for appropriate patients who are starting or restarting LAMICTAL *[see How Supplied/Storage and Handling (16)]*.
It is recommended that LAMICTAL not be restarted in patients who discontinued due to rash associated with prior treatment with lamotrigine, unless the potential benefits clearly outweigh the risks. If the decision is made to restart a patient who has discontinued lamotrigine, the need to restart with the initial dosing recommendations should be assessed. The greater the interval of time since the previous

Table 1. Escalation Regimen for LAMICTAL in Patients Over 12 Years of Age With Epilepsy

	For Patients Taking Valproate*	For Patients Taking AEDs Other Than Carbamazepine, Phenytoin, Phenobarbital, or Primidone†, and Not Taking Valproate	For Patients Taking Carbamazepine, Phenytoin, Phenobarbital, or Primidone† and Not Taking Valproate
Weeks 1 and 2	**25 mg every *other* day**	**25 mg every day**	**50 mg/day**
Weeks 3 and 4	**25 mg every day**	**50 mg/day**	**100 mg/day** (in 2 divided doses)
Weeks 5 onwards to maintenance	Increase by 25 to 50 mg/day every 1 to 2 weeks	Increase by 50 mg/day every 1 to 2 weeks	Increase by 100 mg/day every 1 to 2 weeks.
Usual Maintenance Dose	**100 to 200 mg/day with valproate alone** **100 to 400 mg/day with valproate and other drugs that induce glucuronidation** (in 1 or 2 divided doses)	**225 to 375 mg/day** (in 2 divided doses)	**300 to 500 mg/day** (in 2 divided doses)

* Valproate has been shown to inhibit glucuronidation and decrease the apparent clearance of lamotrigine *[see Drug Interactions (7), Pharmacokinetics (12.3)]*.
† These drugs induce glucuronidation and increase clearance *[see Drug Interactions (7), Pharmacokinetics (12.3)]*. Other drugs which have similar effects include estrogen-containing oral contraceptives and rifampin *[see Drug Interactions (7), Pharmacokinetics (12.3)]*. Dosing recommendations for oral contraceptives can be found in General Dosing Considerations *[see Dosage and Administration (2.1)]*. Patients on rifampin should follow the same dosing titration/maintenance regimen used with drugs that induce glucuronidation and increase clearance.

Table 2. Escalation Regimen for LAMICTAL in Patients 2 to 12 Years of Age With Epilepsy

	For Patients Taking Valproate*	For Patients Taking AEDs Other Than Carbamazepine, Phenytoin, Phenobarbital, or Primidone†, and Not Taking Valproate	For Patients Taking Carbamazepine, Phenytoin, Phenobarbital, or Primidone† and Not Taking Valproate
Weeks 1 and 2	**0.15 mg/kg/day** in 1 or 2 divided doses, rounded down to the nearest whole tablet (see Table 3 for weight based dosing guide)	**0.3 mg/kg/day** in 1 or 2 divided doses, rounded down to the nearest whole tablet	**0.6 mg/kg/day** in 2 divided doses, rounded down to the nearest whole tablet
Weeks 3 and 4	**0.3 mg/kg/day** in 1 or 2 divided doses, rounded down to the nearest whole tablet (see Table 3 for weight based dosing guide)	**0.6 mg/kg/day** in 2 divided doses, rounded down to the nearest whole tablet	**1.2 mg/kg/day** in 2 divided doses, rounded down to the nearest whole tablet
Weeks 5 onwards to maintenance	The dose should be increased every 1 to 2 weeks as follows: calculate 0.3 mg/kg/day, round this amount down to the nearest whole tablet, and add this amount to the previously administered daily dose	The dose should be increased every 1 to 2 weeks as follows: calculate 0.6 mg/kg/day, round this amount down to the nearest whole tablet, and add this amount to the previously administered daily dose	The dose should be increased every 1 to 2 weeks as follows: calculate 1.2 mg/kg/day, round this amount down to the nearest whole tablet, and add this amount to the previously administered daily dose
Usual Maintenance Dose	**1 to 5 mg/kg/day** (maximum 200 mg/day in 1 or 2 divided doses). **1 to 3 mg/kg/day** with valproate alone	**4.5 to 7.5 mg/kg/day** (maximum 300 mg/day in 2 divided doses)	**5 to 15 mg/kg/day** (maximum 400 mg/day in 2 divided doses)
Maintenance dose in patients less than 30 kg	May need to be increased by as much as 50%, based on clinical response	May need to be increased by as much as 50%, based on clinical response	May need to be increased by as much as 50%, based on clinical response

Note: Only whole tablets should be used for dosing.
* Valproate has been shown to inhibit glucuronidation and decrease the apparent clearance of lamotrigine *[see Drug Interactions (7), Pharmacokinetics (12.3)]*.
† These drugs induce glucuronidation and increase clearance *[see Drug Interactions (7), Pharmacokinetics (12.3)]*. Other drugs which have similar effects include estrogen-containing oral contraceptives and rifampin *[see Drug Interactions (7), Pharmacokinetics (12.3)]*. Dosing recommendations for oral contraceptives can be found in General Dosing Considerations *[see Dosage and Administration (2.1)]*. Patients on rifampin should follow the same dosing titration/maintenance regimen used with drugs that induce glucuronidation and increase clearance.

dose, the greater consideration should be given to restarting with the initial dosing recommendations. If a patient has discontinued lamotrigine for a period of more than 5 half-lives, it is recommended that initial dosing recommendations and guidelines be followed. The half-life of lamotrigine is affected by other concomitant medications *[see Clinical Pharmacology (12.3)]*.
LAMICTAL Added to Drugs Known to Induce or Inhibit Glucuronidation: Drugs other than those listed in the Clinical Pharmacology section *[see Clinical Pharmacology (12.3)]* have not been systematically evaluated in combination with lamotrigine. Because lamotrigine is metabolized predominantly by glucuronic acid conjugation, drugs that are known to induce or inhibit glucuronidation may affect the apparent clearance of lamotrigine and doses of LAMICTAL may require adjustment based on clinical response.
Target Plasma Levels for Patients With Epilepsy or Bipolar Disorder: A therapeutic plasma concentration range has not been established for lamotrigine. Dosing of LAMICTAL should be based on therapeutic response *[see Clinical Pharmacology (12.3)]*.
Women Taking Estrogen-Containing Oral Contraceptives: *Starting LAMICTAL in Women Taking Estrogen-Containing Oral Contraceptives:* Although estrogen-containing oral contraceptives have been shown to increase the clearance of lamotrigine *[see Clinical Pharmacology (12.3)]*, no adjustments to the recommended dose-escalation

guidelines for LAMICTAL should be necessary solely based on the use of estrogen-containing oral contraceptives. Therefore, dose escalation should follow the recommended guidelines for initiating adjunctive therapy with LAMICTAL based on the concomitant AED or other concomitant medications (see Table 1 or Table 5). See below for adjustments to maintenance doses of LAMICTAL in women taking estrogen-containing oral contraceptives.

Adjustments to the Maintenance Dose of LAMICTAL In Women Taking Estrogen-Containing Oral Contraceptives:

(1) Taking Estrogen-Containing Oral Contraceptives: For women not taking carbamazepine, phenytoin, phenobarbital, primidone, or rifampin, the maintenance dose of LAMICTAL will in most cases need to be increased, by as much as 2-fold over the recommended target maintenance dose, in order to maintain a consistent lamotrigine plasma level *[see Clinical Pharmacology (12.3)]*.

(2) Starting Estrogen-Containing Oral Contraceptives: In women taking a stable dose of LAMICTAL and not taking carbamazepine, phenytoin, phenobarbital, primidone, or rifampin, the maintenance dose will in most cases need to be increased by as much as 2-fold in order to maintain a consistent lamotrigine plasma level. The dose increases should begin at the same time that the oral contraceptive is introduced and continue, based on clinical response, no more rapidly than 50 to 100 mg/day every week. Dose increases should not exceed the recommended rate (see Table 1 or Table 5) unless lamotrigine plasma levels or clinical response support larger increases. Gradual transient increases in lamotrigine plasma levels may occur during the week of inactive hormonal preparation ("pill-free" week), and these increases will be greater if dose increases are made in the days before or during the week of inactive hormonal preparation. Increased lamotrigine plasma levels could result in additional adverse reactions, such as dizziness, ataxia, and diplopia. If adverse reactions attributable to LAMICTAL consistently occur during the "pill-free" week, dose adjustments to the overall maintenance dose may be necessary. Dose adjustments limited to the "pill-free" week are not recommended. For women taking LAMICTAL in addition to carbamazepine, phenytoin, phenobarbital, primidone, or rifampin, no adjustment should be necessary to the dose of LAMICTAL.

(3) Stopping Estrogen-Containing Oral Contraceptives: For women not taking carbamazepine, phenytoin, phenobarbital, primidone, or rifampin, the maintenance dose of LAMICTAL will in most cases need to be decreased by as much as 50% in order to maintain a consistent lamotrigine plasma level. The decrease in dose of LAMICTAL should not exceed 25% of the total daily dose per week over a 2-week period, unless clinical response or lamotrigine plasma levels indicate otherwise *[see Clinical Pharmacology (12.3)]*. For women taking LAMICTAL in addition to carbamazepine, phenytoin, phenobarbital, primidone, or rifampin, no adjustment to the dose of LAMICTAL should be necessary.

Women and Other Hormonal Contraceptive Preparations or Hormone Replacement Therapy: The effect of other hormonal contraceptive preparations or hormone replacement therapy on the pharmacokinetics of lamotrigine has not been systematically evaluated. It has been reported that ethinylestradiol, not progestogens, increased the clearance of lamotrigine up to 2-fold, and the progestin-only pills had no effect on lamotrigine plasma levels. Therefore, adjustments to the dosage of LAMICTAL in the presence of progestogens alone will likely not be needed.

Patients With Hepatic Impairment: Experience in patients with hepatic impairment is limited. Based on a clinical pharmacology study in 24 patients with mild, moderate, and severe liver impairment *[see Use in Specific Populations (8.6), Clinical Pharmacology (12.3)]*, the following general recommendations can be made. No dosage adjustment is needed in patients with mild liver impairment. Initial, escalation, and maintenance doses should generally be reduced by approximately 25% in patients with moderate and severe liver impairment without ascites and 50% in patients with severe liver impairment with ascites. Escalation and maintenance doses may be adjusted according to clinical response.

Patients With Renal Impairment: Initial doses of LAMICTAL should be based on patients' concomitant medications (see Tables 1-3 or Table 5); reduced maintenance doses may be effective for patients with significant renal impairment *[see Use in Specific Populations (8.7), Clinical Pharmacology (12.3)]*. Few patients with severe renal impairment have been evaluated during chronic treatment with LAMICTAL. Because there is inadequate experience in this population, LAMICTAL should be used with caution in these patients.

Discontinuation Strategy: *Epilepsy:* For patients receiving LAMICTAL in combination with other AEDs, a re-evaluation of all AEDs in the regimen should be considered if a change in seizure control or an appearance or worsening of adverse reactions is observed.

If a decision is made to discontinue therapy with LAMICTAL, a step-wise reduction of dose over at least 2 weeks (approximately 50% per week) is recommended unless safety concerns require a more rapid withdrawal *[see Warnings and Precautions (5.9)]*.

Discontinuing carbamazepine, phenytoin, phenobarbital, or primidone should prolong the half-life of lamotrigine; discontinuing valproate should shorten the half-life of lamotrigine.

Bipolar Disorder: In the controlled clinical trials, there was no increase in the incidence, type, or severity of adverse reactions following abrupt termination of LAMICTAL. In clinical trials in patients with Bipolar Disorder, 2 patients experienced seizures shortly after abrupt withdrawal of LAMICTAL. However, there were confounding factors that may have contributed to the occurrence of seizures in these bipolar patients. Discontinuation of LAMICTAL should involve a step-wise reduction of dose over at least 2 weeks (approximately 50% per week) unless safety concerns require a more rapid withdrawal *[see Warnings and Precautions (5.9)]*.

Table 3. The Initial Weight-Based Dosing Guide for Patients 2 to 12 Years Taking Valproate (Weeks 1 to 4) With Epilepsy

If the patient's weight is		Give this daily dose, using the most appropriate combination of LAMICTAL 2-mg and 5-mg tablets	
Greater than	And less than	Weeks 1 and 2	Weeks 3 and 4
6.7 kg	14 kg	2 mg every *other* day	2 mg every day
14.1 kg	27 kg	2 mg every day	4 mg every day
27.1 kg	34 kg	4 mg every day	8 mg every day
34.1 kg	40 kg	5 mg every day	10 mg every day

2.2 Epilepsy—Adjunctive Therapy

This section provides specific dosing recommendations for patients greater than 12 years of age and patients 2 to 12 years of age. Within each of these age-groups, specific dosing recommendations are provided depending upon concomitant AED or other concomitant medications (Table 1 for patients greater than 12 years of age and Table 2 for patients 2 to 12 years of age). A weight-based dosing guide for patients 2 to 12 years of age on concomitant valproate is provided in Table 3.

Patients Over 12 Years of Age: Recommended dosing guidelines are summarized in Table 1.

[See table 1 at top of previous page]

Patients 2 to 12 Years of Age: Recommended dosing guidelines are summarized in Table 2.

Smaller starting doses and slower dose escalations than those used in clinical trials are recommended because of the suggestion that the risk of rash may be decreased by smaller starting doses and slower dose escalations. Therefore, maintenance doses will take longer to reach in clinical practice than in clinical trials. It may take several weeks to months to achieve an individualized maintenance dose. Maintenance doses in patients weighing less than 30 kg, regardless of age or concomitant AED, may need to be increased as much as 50%, based on clinical response.

The smallest available strength of LAMICTAL Chewable Dispersible Tablets is 2 mg, and only whole tablets should be administered. If the calculated dose cannot be achieved using whole tablets, the dose should be rounded down to the nearest whole tablet *[see How Supplied/Storage and Handling (16) and Medication Guide]*.

[See table 2 at top of previous page]

[See table 3 above]

Usual Adjunctive Maintenance Dose for Epilepsy: The usual maintenance doses identified in Tables 1 and 2 are derived from dosing regimens employed in the placebo-controlled adjunctive studies in which the efficacy of LAMICTAL was established. In patients receiving multidrug regimens employing carbamazepine, phenytoin, phenobarbital, or primidone **without valproate**, maintenance doses of adjunctive LAMICTAL as high as 700 mg/day have been used. In patients receiving **valproate alone**, maintenance doses of adjunctive LAMICTAL as high as 200 mg/day have been used. The advantage of using doses above those recommended in Tables 1 through 4 has not been established in controlled trials.

2.3 Epilepsy—Conversion From Adjunctive Therapy to Monotherapy

The goal of the transition regimen is to effect the conversion to monotherapy with LAMICTAL under conditions that ensure adequate seizure control while mitigating the risk of serious rash associated with the rapid titration of LAMICTAL.

The recommended maintenance dose of LAMICTAL as monotherapy is 500 mg/day given in 2 divided doses.

To avoid an increased risk of rash, the recommended initial dose and subsequent dose escalations of LAMICTAL should not be exceeded *[see Boxed Warning]*.

Conversion From Adjunctive Therapy With Carbamazepine, Phenytoin, Phenobarbital, or Primidone to Monotherapy With LAMICTAL: After achieving a dose of 500 mg/day of LAMICTAL according to the guidelines in Table 1, the concomitant AED should be withdrawn by 20% decrements each week over a 4-week period. The regimen for the withdrawal of the concomitant AED is based on experience gained in the controlled monotherapy clinical trial.

Conversion from Adjunctive Therapy With Valproate to Monotherapy With LAMICTAL: The conversion regimen involves 4 steps outlined in Table 4.

Table 4. Conversion From Adjunctive Therapy With Valproate to Monotherapy With LAMICTAL in Patients ≥16 Years of Age with Epilepsy

	LAMICTAL	Valproate
Step 1	Achieve a dose of 200 mg/day according to guidelines in Table 1 (if not already on 200 mg/day).	Maintain previous stable dose.
Step 2	Maintain at 200 mg/day.	Decrease to 500 mg/day by decrements no greater than 500 mg/day/week and then maintain the dose of 500 mg/day for 1 week.
Step 3	Increase to 300 mg/day and maintain for 1 week.	Simultaneously decrease to 250 mg/day and maintain for 1 week.
Step 4	Increase by 100 mg/day every week to achieve maintenance dose of 500 mg/day.	Discontinue.

Conversion from Adjunctive Therapy With AEDs Other Than Carbamazepine, Phenytoin, Phenobarbital, Primidone, or Valproate to Monotherapy With LAMICTAL: No specific dosing guidelines can be provided for conversion to monotherapy with LAMICTAL with AEDs other than carbamazepine, phenobarbital, phenytoin, primidone, or valproate.

2.4 Bipolar Disorder

The goal of maintenance treatment with LAMICTAL is to delay the time to occurrence of mood episodes (depression, mania, hypomania, mixed episodes) in patients treated for acute mood episodes with standard therapy. The target dose of LAMICTAL is 200 mg/day (100 mg/day in patients taking valproate, which decreases the apparent clearance of lamotrigine, and 400 mg/day in patients not taking valproate and taking either carbamazepine, phenytoin, phenobarbital, primidone, or rifampin, which increase the apparent clearance of lamotrigine). In the clinical trials, doses up to 400 mg/day as monotherapy were evaluated; however, no additional benefit was seen at 400 mg/day compared with 200 mg/day *[see Clinical Studies (14.2)]*. Accordingly, doses above 200 mg/day are not recommended. Treatment with LAMICTAL is introduced, based on concurrent medications, according to the regimen outlined in Table 5. If other psychotropic medications are withdrawn following stabilization, the dose of LAMICTAL should be adjusted. For patients discontinuing valproate, the dose of LAMICTAL should be doubled over a 2-week period in equal weekly increments (see Table 6). For patients discontinuing carbamazepine, phenytoin, phenobarbital, primidone, or rifampin, the dose of LAMICTAL should remain constant

Continued on next page

Lamictal—Cont.

for the first week and then should be decreased by half over a 2-week period in equal weekly decrements (see Table 6). The dose of LAMICTAL may then be further adjusted to the target dose (200 mg) as clinically indicated.

If other drugs are subsequently introduced, the dose of LAMICTAL may need to be adjusted. In particular, the introduction of valproate requires reduction in the dose of LAMICTAL *[see Drug Interactions (7), Clinical Pharmacology (12.3)]*.

To avoid an increased risk of rash, the recommended initial dose and subsequent dose escalations of LAMICTAL should not be exceeded *[see Boxed Warning]*.

[See table 5 above]

[See table 6 above]

The benefit of continuing treatment in patients who had been stabilized in an 8- to 16-week open-label phase with LAMICTAL was established in 2 randomized, placebo-controlled clinical maintenance trials *[see Clinical Studies (14.2)]*. However, the optimal duration of treatment with LAMICTAL has not been established. Thus, patients should be periodically reassessed to determine the need for maintenance treatment.

2.5 Administration of LAMICTAL Chewable Dispersible Tablets

LAMICTAL Chewable Dispersible Tablets may be swallowed whole, chewed, or dispersed in water or diluted fruit juice. If the tablets are chewed, consume a small amount of water or diluted fruit juice to aid in swallowing.

To disperse LAMICTAL Chewable Dispersible Tablets, add the tablets to a small amount of liquid (1 teaspoon, or enough to cover the medication). Approximately 1 minute later, when the tablets are completely dispersed, swirl the solution and consume the entire quantity immediately. *No attempt should be made to administer partial quantities of the dispersed tablets.*

2.6 Administration of LAMICTAL ODT Orally Disintegrating Tablets

LAMICTAL ODT Orally Disintegrating Tablets should be placed onto the tongue and moved around in the mouth. The tablet will disintegrate rapidly, can be swallowed with or without water, and can be taken with or without food.

3 DOSAGE FORMS AND STRENGTHS

3.1 Tablets

25 mg, white, scored, shield-shaped tablets debossed with "LAMICTAL" and "25"

100 mg, peach, scored, shield-shaped tablets debossed with "LAMICTAL" and "100"

150 mg, cream, scored, shield-shaped tablets debossed with "LAMICTAL" and "150"

200 mg, blue, scored, shield-shaped tablets debossed with "LAMICTAL" and "200"

3.2 Chewable Dispersible Tablets

2 mg, white to off-white, round tablets debossed with "LTG" over "2"

5 mg, white to off-white, caplet-shaped tablets debossed with "GX CL2"

25 mg, white, super elliptical-shaped tablets debossed with "GX CL5"

3.3 Orally Disintegrating Tablets

25 mg, white to off-white, round, flat-faced, radius edge, tablets debossed with "LMT" on one side and "25" on the other side.

50 mg, white to off-white, round, flat-faced, radius edge, tablets debossed with "LMT" on one side and "50" on the other side.

100 mg, white to off-white, round, flat-faced, radius edge, tablets debossed with "LAMICTAL" on one side and "100" on the other side.

200 mg, white to off-white, round, flat-faced, radius edge, tablets debossed with "LAMICTAL" on one side and "200" on the other side.

3.4 Potential Medication Errors

Patients should be strongly advised to visually inspect their tablets to verify that they are receiving LAMICTAL as well as the correct formulation of LAMICTAL each time they fill their prescription. Depictions of the LAMICTAL Tablets, Chewable Dispersible Tablets, and Orally Disintegrating Tablets can be found in the Medication Guide that accompanies the product.

4 CONTRAINDICATIONS

LAMICTAL is contraindicated in patients who have demonstrated hypersensitivity to the drug or its ingredients *[see Boxed Warning]*.

5 WARNINGS AND PRECAUTIONS

5.1 Serious Skin Rashes *[see Boxed Warning]*

Pediatric Population: The incidence of serious rash associated with hospitalization and discontinuation of LAMICTAL in a prospectively followed cohort of pediatric patients (2 to 16 years of age) with epilepsy receiving adjunctive therapy was approximately 0.8% (16 of 1,983). When 14 of these cases were reviewed by 3 expert dermatologists, there was considerable disagreement as to their proper classification. To illustrate, one dermatologist considered none of the cases to be Stevens-Johnson syndrome; another assigned 7 of the 14 to this diagnosis. There was 1 rash-related death in this 1,983-patient cohort. Additionally, there have been rare cases of toxic epidermal necrolysis with and without permanent sequelae and/or death in US and foreign postmarketing experience.

There is evidence that the inclusion of valproate in a multidrug regimen increases the risk of serious, potentially life-threatening rash in pediatric patients. In pediatric patients who used valproate concomitantly, 1.2% (6 of 482) experienced a serious rash compared with 0.6% (6 of 952) patients not taking valproate.

Adult Population: Serious rash associated with hospitalization and discontinuation of LAMICTAL occurred in 0.3% (11 of 3,348) of adult patients who received LAMICTAL in premarketing clinical trials of epilepsy. In the bipolar and other mood disorders clinical trials, the rate of serious rash was 0.08% (1 of 1,233) of adult patients who received LAMICTAL as initial monotherapy and 0.13% (2 of 1,538) of adult patients who received LAMICTAL as adjunctive therapy. No fatalities occurred among these individuals. However, in worldwide postmarketing experience, rare cases of rash-related death have been reported, but their numbers are too few to permit a precise estimate of the rate.

Among the rashes leading to hospitalization were Stevens-Johnson syndrome, toxic epidermal necrolysis, angioedema, and a rash associated with a variable number of the following systemic manifestations: fever, lymphadenopathy, facial swelling, and hematologic and hepatologic abnormalities.

There is evidence that the inclusion of valproate in a multidrug regimen increases the risk of serious, potentially life-threatening rash in adults. Specifically, of 584 patients administered LAMICTAL with valproate in epilepsy clinical trials, 6 (1%) were hospitalized in association with rash; in contrast, 4 (0.16%) of 2,398 clinical trial patients and volunteers administered LAMICTAL in the absence of valproate were hospitalized.

Patients With History of Allergy or Rash to Other AEDs: The risk of nonserious rash may be increased when the recommended initial dose and/or the rate of dose escalation of LAMICTAL is exceeded and in patients with a history of allergy or rash to other AEDs.

5.2 Hypersensitivity Reactions

Hypersensitivity reactions, some fatal or life-threatening, have also occurred. Some of these reactions have included clinical features of multiorgan failure/dysfunction, including hepatic abnormalities and evidence of disseminated intravascular coagulation. It is important to note that early manifestations of hypersensitivity (e.g., fever, lymphadenopathy) may be present even though a rash is not evident. If such signs or symptoms are present, the patient should be evaluated immediately. LAMICTAL should be discontinued if an alternative etiology for the signs or symptoms cannot be established.

Prior to initiation of treatment with LAMICTAL, the patient should be instructed that a rash or other signs or symptoms of hypersensitivity (e.g., fever, lymphadenopathy) may herald a serious medical event and that the patient should report any such occurrence to a physician immediately.

5.3 Acute Multiorgan Failure

Multiorgan failure, which in some cases has been fatal or irreversible, has been observed in patients receiving LAMICTAL. Fatalities associated with multiorgan failure and various degrees of hepatic failure have been reported in 2 of 3,796 adult patients and 4 of 2,435 pediatric patients who received LAMICTAL in epilepsy clinical trials. No such fatalities have been reported in bipolar patients in clinical trials. Rare fatalities from multiorgan failure have also been reported in compassionate plea and postmarketing use. The majority of these deaths occurred in association with other serious medical events, including status epilepticus and overwhelming sepsis, and hantavirus, making it difficult to identify the initial cause.

Additionally, 3 patients (a 45-year-old woman, a 3.5-year-old boy, and an 11-year-old girl) developed multiorgan dysfunction and disseminated intravascular coagulation 9 to 14 days after LAMICTAL was added to their AED regimens. Rash and elevated transaminases were also present in all patients and rhabdomyolysis was noted in 2 patients. Both

Table 5. Escalation Regimen for LAMICTAL for Patients With Bipolar Disorder*

	For Patients Taking Valproate‡	For Patients Not Taking Carbamazepine, Phenytoin, Phenobarbital, Primidone, or Rifampin† and Not Taking Valproate‡	For Patients Taking Carbamazepine, Phenytoin, Phenobarbital, Primidone, or Rifampin† and Not Taking Valproate‡
Weeks 1 and 2	25 mg every *other* day	25 mg daily	50 mg daily
Weeks 3 and 4	25 mg daily	50 mg daily	100 mg daily, in divided doses
Week 5	50 mg daily	100 mg daily	200 mg daily, in divided doses
Week 6	100 mg daily	200 mg daily	300 mg daily, in divided doses
Week 7	100 mg daily	200 mg daily	up to 400 mg daily, in 12 divided doses

* See *Drug Interactions (7)* and *Clinical Pharmacology (12.3)* for a description of known drug interactions.

† Carbamazepine, phenytoin, phenobarbital, primidone, and rifampin have been shown to increase the apparent clearance of lamotrigine.

‡ Valproate has been shown to decrease the apparent clearance of lamotrigine.

Table 6. Dosage Adjustments to LAMICTAL for Patients With Bipolar Disorder Following Discontinuation of Psychotropic Medications*

	Discontinuation of Psychotropic Drugs (excluding Carbamazepine, Phenytoin, Phenobarbital, Primidone, Rifampin†, or Valproate‡)	After Discontinuation of Valproate‡	After Discontinuation of Carbamazepine, Phenytoin, Phenobarbital, Primidone, or Rifampin†
		Current dose of LAMICTAL (mg/day) 100	Current dose of LAMICTAL (mg/day) 400
Week 1	Maintain current dose of LAMICTAL	150	400
Week 2	Maintain current dose of LAMICTAL	200	300
Week 3 onward	Maintain current dose of LAMICTAL	200	200

* See *Drug Interactions (7)* and *Clinical Pharmacology (12.3)* for a description of known drug interactions.

† Carbamazepine, phenytoin, phenobarbital, primidone, and rifampin have been shown to increase the apparent clearance of lamotrigine.

‡ Valproate has been shown to decrease the apparent clearance of lamotrigine.

pediatric patients were receiving concomitant therapy with valproate, while the adult patient was being treated with carbamazepine and clonazepam. All patients subsequently recovered with supportive care after treatment with LAMICTAL was discontinued.

5.4 Blood Dyscrasias

There have been reports of blood dyscrasias that may or may not be associated with the hypersensitivity syndrome. These have included neutropenia, leukopenia, anemia, thrombocytopenia, pancytopenia, and, rarely, aplastic anemia and pure red cell aplasia.

5.5 Suicidal Behavior and Ideation

Antiepileptic drugs (AEDs), including LAMICTAL, increase the risk of suicidal thoughts or behavior in patients taking these drugs for any indication. Patients treated with any AED for any indication should be monitored for the emergence or worsening of depression, suicidal thoughts or behavior, and/or any unusual changes in mood or behavior.

Pooled analyses of 199 placebo-controlled clinical trials (mono- and adjunctive therapy) of 11 different AEDs showed that patients randomized to one of the AEDs had approximately twice the risk (adjusted Relative Risk 1.8, 95% CI:1.2, 2.7) of suicidal thinking or behavior compared to patients randomized to placebo. In these trials, which had a median treatment duration of 12 weeks, the estimated incidence of suicidal behavior or ideation among 27,863 AED-treated patients was 0.43%, compared to 0.24% among 16,029 placebo-treated patients, representing an increase of approximately 1 case of suicidal thinking or behavior for every 530 patients treated. There were 4 suicides in drug-treated patients in the trials and none in placebo-treated patients, but the number of events is too small to allow any conclusion about drug effect on suicide.

The increased risk of suicidal thoughts or behavior with AEDs was observed as early as 1 week after starting treatment with AEDs and persisted for the duration of treatment assessed. Because most trials included in the analysis did not extend beyond 24 weeks, the risk of suicidal thoughts or behavior beyond 24 weeks could not be assessed.

The risk of suicidal thoughts or behavior was generally consistent among drugs in the data analyzed. The finding of increased risk with AEDs of varying mechanism of action and across a range of indications suggests that the risk applies to all AEDs used for any indication. The risk did not vary substantially by age (5 to 100 years) in the clinical trials analyzed.

Table 7 shows absolute and relative risk by indication for all evaluated AEDs.

Table 7. Risk by Indication for Antiepileptic Drugs in the Pooled Analysis

Indication	Placebo Patients With Events Per 1,000 Patients	Drug Patients With Events Per 1,000 Patients	Relative Risk: Incidence of Events in Drug Patients/ Incidence in Placebo Patients	Risk Difference: Additional Drug Patients With Events Per 1,000 Patients
Epilepsy	1.0	3.4	3.5	2.4
Psychiatric	5.7	8.5	1.5	2.9
Other	1.0	1.8	1.9	0.9
Total	2.4	4.3	1.8	1.9

[See table 7 above]

The relative risk for suicidal thoughts or behavior was higher in clinical trials for epilepsy than in clinical trials for psychiatric or other conditions, but the absolute risk differences were similar for the epilepsy and psychiatric indications.

Anyone considering prescribing LAMICTAL or any other AED must balance the risk of suicidal thoughts or behavior with the risk of untreated illness. Epilepsy and many other illnesses for which AEDs are prescribed are themselves associated with morbidity and mortality and an increased risk of suicidal thoughts and behavior. Should suicidal thoughts and behavior emerge during treatment, the prescriber needs to consider whether the emergence of these symptoms in any given patient may be related to the illness being treated.

Patients, their caregivers, and families should be informed that AEDs increase the risk of suicidal thoughts and behavior and should be advised of the need to be alert for the emergence or worsening of the signs and symptoms of depression, any unusual changes in mood or behavior, or the emergence of suicidal thoughts, behavior, or thoughts about self-harm. Behaviors of concern should be reported immediately to healthcare providers.

5.6 Use in Patients With Bipolar Disorder

Acute Treatment of Mood Episodes: Safety and effectiveness of LAMICTAL in the acute treatment of mood episodes have not been established.

Children and Adolescents (less than 18 years of age): Safety and effectiveness of LAMICTAL in patients below the age of 18 years with mood disorders have not been established *[see Suicidal Behavior and Ideation (5.5)]*.

Clinical Worsening and Suicide Risk Associated With Bipolar Disorder: Patients with bipolar disorder may experience worsening of their depressive symptoms and/or the emergence of suicidal ideation and behaviors (suicidality) whether or not they are taking medications for bipolar disorder. Patients should be closely monitored for clinical worsening (including development of new symptoms) and suicidality, especially at the beginning of a course of treatment, or at the time of dose changes.

In addition, patients with a history of suicidal behavior or thoughts, those patients exhibiting a significant degree of suicidal ideation prior to commencement of treatment, and young adults are at an increased risk of suicidal thoughts or suicide attempts, and should receive careful monitoring during treatment *[see Suicidal Behavior and Ideation (5.5)]*. Consideration should be given to changing the therapeutic regimen, including possibly discontinuing the medication, in patients who experience clinical worsening (including development of new symptoms) and/or the emergence of suicidal ideation/behavior especially if these symptoms are severe, abrupt in onset, or were not part of the patient's presenting symptoms.

Prescriptions for LAMICTAL should be written for the smallest quantity of tablets consistent with good patient management in order to reduce the risk of overdose. Overdoses have been reported for LAMICTAL, some of which have been fatal *[see Overdosage (10.1)]*.

5.7 Potential Medication Errors

Medication errors involving LAMICTAL have occurred. In particular, the name LAMICTAL or lamotrigine can be confused with the names of other commonly used medications. Medication errors may also occur between the different formulations of LAMICTAL. To reduce the potential of medication errors, write and say LAMICTAL clearly. Depictions of the LAMICTAL Tablets, Chewable Dispersible Tablets, and Orally Disintegrating Tablets can be found in the Medication Guide that accompanies the product to highlight the distinctive markings, colors, and shapes that serve to identify the different presentations of the drug and thus may help reduce the risk of medication errors. To avoid the medication error of using the wrong drug or formulation, patients should be strongly advised to visually inspect their tablets to verify that they are LAMICTAL, as well as the correct formulation of LAMICTAL, each time they fill their prescription.

5.8 Concomitant Use With Oral Contraceptives

Some estrogen-containing oral contraceptives have been shown to decrease serum concentrations of lamotrigine *[see Clinical Pharmacology (12.3)]*. **Dosage adjustments will be necessary in most patients who start or stop estrogen-containing oral contraceptives while taking LAMICTAL** *[see Dosage and Administration (2.1)]*. During the week of inactive hormone preparation ("pill-free" week) of oral contraceptive therapy, plasma lamotrigine levels are expected to rise, as much as doubling at the end of the week. Adverse reactions consistent with elevated levels of lamotrigine, such as dizziness, ataxia, and diplopia, could occur.

5.9 Withdrawal Seizures

As with other AEDs, LAMICTAL should not be abruptly discontinued. In patients with epilepsy there is a possibility of increasing seizure frequency. In clinical trials in patients with Bipolar Disorder, 2 patients experienced seizures shortly after abrupt withdrawal of LAMICTAL. However, there were confounding factors that may have contributed to the occurrence of seizures in these bipolar patients. Unless safety concerns require a more rapid withdrawal, the dose of LAMICTAL should be tapered over a period of at least 2 weeks (approximately 50% reduction per week) *[see Dosage and Administration (2.1)]*.

5.10 Status Epilepticus

Valid estimates of the incidence of treatment-emergent status epilepticus among patients treated with LAMICTAL are difficult to obtain because reporters participating in clinical trials did not all employ identical rules for identifying cases. At a minimum, 7 of 2,343 adult patients had episodes that could unequivocally be described as status epilepticus. In addition, a number of reports of variably defined episodes of seizure exacerbation (e.g., seizure clusters, seizure flurries, etc.) were made.

5.11 Sudden Unexplained Death in Epilepsy (SUDEP)

During the premarketing development of LAMICTAL, 20 sudden and unexplained deaths were recorded among a cohort of 4,700 patients with epilepsy (5,747 patient-years of exposure).

Some of these could represent seizure-related deaths in which the seizure was not observed, e.g., at night. This represents an incidence of 0.0035 deaths per patient-year. Although this rate exceeds that expected in a healthy population matched for age and sex, it is within the range of estimates for the incidence of sudden unexplained deaths in patients with epilepsy not receiving LAMICTAL (ranging from 0.0005 for the general population of patients with epilepsy, to 0.004 for a recently studied clinical trial population similar to that in the clinical development program for LAMICTAL, to 0.005 for patients with refractory epilepsy). Consequently, whether these figures are reassuring or suggest concern depends on the comparability of the populations reported upon to the cohort receiving LAMICTAL and the accuracy of the estimates provided. Probably most reassuring is the similarity of estimated SUDEP rates in patients receiving LAMICTAL and those receiving other AEDs, chemically unrelated to each other, that underwent clinical testing in similar populations. Importantly, that drug is chemically unrelated to LAMICTAL. This evidence suggests, although it certainly does not prove, that the high SUDEP rates reflect population rates, not a drug effect.

5.12 Addition of LAMICTAL to a Multidrug Regimen That Includes Valproate

Because valproate reduces the clearance of lamotrigine, the dosage of lamotrigine in the presence of valproate is less than half of that required in its absence.

5.13 Binding in the Eye and Other Melanin-Containing Tissues

Because lamotrigine binds to melanin, it could accumulate in melanin-rich tissues over time. This raises the possibility that lamotrigine may cause toxicity in these tissues after extended use. Although ophthalmological testing was performed in one controlled clinical trial, the testing was inadequate to exclude subtle effects or injury occurring after long-term exposure. Moreover, the capacity of available tests to detect potentially adverse consequences, if any, of lamotrigine's binding to melanin is unknown *[see Clinical Pharmacology (12.2)]*.

Accordingly, although there are no specific recommendations for periodic ophthalmological monitoring, prescribers should be aware of the possibility of long-term ophthalmologic effects.

5.14 Laboratory Tests

The value of monitoring plasma concentrations of lamotrigine in patients treated with LAMICTAL has not been established. Because of the possible pharmacokinetic interactions between lamotrigine and other drugs including AEDs (see Table 15), monitoring of the plasma levels of lamotrigine and concomitant drugs may be indicated, particularly during dosage adjustments. In general, clinical judgment should be exercised regarding monitoring of plasma levels of lamotrigine and other drugs and whether or not dosage adjustments are necessary.

6 ADVERSE REACTIONS

The following adverse reactions are described in more detail in the *Warnings and Precautions* section of the label:

- Serious skin rashes *[see Warnings and Precautions (5.1)]*
- Hypersensitivity reactions *[see Warnings and Precautions (5.2)]*
- Acute multiorgan failure *[see Warnings and Precautions (5.3)]*
- Blood dyscrasias *[see Warnings and Precautions (5.4)]*
- Suicidal behavior and ideation *[see Warnings and Precautions (5.5)]*
- Withdrawal seizures *[see Warnings and Precautions (5.9)]*
- Status epilepticus *[see Warnings and Precautions (5.10)]*
- Sudden unexplained death in epilepsy *[see Warnings and Precautions (5.11)]*

Continued on next page

Product information on these pages is effective as of June 2009. Further information is available at 1-888-825-5249 or www.gsk.com.

Lamictal—Cont.

6.1 Clinical Trials

Because clinical trials are conducted under widely varying conditions, adverse reaction rates observed in the clinical trials of a drug cannot be directly compared with rates in the clinical trials of another drug and may not reflect the rates observed in practice.

LAMICTAL has been evaluated for safety in patients with epilepsy and in patients with Bipolar I Disorder. Adverse reactions reported for each of these patient populations are provided below. Excluded are adverse reactions considered too general to be informative and those not reasonably attributable to the use of the drug.

Epilepsy: *Most Common Adverse Reactions in All Clinical Studies: Adjunctive Therapy in Adults With Epilepsy:* The most commonly observed (≥5% for LAMICTAL and more common on drug than placebo) adverse reactions seen in association with LAMICTAL during adjunctive therapy in adults and not seen at an equivalent frequency among placebo-treated patients were: dizziness, ataxia, somnolence, headache, diplopia, blurred vision, nausea, vomiting, and rash. Dizziness, diplopia, ataxia, blurred vision, nausea, and vomiting were dose-related. Dizziness, diplopia, ataxia, and blurred vision occurred more commonly in patients receiving carbamazepine with LAMICTAL than in patients receiving other AEDs with LAMICTAL. Clinical data suggest a higher incidence of rash, including serious rash, in patients receiving concomitant valproate than in patients not receiving valproate *[see Warnings and Precautions (5.1)]*.

Approximately 11% of the 3,378 adult patients who received LAMICTAL as adjunctive therapy in premarketing clinical trials discontinued treatment because of an adverse reaction. The adverse reactions most commonly associated with discontinuation were rash (3.0%), dizziness (2.8%), and headache (2.5%).

In a dose-response study in adults, the rate of discontinuation of LAMICTAL for dizziness, ataxia, diplopia, blurred vision, nausea, and vomiting was dose-related.

Monotherapy in Adults With Epilepsy: The most commonly observed (≥5% for LAMICTAL and more common on drug than placebo) adverse reactions seen in association with the use of LAMICTAL during the monotherapy phase of the controlled trial in adults not seen at an equivalent rate in the control group were vomiting, coordination abnormality, dyspepsia, nausea, dizziness, rhinitis, anxiety, insomnia, infection, pain, weight decrease, chest pain, and dysmenorrhea. The most commonly observed (≥5% for LAMICTAL and more common on drug than placebo) adverse reactions associated with the use of LAMICTAL during the conversion to monotherapy (add-on) period, not seen at an equivalent frequency among low-dose valproate-treated patients, were dizziness, headache, nausea, asthenia, coordination abnormality, vomiting, rash, somnolence, diplopia, ataxia, accidental injury, tremor, blurred vision, insomnia, nystagmus, diarrhea, lymphadenopathy, pruritus, and sinusitis.

Approximately 10% of the 420 adult patients who received LAMICTAL as monotherapy in premarketing clinical trials discontinued treatment because of an adverse reaction. The adverse reactions most commonly associated with discontinuation were rash (4.5%), headache (3.1%), and asthenia (2.4%).

Adjunctive Therapy in Pediatric Patients With Epilepsy: The most commonly observed (≥5% for LAMICTAL and more common on drug than placebo) adverse reactions seen in association with the use of LAMICTAL as adjunctive treatment in pediatric patients 2 to 16 years of age and not seen at an equivalent rate in the control group were infection, vomiting, rash, fever, somnolence, accidental injury, dizziness, diarrhea, abdominal pain, nausea, ataxia, tremor, asthenia, bronchitis, flu syndrome, and diplopia.

In 339 patients 2 to 16 years of age with partial seizures or generalized seizures of Lennox-Gastaut syndrome, 4.2% of patients on LAMICTAL and 2.9% of patients on placebo discontinued due to adverse reactions. The most commonly reported adverse reaction that led to discontinuation of LAMICTAL was rash.

Approximately 11.5% of the 1,081 pediatric patients 2 to 16 years of age who received LAMICTAL as adjunctive therapy in premarketing clinical trials discontinued treatment because of an adverse reaction. The adverse reactions most commonly associated with discontinuation were rash (4.4%), reaction aggravated (1.7%), and ataxia (0.6%).

Controlled Adjunctive Clinical Studies in Adults With Epilepsy: Table 8 lists treatment-emergent adverse reactions that occurred in at least 2% of adult patients with epilepsy treated with LAMICTAL in placebo-controlled trials and were numerically more common in the patients treated with LAMICTAL. In these studies, either LAMICTAL or placebo was added to the patient's current AED therapy. Adverse reactions were usually mild to moderate in intensity.

Table 8. Treatment-Emergent Adverse Reaction Incidence in Placebo-Controlled Adjunctive Trials in Adult Patients With Epilepsy* (Adverse reactions in at least 2% of patients treated with LAMICTAL and numerically more frequent than in the placebo group.)

Body System/ Adverse Experience	Percent of Patients Receiving Adjunctive LAMICTAL (n = 711)	Percent of Patients Receiving Adjunctive Placebo (n = 419)
Body as a whole		
Headache	29	19
Flu syndrome	7	6
Fever	6	4
Abdominal pain	5	4
Neck pain	2	1
Reaction aggravated (seizure exacerbation)	2	1
Digestive		
Nausea	19	10
Vomiting	9	4
Diarrhea	6	4
Dyspepsia	5	2
Constipation	4	3
Anorexia	2	1
Musculoskeletal		
Arthralgia	2	0
Nervous		
Dizziness	38	13
Ataxia	22	6
Somnolence	14	7
Incoordination	6	2
Insomnia	6	2
Tremor	4	1
Depression	4	3
Anxiety	4	3
Convulsion	3	1
Irritability	3	2
Speech disorder	3	0
Concentration disturbance	2	1
Respiratory		
Rhinitis	14	9
Pharyngitis	10	9
Cough increased	8	6
Skin and appendages		
Rash	10	5
Pruritus	3	2
Special senses		
Diplopia	28	7
Blurred vision	16	5
Vision abnormality	3	1
Urogenital		
Female patients only	(n = 365)	(n = 207)
Dysmenorrhea	7	6
Vaginitis	4	1
Amenorrhea	2	1

*Patients in these adjunctive studies were receiving 1 to 3 of the following concomitant AEDs (carbamazepine, phenytoin, phenobarbital, or primidone) in addition to LAMICTAL or placebo. Patients may have reported multiple adverse reactions during the study or at discontinuation; thus, patients may be included in more than one category.

In a randomized, parallel study comparing placebo and 300 and 500 mg/day of LAMICTAL, some of the more common drug-related adverse reactions were dose-related (see Table 9).

Table 9. Dose-Related Adverse Reactions From a Randomized, Placebo-Controlled Adjunctive Trial in Adults With Epilepsy

Adverse Reactions	Percent of Patients Experiencing Adverse Reactions: Placebo (n = 73)	LAMICTAL 300 mg (n = 71)	LAMICTAL 500 mg (n = 72)
Ataxia	10	10	28*†
Blurred vision	10	11	25*†
Diplopia	8	24*	49*†
Dizziness	27	31	54*†
Nausea	11	18	25*
Vomiting	4	11	18*

* Significantly greater than placebo group ($p<0.05$).
† Significantly greater than group receiving LAMICTAL 300 mg ($p<0.05$).

The overall adverse reaction profile for LAMICTAL was similar between females and males, and was independent of age. Because the largest non-Caucasian racial subgroup was only 6% of patients exposed to LAMICTAL in placebo-controlled trials, there are insufficient data to support a statement regarding the distribution of adverse reaction reports by race. Generally, females receiving either LAMICTAL as adjunctive therapy or placebo were more likely to report adverse reactions than males. The only adverse reaction for which the reports on LAMICTAL were greater than 10% more frequent in females than males (without a corresponding difference by gender on placebo) was dizziness (difference = 16.5%). There was little difference between females and males in the rates of discontinuation of LAMICTAL for individual adverse reactions.

Controlled Monotherapy Trial in Adults With Partial Seizures: Table 10 lists treatment-emergent adverse reactions that occurred in at least 5% of patients with epilepsy treated with monotherapy with LAMICTAL in a double-blind trial following discontinuation of either concomitant carbamazepine or phenytoin not seen at an equivalent frequency in the control group.

Table 10. Treatment-Emergent Adverse Reaction Incidence in Adults With Partial Seizures in a Controlled Monotherapy Trial* (Adverse reactions in at least 5% of patients treated with LAMICTAL and numerically more frequent than in the valproate group.)

Body System/ Adverse Experience	Percent of Patients Receiving LAMICTAL as Monotherapy† (n = 43)	Percent of Patients Receiving Low-Dose Valproate‡ Monotherapy (n = 44)
Body as a whole		
Pain	5	0
Infection	5	2
Chest pain	5	2
Digestive		
Vomiting	9	0
Dyspepsia	7	2
Nausea	7	2
Metabolic and nutritional		
Weight decrease	5	2
Nervous		
Coordination abnormality	7	0
Dizziness	7	0
Anxiety	5	0
Insomnia	5	2
Respiratory		
Rhinitis	7	2
Urogenital (female patients only)	(n = 21)	(n = 28)
Dysmenorrhea	5	0

* Patients in these studies were converted to LAMICTAL or valproate monotherapy from adjunctive therapy with carbamazepine or phenytoin. Patients may have reported multiple adverse reactions during the study; thus, patients may be included in more than one category.
† Up to 500 mg/day.
‡ 1,000 mg/day.

Adverse reactions that occurred with a frequency of less than 5% and greater than 2% of patients receiving LAMICTAL and numerically more frequent than placebo were:

Body as a Whole: Asthenia, fever.

Digestive: Anorexia, dry mouth, rectal hemorrhage, peptic ulcer.

Metabolic and Nutritional: Peripheral edema.

Nervous System: Amnesia, ataxia, depression, hypesthesia, libido increase, decreased reflexes, increased reflexes, nystagmus, irritability, suicidal ideation.

Respiratory: Epistaxis, bronchitis, dyspnea.

Skin and Appendages: Contact dermatitis, dry skin, sweating.

Special Senses: Vision abnormality.

Incidence in Controlled Adjunctive Trials in Pediatric Patients With Epilepsy: Table 11 lists adverse reactions that occurred in at least 2% of 339 pediatric patients with partial seizures or generalized seizures of Lennox-Gastaut syndrome, who received LAMICTAL up to 15 mg/kg/day or a maximum of 750 mg/day. Reported adverse reactions were classified using COSTART terminology.

Table 11. Treatment-Emergent Adverse Reaction Incidence in Placebo-Controlled Adjunctive Trials in Pediatric Patients With Epilepsy (Adverse reactions in at least 2% of patients treated with LAMICTAL and numerically more frequent than in the placebo group.)

Body System/ Adverse Experience	Percent of Patients Receiving LAMICTAL (n = 168)	Percent of Patients Receiving Placebo (n = 171)
Body as a whole		
Infection	20	17
Fever	15	14
Accidental injury	14	12
Abdominal pain	10	5
Asthenia	8	4
Flu syndrome	7	6
Pain	5	4
Facial edema	2	1
Photosensitivity	2	0
Cardiovascular		
Hemorrhage	2	1
Digestive		
Vomiting	20	16
Diarrhea	11	9
Nausea	10	2
Constipation	4	2
Dyspepsia	2	1
Hemic and lymphatic		
Lymphadenopathy	2	1
Metabolic and nutritional		
Edema	2	0
Nervous system		
Somnolence	17	15
Dizziness	14	4
Ataxia	11	3
Tremor	10	1
Emotional lability	4	2
Gait abnormality	4	2
Thinking abnormality	3	2
Convulsions	2	1
Nervousness	2	1
Vertigo	2	1
Respiratory		
Pharyngitis	14	11
Bronchitis	7	5
Increased cough	7	6
Sinusitis	2	1
Bronchospasm	2	1
Skin		
Rash	14	12
Eczema	2	1
Pruritus	2	1
Special senses		
Diplopia	5	1
Blurred vision	4	1
Visual abnormality	2	0
Urogenital		
Male and female patients		
Urinary tract infection	3	0

Bipolar Disorder: The most commonly observed (≥5%) treatment-emergent adverse reactions seen in association with the use of LAMICTAL as monotherapy (100 to 400 mg/day) in adult patients (≥18 years of age) with Bipolar Disorder in the 2 double-blind, placebo-controlled trials of 18 months' duration, and numerically more frequent than in placebo-treated patients are included in Table 12. Adverse reactions that occurred in at least 5% of patients and were numerically more common during the dose-escalation phase of LAMICTAL in these trials (when patients may have been receiving concomitant medications) compared with the monotherapy phase were: headache (25%), rash (11%), dizziness (10%), diarrhea (8%), dream abnormality (6%), and pruritus (6%).

During the monotherapy phase of the double-blind, placebo-controlled trials of 18 months' duration, 13% of 227 patients who received LAMICTAL (100 to 400 mg/day), 16% of 190 patients who received placebo, and 23% of 166 patients who received lithium discontinued therapy because of an adverse reaction. The adverse reactions which most commonly led to discontinuation of LAMICTAL were rash (3%) and mania/hypomania/mixed mood adverse reactions (2%). Approximately 16% of 2,401 patients who received LAMICTAL (50 to 500 mg/day) for Bipolar Disorder in premarketing trials discontinued therapy because of an adverse reaction; most commonly due to rash (5%) and mania/hypomania/mixed mood adverse reactions (2%).

The overall adverse reaction profile for LAMICTAL was similar between females and males, between elderly and nonelderly patients, and among racial groups.

Table 12. Treatment-Emergent Adverse Reaction Incidence in 2 Placebo-Controlled Trials in Adults With Bipolar I Disorder* (Adverse reactions in at least 5% of patients treated with LAMICTAL as monotherapy and numerically more frequent than in the placebo group.)

Body System/ Adverse Experience	Percent of Patients Receiving LAMICTAL (n = 227)	Percent of Patients Receiving Placebo (n = 190)
General		
Back pain	8	6
Fatigue	8	5
Abdominal pain	6	3
Digestive		
Nausea	14	11
Constipation	5	2
Vomiting	5	2
Nervous System		
Insomnia	10	6
Somnolence	9	7
Xerostomia (dry mouth)	6	4
Respiratory		
Rhinitis	7	4
Exacerbation of cough	5	3
Pharyngitis	5	4
Skin		
Rash (nonserious)†	7	5

* Patients in these studies were converted to LAMICTAL (100 to 400 mg/day) or placebo monotherapy from add-on therapy with other psychotropic medications. Patients may have reported multiple adverse reactions during the study; thus, patients may be included in more than one category.

† In the overall bipolar and other mood disorders clinical trials, the rate of serious rash was 0.08% (1 of 1,233) of adult patients who received LAMICTAL as initial monotherapy and 0.13% (2 of 1,538) of adult patients who received LAMICTAL as adjunctive therapy *[see Warnings and Precautions (5.1)]*.

These adverse reactions were usually mild to moderate in intensity. Other reactions that occurred in 5% or more patients but equally or more frequently in the placebo group included: dizziness, mania, headache, infection, influenza, pain, accidental injury, diarrhea, and dyspepsia.

Adverse reactions that occurred with a frequency of less than 5% and greater than 1% of patients receiving LAMICTAL and numerically more frequent than placebo were:

General: Fever, neck pain.
Cardiovascular: Migraine.
Digestive: Flatulence
Metabolic and Nutritional: Weight gain, edema.
Musculoskeletal: Arthralgia, myalgia.
Nervous System: Amnesia, depression, agitation, emotional lability, dyspraxia, abnormal thoughts, dream abnormality, hypoesthesia.
Respiratory: Sinusitis.
Urogenital: Urinary frequency.

Adverse Reactions Following Abrupt Discontinuation: In the 2 maintenance trials, there was no increase in the incidence, severity or type of adverse reactions in Bipolar Disorder patients after abruptly terminating therapy with LAMICTAL. In clinical trials in patients with Bipolar Disorder, 2 patients experienced seizures shortly after abrupt withdrawal of LAMICTAL. However, there were confounding factors that may have contributed to the occurrence of seizures in these bipolar patients *[see Warnings and Precautions (5.9)]*.

Mania/Hypomania/Mixed Episodes: During the double-blind, placebo-controlled clinical trials in Bipolar I Disorder in which patients were converted to monotherapy with LAMICTAL (100 to 400 mg/day) from other psychotropic medications and followed for up to 18 months, the rates of manic or hypomanic or mixed mood episodes reported as adverse reactions were 5% for patients treated with LAMICTAL (n = 227), 4% for patients treated with lithium (n = 166), and 7% for patients treated with placebo (n = 190). In all bipolar controlled trials combined, adverse reactions of mania (including hypomania and mixed mood episodes) were reported in 5% of patients treated with LAMICTAL (n = 956), 3% of patients treated with lithium (n = 280), and 4% of patients treated with placebo (n = 803).

6.2 Other Adverse Reactions Observed in All Clinical Trials

LAMICTAL has been administered to 6,694 individuals for whom complete adverse reaction data was captured during all clinical trials, only some of which were placebo controlled. During these trials, all adverse reactions were recorded by the clinical investigators using terminology of their own choosing. To provide a meaningful estimate of the proportion of individuals having adverse reactions, similar types of adverse reactions were grouped into a smaller number of standardized categories using modified COSTART dictionary terminology. The frequencies presented represent the proportion of the 6,694 individuals exposed to LAMICTAL who experienced an event of the type cited on at least one occasion while receiving LAMICTAL. All reported adverse reactions are included except those already listed in the previous tables or elsewhere in the labeling, those too general to be informative, and those not reasonably associated with the use of the drug.

Adverse reactions are further classified within body system categories and enumerated in order of decreasing frequency using the following definitions: *frequent* adverse reactions are defined as those occurring in at least 1/100 patients; *infrequent* adverse reactions are those occurring in 1/100 to 1/1,000 patients; *rare* adverse reactions are those occurring in fewer than 1/1,000 patients.

Body as a Whole: *Infrequent:* Allergic reaction, chills, and malaise.

Cardiovascular System: *Infrequent:* Flushing, hot flashes, hypertension, palpitations, postural hypotension, syncope, tachycardia, and vasodilation.

Dermatological: *Infrequent:* Acne, alopecia, hirsutism, maculopapular rash, skin discoloration, and urticaria. *Rare:* Angioedema, erythema, exfoliative dermatitis, fungal dermatitis, herpes zoster, leukoderma, multiforme erythema, petechial rash, pustular rash, Stevens-Johnson syndrome, and vesiculobullous rash.

Digestive System: *Infrequent:* Dysphagia, eructation, gastritis, gingivitis, increased appetite, increased salivation, liver function tests abnormal, and mouth ulceration. *Rare:* Gastrointestinal hemorrhage, glossitis, gum hemorrhage, gum hyperplasia, hematemesis, hemorrhagic colitis, hepatitis, melena, stomach ulcer, stomatitis, and tongue edema.

Endocrine System: *Rare:* Goiter and hypothyroidism.

Hematologic and Lymphatic System: *Infrequent:* Ecchymosis and leukopenia. *Rare:* Anemia, eosinophilia, fibrin decrease, fibrinogen decrease, iron deficiency anemia, leukocytosis, lymphocytosis, macrocytic anemia, petechia, and thrombocytopenia.

Metabolic and Nutritional Disorders: *Infrequent:* Aspartate transaminase increased. *Rare:* Alcohol intolerance, alkaline phosphatase increase, alanine transaminase increase, bilirubinemia, general edema, gamma glutamyl transpeptidase increase, and hyperglycemia.

Musculoskeletal System: *Infrequent:* Arthritis, leg cramps, myasthenia, and twitching. *Rare:* Bursitis, muscle atrophy, pathological fracture, and tendinous contracture.

Nervous System: *Frequent:* Confusion and paresthesia. *Infrequent:* Akathisia, apathy, aphasia, CNS depression, depersonalization, dysarthria, dyskinesia, euphoria, hallucinations, hostility, hyperkinesia, hypertonia, libido decreased, memory decrease, mind racing, movement disorder, myoclonus, panic attack, paranoid reaction, personality disorder, psychosis, sleep disorder, stupor, and suicidal ideation. *Rare:* Choreoathetosis, delirium, delusions, dysphoria, dystonia, extrapyramidal syndrome, faintness, grand mal convulsions, hemiplegia, hyperalgesia, hyperesthesia, hypokinesia, hypotonia, manic depression reaction, muscle spasm, neuralgia, neurosis, paralysis, and peripheral neuritis.

Respiratory System: *Infrequent:* Yawn. *Rare:* Hiccup and hyperventilation.

Special Senses: *Frequent:* Amblyopia. *Infrequent:* Abnormality of accommodation, conjunctivitis, dry eyes, ear pain, photophobia, taste perversion, and tinnitus. *Rare:* Deafness, lacrimation disorder, oscillopsia, parosmia, ptosis, strabismus, taste loss, uveitis, and visual field defect.

Continued on next page

Product information on these pages is effective as of June 2009. Further information is available at 1-888-825-5249 or www.gsk.com.

Lamictal—Cont.

Urogenital System: *Infrequent:* Abnormal ejaculation, hematuria, impotence, menorrhagia, polyuria, and urinary incontinence. *Rare:* Acute kidney failure, anorgasmia, breast abscess, breast neoplasm, creatinine increase, cystitis, dysuria, epididymitis, female lactation, kidney failure, kidney pain, nocturia, urinary retention, and urinary urgency.

6.3 Postmarketing Experience

The following adverse events (not listed above in clinical trials or other sections of the prescribing information) have been identified during postapproval use of LAMICTAL. Because these events are reported voluntarily from a population of uncertain size, it is not always possible to reliably estimate their frequency or establish a causal relationship to drug exposure.

Blood and Lymphatic: Agranulocytosis, hemolytic anemia.
Gastrointestinal: Esophagitis.
Hepatobiliary Tract and Pancreas: Pancreatitis.
Immunologic: Lupus-like reaction, vasculitis.
Lower Respiratory: Apnea.
Musculoskeletal: Rhabdomyolysis has been observed in patients experiencing hypersensitivity reactions.
Neurology: Exacerbation of Parkinsonian symptoms in patients with pre-existing Parkinson's disease, tics.
Non-site Specific: Progressive immunosuppression.

7 DRUG INTERACTIONS

Significant drug interactions with lamotrigine are summarized in Table 13. Additional details of these drug interaction studies are provided in the Clinical Pharmacology subsection *[see Clinical Pharmacology (12.3)]*.

Table 13. Established and Other Potentially Significant Drug Interactions

Concomitant Drug	Effect on Concentration of Lamotrigine or Concomitant Drug	Clinical Comment
Estrogen-containing oral contraceptive preparation containing 30 mcg ethinylestradiol and 150 mcg levonorgestrel	↓ lamotrigine	Decreased lamotrigine levels approximately 50%.
	↓ levonorgestrel	Decrease in levonorgestrel component by 19%.
Carbamazepine (CBZ) and CBZ epoxide	↓ lamotrigine	Addition of carbamazepine decreases lamotrigine concentration approximately 40%.
	? CBZ epoxide	May increase CBZ epoxide levels.
Phenobarbital/ Primidone	↓ lamotrigine	Decreased lamotrigine concentration approximately 40%.
Phenytoin (PHT)	↓ lamotrigine	Decreased lamotrigine concentration approximately 40%.
Rifampin	↓ lamotrigine	Decreased lamotrigine AUC approximately 40%.
Valproate	↑ lamotrigine	Increased lamotrigine concentrations slightly more than 2-fold.
	? valproate	Decreased valproate concentrations an average of 25% over a 3-week period then stabilized in healthy volunteers; no change in controlled clinical trials in epilepsy patients.

↓ = Decreased (induces lamotrigine glucuronidation).
↑ = Increased (inhibits lamotrigine glucuronidation).
? = Conflicting data.

8 USE IN SPECIFIC POPULATIONS

8.1 Pregnancy

Teratogenic Effects: Pregnancy Category C. No evidence of teratogenicity was found in mice, rats, or rabbits when lamotrigine was orally administered to pregnant animals during the period of organogenesis at doses up to 1.2, 0.5, and 1.1 times, respectively, on a mg/m^2 basis, the highest usual human maintenance dose (i.e., 500 mg/day). However, maternal toxicity and secondary fetal toxicity producing reduced fetal weight and/or delayed ossification were seen in mice and rats, but not in rabbits at these doses. Teratology studies were also conducted using bolus intravenous administration of the isethionate salt of lamotrigine in rats and rabbits. In rat dams administered an intravenous dose at 0.6 times the highest usual human maintenance dose, the incidence of intrauterine death without signs of teratogenicity was increased.

A behavioral teratology study was conducted in rats dosed during the period of organogenesis. At day 21 postpartum, offspring of dams receiving 5 mg/kg/day or higher displayed a significantly longer latent period for open field exploration and a lower frequency of rearing. In a swimming maze test performed on days 39 to 44 postpartum, time to completion was increased in offspring of dams receiving 25 mg/kg/day. These doses represent 0.1 and 0.5 times the clinical dose on a mg/m^2 basis, respectively.

Lamotrigine did not affect fertility, teratogenesis, or postnatal development when rats were dosed prior to and during mating, and throughout gestation and lactation at doses equivalent to 0.4 times the highest usual human maintenance dose on a mg/m^2 basis.

When pregnant rats were orally dosed at 0.1, 0.14, or 0.3 times the highest human maintenance dose (on a mg/m^2 basis) during the latter part of gestation (days 15 to 20), maternal toxicity and fetal death were seen. In dams, food consumption and weight gain were reduced, and the gestation period was slightly prolonged (22.6 vs. 22.0 days in the control group). Stillborn pups were found in all 3 drug-treated groups with the highest number in the high-dose group. Postnatal death was also seen, but only in the 2 highest doses, and occurred between days 1 and 20. Some of these deaths appear to be drug-related and not secondary to the maternal toxicity. A no-observed-effect level (NOEL) could not be determined for this study.

Although lamotrigine was not found to be teratogenic in the above studies, lamotrigine decreases fetal folate concentrations in rats, an effect known to be associated with teratogenesis in animals and humans. There are no adequate and well-controlled studies in pregnant women. Because animal reproduction studies are not always predictive of human response, this drug should be used during pregnancy only if the potential benefit justifies the potential risk to the fetus.

Non-Teratogenic Effects: As with other AEDs, physiological changes during pregnancy may affect lamotrigine concentrations and/or therapeutic effect. There have been reports of decreased lamotrigine concentrations during pregnancy and restoration of pre-partum concentrations after delivery. Dosage adjustments may be necessary to maintain clinical response.

Pregnancy Exposure Registry: To provide information regarding the effects of in utero exposure to LAMICTAL, physicians are advised to recommend that pregnant patients taking LAMICTAL enroll in the North American Antiepileptic Drug (NAAED) Pregnancy Registry. This can be done by calling the toll free number 1-888-233-2334, and must be done by patients themselves. Information on the registry can also be found at the website http://www.aedpregnancyregistry.org/.

Physicians are also encouraged to register patients in the Lamotrigine Pregnancy Registry; enrollment in this registry must be done prior to any prenatal diagnostic tests and **before fetal outcome is known.** Physicians can obtain information by calling the Lamotrigine Pregnancy Registry at 1-800-336-2176 (toll-free).

8.2 Labor and Delivery

The effect of LAMICTAL on labor and delivery in humans is unknown.

8.3 Nursing Mothers

Preliminary data indicate that lamotrigine passes into human milk. Because the effects on the infant exposed to lamotrigine by this route are unknown, breastfeeding while taking LAMICTAL is not recommended.

8.4 Pediatric Use

LAMICTAL is indicated for adjunctive therapy in patients ≥2 years of age for partial seizures, the generalized seizures of Lennox-Gastaut syndrome, and primary generalized tonic-clonic seizures.

Safety and efficacy of LAMICTAL, used as adjunctive treatment for partial seizures, were not demonstrated in a small randomized, double-blind, placebo-controlled, withdrawal study in very young pediatric patients (1 to 24 months). LAMICTAL was associated with an increased risk for infectious adverse reactions (LAMICTAL 37%, Placebo 5%), and respiratory adverse reactions (LAMICTAL 26%, Placebo 5%). Infectious adverse reactions included: bronchiolitis, bronchitis, ear infection, eye infection, otitis externa, pharyngitis, urinary tract infection, and viral infection. Respiratory adverse reactions included nasal congestion, cough, and apnea.

Safety and effectiveness in patients below the age of 18 years with Bipolar Disorder has not been established.

8.5 Geriatric Use

Clinical studies of LAMICTAL for epilepsy and in Bipolar Disorder did not include sufficient numbers of subjects 65 years of age and over to determine whether they respond differently from younger subjects or exhibit a different safety profile than that of younger patients. In general, dose selection for an elderly patient should be cautious, usually starting at the low end of the dosing range, reflecting the greater frequency of decreased hepatic, renal, or cardiac function, and of concomitant disease or other drug therapy.

8.6 Patients With Hepatic Impairment

Experience in patients with hepatic impairment is limited. Based on a clinical pharmacology study in 24 patients with mild, moderate, and severe liver impairment *[see Clinical Pharmacology (12.3)]*, the following general recommendations can be made. No dosage adjustment is needed in patients with mild liver impairment. Initial, escalation, and maintenance doses should generally be reduced by approximately 25% in patients with moderate and severe liver impairment without ascites and 50% in patients with severe liver impairment with ascites. Escalation and maintenance doses may be adjusted according to clinical response *[see Dosage and Administration (2.1)]*.

8.7 Patients With Renal Impairment

Lamotrigine is metabolized mainly by glucuronic acid conjugation, with the majority of the metabolites being recovered in the urine. In a small study comparing a single dose of lamotrigine in patients with varying degrees of renal impairment with healthy volunteers, the plasma half-life of lamotrigine was significantly longer in the patients with renal impairment *[see Clinical Pharmacology (12.3)]*.

Initial doses of LAMICTAL should be based on patients' AED regimen; reduced maintenance doses may be effective for patients with significant renal impairment. Few patients with severe renal impairment have been evaluated during chronic treatment with LAMICTAL. Because there is inadequate experience in this population, LAMICTAL should be used with caution in these patients *[see Dosage and Administration (2.1)]*.

10 OVERDOSAGE

10.1 Human Overdose Experience

Overdoses involving quantities up to 15 g have been reported for LAMICTAL, some of which have been fatal. Overdose has resulted in ataxia, nystagmus, increased seizures, decreased level of consciousness, coma, and intraventricular conduction delay.

10.2 Management of Overdose

There are no specific antidotes for lamotrigine. Following a suspected overdose, hospitalization of the patient is advised. General supportive care is indicated, including frequent monitoring of vital signs and close observation of the patient. If indicated, emesis should be induced or gastric lavage should be performed; usual precautions should be taken to protect the airway. It should be kept in mind that lamotrigine is rapidly absorbed *[see Clinical Pharmacology (12.3)]*. It is uncertain whether hemodialysis is an effective means of removing lamotrigine from the blood. In 6 renal failure patients, about 20% of the amount of lamotrigine in the body was removed by hemodialysis during a 4-hour session. A Poison Control Center should be contacted for information on the management of overdosage of LAMICTAL.

11 DESCRIPTION

LAMICTAL (lamotrigine), an AED of the phenyltriazine class, is chemically unrelated to existing AEDs. Its chemical name is 3,5-diamino-6-(2,3-dichlorophenyl)-*as*-triazine, its molecular formula is $C_9H_7N_5Cl_2$, and its molecular weight is 256.09. Lamotrigine is a white to pale cream-colored powder and has a pK_a of 5.7. Lamotrigine is very slightly soluble in water (0.17 mg/mL at 25°C) and slightly soluble in 0.1 M HCl (4.1 mg/mL at 25°C). The structural formula is:

LAMICTAL Tablets are supplied for oral administration as 25 mg (white), 100 mg (peach), 150 mg (cream), and 200 mg (blue) tablets. Each tablet contains the labeled amount of lamotrigine and the following inactive ingredients: lactose; magnesium stearate; microcrystalline cellulose; povidone; sodium starch glycolate; FD&C Yellow No. 6 Lake (100 mg tablet only); ferric oxide, yellow (150 mg tablet only); and FD&C Blue No. 2 Lake (200 mg tablet only).

LAMICTAL Chewable Dispersible Tablets are supplied for oral administration. The tablets contain 2 mg (white), 5 mg (white), or 25 mg (white) of lamotrigine and the following inactive ingredients: blackcurrant flavor, calcium carbonate, low-substituted hydroxypropylcellulose, magnesium aluminum silicate, magnesium stearate, povidone, saccharin sodium, and sodium starch glycolate.
LAMICTAL ODT Orally Disintegrating Tablets are supplied for oral administration. The tablets contain 25 mg (white to off-white), 50 mg (white to off-white), 100 mg (white to off-white), or 200 mg (white to off-white) of lamotrigine and the following inactive ingredients: artificial cherry flavor, crospovidone, ethylcellulose, magnesium stearate, mannitol, polyethylene, and sucralose.
LAMICTAL ODT Orally Disintegrating Tablets are formulated using technologies (Microcaps®* and AdvaTab®*) designed to mask the bitter taste of lamotrigine and achieve a rapid dissolution profile. Tablet characteristics including flavor, mouth-feel, after-taste, and ease of use were rated as favorable in a study of 108 healthy volunteers.

12 CLINICAL PHARMACOLOGY

12.1 Mechanism of Action

The precise mechanism(s) by which lamotrigine exerts its anticonvulsant action are unknown. In animal models designed to detect anticonvulsant activity, lamotrigine was effective in preventing seizure spread in the maximum electroshock (MES) and pentylenetetrazol (scMet) tests, and prevented seizures in the visually and electrically evoked after-discharge (EEAD) tests for antiepileptic activity. Lamotrigine also displayed inhibitory properties in the kindling model in rats both during kindling development and in the fully kindled state. The relevance of these models to human epilepsy, however, is not known.
One proposed mechanism of action of lamotrigine, the relevance of which remains to be established in humans, involves an effect on sodium channels. In vitro pharmacological studies suggest that lamotrigine inhibits voltage-sensitive sodium channels, thereby stabilizing neuronal membranes and consequently modulating presynaptic transmitter release of excitatory amino acids (e.g., glutamate and aspartate).
Although the relevance for human use is unknown, the following data characterize the performance of lamotrigine in receptor binding assays. Lamotrigine had a weak inhibitory effect on the serotonin 5-HT_3 receptor (IC_{50} = 18 μM). It does not exhibit high affinity binding (IC_{50}>100 μM) to the following neurotransmitter receptors: adenosine A_1 and A_2; adrenergic α_1, α_2, and β; dopamine D_1 and D_2; γ-aminobutyric acid (GABA) A and B; histamine H_1; kappa opioid; muscarinic acetylcholine; and serotonin 5-HT_2. Studies have failed to detect an effect of lamotrigine on dihydropyridine-sensitive calcium channels. It had weak effects at sigma opioid receptors (IC_{50} = 145 μM). Lamotrigine did not inhibit the uptake of norepinephrine, dopamine, or serotonin, (IC_{50}>200 μM) when tested in rat synaptosomes and/or human platelets in vitro.
Effect of Lamotrigine on N-Methyl d-Aspartate-Receptor Mediated Activity: Lamotrigine did not inhibit N-methyl d-aspartate (NMDA)-induced depolarizations in rat cortical slices or NMDA-induced cyclic GMP formation in immature rat cerebellum, nor did lamotrigine displace compounds that are either competitive or noncompetitive ligands at this glutamate receptor complex (CNQX, CGS, TCHP). The IC_{50} for lamotrigine effects on NMDA-induced currents (in the presence of 3 μM of glycine) in cultured hippocampal neurons exceeded 100 μM.
The mechanisms by which lamotrigine exerts its therapeutic action in Bipolar Disorder have not been established.

12.2 Pharmacodynamics

Folate Metabolism: In vitro, lamotrigine inhibited dihydrofolate reductase, the enzyme that catalyzes the reduction of dihydrofolate to tetrahydrofolate. Inhibition of this enzyme may interfere with the biosynthesis of nucleic acids and proteins. When oral daily doses of lamotrigine were given to pregnant rats during organogenesis, fetal, placental, and maternal folate concentrations were reduced. Significantly reduced concentrations of folate are associated with teratogenesis *[see Use in Specific Populations (8.1)]*. Folate concentrations were also reduced in male rats given repeated oral doses of lamotrigine. Reduced concentrations were partially returned to normal when supplemented with folinic acid.
Accumulation in Kidneys: Lamotrigine accumulated in the kidney of the male rat, causing chronic progressive nephrosis, necrosis, and mineralization. These findings are attributed to α-2 microglobulin, a species- and sex-specific protein that has not been detected in humans or other animal species.
Melanin Binding: Lamotrigine binds to melanin-containing tissues, e.g., in the eye and pigmented skin. It has been found in the uveal tract up to 52 weeks after a single dose in rodents.

Table 14. Mean* Pharmacokinetic Parameters in Healthy Volunteers and Adult Patients With Epilepsy

Adult Study Population	Number of Subjects	T_{max}: Time of Maximum Plasma Concentration (hr)	$t_{1/2}$: Elimination Half-life (hr)	Cl/F: Apparent Plasma Clearance (mL/min/kg)
Healthy volunteers taking no other medications:				
Single-dose LAMICTAL	179	2.2 (0.25-12.0)	32.8 (14.0-103.0)	0.44 (0.12-1.10)
Multiple-dose LAMICTAL	36	1.7 (0.5-4.0)	25.4 (11.6-61.6)	0.58 (0.24-1.15)
Healthy volunteers taking valproate:				
Single-dose LAMICTAL	6	1.8 (1.0-4.0)	48.3 (31.5-88.6)	0.30 (0.14-0.42)
Multiple-dose LAMICTAL	18	1.9 (0.5-3.5)	70.3 (41.9-113.5)	0.18 (0.12-0.33)
Patients with epilepsy taking valproate only:				
Single-dose LAMICTAL	4	4.8 (1.8-8.4)	58.8 (30.5-88.8)	0.28 (0.16-0.40)
Patients with epilepsy taking carbamazepine, phenytoin, phenobarbital, or primidone† plus valproate:				
Single-dose LAMICTAL	25	3.8 (1.0-10.0)	27.2 (11.2-51.6)	0.53 (0.27-1.04)
Patients with epilepsy taking carbamazepine, phenytoin, phenobarbital, or primidone†:				
Single-dose LAMICTAL	24	2.3 (0.5-5.0)	14.4 (6.4-30.4)	1.10 (0.51-2.22)
Multiple-dose LAMICTAL	17	2.0 (0.75-5.93)	12.6 (7.5-23.1)	1.21 (0.66-1.82)

* The majority of parameter means determined in each study had coefficients of variation between 20% and 40% for half-life and Cl/F and between 30% and 70% for T_{max}. The overall mean values were calculated from individual study means that were weighted based on the number of volunteers/patients in each study. The numbers in parentheses below each parameter mean represent the range of individual volunteer/patient values across studies.
† Carbamazepine, phenobarbital, phenytoin, and primidone have been shown to increase the apparent clearance of lamotrigine. Estrogen-containing oral contraceptives and rifampin have also been shown to increase the apparent clearance of lamotrigine *[see Drug Interactions (7)]*.

Cardiovascular: In dogs, lamotrigine is extensively metabolized to a 2-N-methyl metabolite. This metabolite causes dose-dependent prolongations of the PR interval, widening of the QRS complex, and, at higher doses, complete AV conduction block. Similar cardiovascular effects are not anticipated in humans because only trace amounts of the 2-N-methyl metabolite (<0.6% of lamotrigine dose) have been found in human urine *[see Clinical Pharmacology (12.3)]*. However, it is conceivable that plasma concentrations of this metabolite could be increased in patients with a reduced capacity to glucuronidate lamotrigine (e.g., in patients with liver disease).

12.3 Pharmacokinetics

The pharmacokinetics of lamotrigine have been studied in patients with epilepsy, healthy young and elderly volunteers, and volunteers with chronic renal failure. Lamotrigine pharmacokinetic parameters for adult and pediatric patients and healthy normal volunteers are summarized in Tables 14 and 16.
[See table 14 above]
Absorption: Lamotrigine is rapidly and completely absorbed after oral administration with negligible first-pass metabolism (absolute bioavailability is 98%). The bioavailability is not affected by food. Peak plasma concentrations occur anywhere from 1.4 to 4.8 hours following drug administration. The lamotrigine chewable/dispersible tablets were found to be equivalent, whether they were administered as dispersed in water, chewed and swallowed, or swallowed as whole, to the lamotrigine compressed tablets in terms of rate and extent of absorption. In terms of rate and extent of absorption, lamotrigine orally disintegrating tablets whether disintegrated in the mouth or swallowed whole with water were equivalent to the lamotrigine compressed tablets swallowed with water.
Dose Proportionality: In healthy volunteers not receiving any other medications and given single doses, the plasma concentrations of lamotrigine increased in direct proportion to the dose administered over the range of 50 to 400 mg. In 2 small studies (n = 7 and 8) of patients with epilepsy who were maintained on other AEDs, there also was a linear relationship between dose and lamotrigine plasma concentrations at steady state following doses of 50 to 350 mg twice daily.
Distribution: Estimates of the mean apparent volume of distribution (Vd/F) of lamotrigine following oral administration ranged from 0.9 to 1.3 L/kg. Vd/F is independent of dose and is similar following single and multiple doses in both patients with epilepsy and in healthy volunteers.
Protein Binding: Data from in vitro studies indicate that lamotrigine is approximately 55% bound to human plasma proteins at plasma lamotrigine concentrations from 1 to 10 mcg/mL (10 mcg/mL is 4 to 6 times the trough plasma concentration observed in the controlled efficacy trials). Because lamotrigine is not highly bound to plasma proteins, clinically significant interactions with other drugs through competition for protein binding sites are unlikely. The binding of lamotrigine to plasma proteins did not change in the presence of therapeutic concentrations of phenytoin, phenobarbital, or valproate. Lamotrigine did not displace other AEDs (carbamazepine, phenytoin, phenobarbital) from protein binding sites.
Metabolism: Lamotrigine is metabolized predominantly by glucuronic acid conjugation; the major metabolite is an inactive 2-N-glucuronide conjugate. After oral administration of 240 mg of ^{14}C-lamotrigine (15 μCi) to 6 healthy volunteers, 94% was recovered in the urine and 2% was recovered in the feces. The radioactivity in the urine consisted of unchanged lamotrigine (10%), the 2-N-glucuronide (76%), a 5-N-glucuronide (10%), a 2-N-methyl metabolite (0.14%), and other unidentified minor metabolites (4%).
Enzyme Induction: The effects of lamotrigine on the induction of specific families of mixed-function oxidase isozymes have not been systematically evaluated.
Following multiple administrations (150 mg twice daily) to normal volunteers taking no other medications, lamotrigine induced its own metabolism, resulting in a 25% decrease in $t_{1/2}$ and a 37% increase in Cl/F at steady state compared with values obtained in the same volunteers following a single dose. Evidence gathered from other sources suggests that self-induction by lamotrigine may not occur when lamotrigine is given as adjunctive therapy in patients receiving carbamazepine, phenytoin, phenobarbital, primidone, or rifampin.
Elimination: The elimination half-life and apparent clearance of lamotrigine following administration of LAMICTAL to adult patients with epilepsy and healthy volunteers is summarized in Table 14. Half-life and apparent oral clearance vary depending on concomitant AEDs.
Drug Interactions: The apparent clearance of lamotrigine is affected by the coadministration of certain medications *[see Warnings and Precautions (5.8, 5.12), Drug Interactions (7)]*.
The net effects of drug interactions with LAMICTAL are summarized in Tables 13 and 15, followed by details of the drug interaction studies below.

Continued on next page

Product information on these pages is effective as of June 2009. Further information is available at 1-888-825-5249 or www.gsk.com.

Lamictal—Cont.

Table 15. Summary of Drug Interactions With LAMICTAL

Drug	Drug Plasma Concentration With Adjunctive LAMICTAL*	Lamotrigine Plasma Concentration With Adjunctive Drugs†
Oral contraceptives (e.g., ethinylestradiol/ levonorgestrel)‡	↔§	↓
Bupropion	Not assessed	↔
Carbamazepine (CBZ)	↔	↓
CBZ epoxide‖	?	
Felbamate	Not assessed	↔
Gabapentin	Not assessed	↔
Levetiracetam	↔	↔
Lithium	↔	Not assessed
Olanzapine	↔	↔¶
Oxcarbazepine	↔	↔
10-monohydroxy oxcarbazepine metabolite#	↔	
Phenobarbital/ primidone	↔	↓
Phenytoin (PHT)	↔	↓
Pregabalin	↔	↔
Rifampin	Not assessed	↓
Topiramate	↔**	↔
Valproate	↓	↑
Valproate + PHT and/or CBZ	Not assessed	↔
Zonisamide	Not assessed	↔

* From adjunctive clinical trials and volunteer studies.
† Net effects were estimated by comparing the mean clearance values obtained in adjunctive clinical trials and volunteers studies.
‡ The effect of other hormonal contraceptive preparations or hormone replacement therapy on the pharmacokinetics of lamotrigine has not been systematically evaluated in clinical trials, although the effect may be similar to that seen with the ethinylestradiol/levonorgestrel combinations.
§ Modest decrease in levonorgestrel.
‖ Not administered, but an active metabolite of carbamazepine.
¶ Slight decrease, not expected to be clinically relevant.
\# Not administered, but an active metabolite of oxcarbazepine.
** Slight increase not expected to be clinically relevant.
↔= No significant effect.
? = Conflicting data.

Estrogen-Containing Oral Contraceptives: In 16 female volunteers, an oral contraceptive preparation containing 30 mcg ethinylestradiol and 150 mcg levonorgestrel increased the apparent clearance of lamotrigine (300 mg/day) by approximately 2-fold with a mean decrease in AUC of 52% and in C_{max} of 39%. In this study, trough serum lamotrigine concentrations gradually increased and were approximately 2-fold higher on average at the end of the week of the inactive hormone preparation compared with trough lamotrigine concentrations at the end of the active hormone cycle.

Gradual transient increases in lamotrigine plasma levels (approximate 2-fold increase) occurred during the week of inactive hormone preparation ("pill-free" week) for women not also taking a drug that increased the clearance of lamotrigine (carbamazepine, phenytoin, phenobarbital, primidone, or rifampin). The increase in lamotrigine plasma levels will be greater if the dose of LAMICTAL is increased in the few days before or during the "pill-free" week. Increases in lamotrigine plasma levels could result in dose-dependent adverse effects.

In the same study, coadministration of LAMICTAL (300 mg/day) in 16 female volunteers did not affect the pharmacokinetics of the ethinylestradiol component of the oral contraceptive preparation. There were mean decreases in the AUC and C_{max} of the levonorgestrel component of 19% and 12%, respectively. Measurement of serum progesterone indicated that there was no hormonal evidence of ovulation in any of the 16 volunteers, although measurement of serum FSH, LH, and estradiol indicated that there was some loss of suppression of the hypothalamic-pituitary-ovarian axis.

The effects of doses of LAMICTAL other than 300 mg/day have not been systematically evaluated in controlled clinical trials.

The clinical significance of the observed hormonal changes on ovulatory activity is unknown. However, the possibility of decreased contraceptive efficacy in some patients cannot be excluded. Therefore, patients should be instructed to promptly report changes in their menstrual pattern (e.g., break-through bleeding).

Dosage adjustments may be necessary for women receiving estrogen-containing oral contraceptive preparations *[see Dosage and Administration (2.1)]*.

Other Hormonal Contraceptives or Hormone Replacement Therapy: The effect of other hormonal contraceptive preparations or hormone replacement therapy on the pharmacokinetics of lamotrigine has not been systematically evaluated. It has been reported that ethinylestradiol, not progestogens, increased the clearance of lamotrigine up to 2-fold, and the progestin-only pills had no effect on lamotrigine plasma levels. Therefore, adjustments to the dosage of LAMICTAL in the presence of progestogens alone will likely not be needed.

Bupropion: The pharmacokinetics of a 100-mg single dose of LAMICTAL in healthy volunteers (n = 12) were not changed by coadministration of bupropion sustained-release formulation (150 mg twice a day) starting 11 days before LAMICTAL.

Carbamazepine: LAMICTAL has no appreciable effect on steady-state carbamazepine plasma concentration. Limited clinical data suggest there is a higher incidence of dizziness, diplopia, ataxia, and blurred vision in patients receiving carbamazepine with lamotrigine than in patients receiving other AEDs with lamotrigine *[see Adverse Reactions (6.1)]*. The mechanism of this interaction is unclear. The effect of lamotrigine on plasma concentrations of carbamazepine-epoxide is unclear. In a small subset of patients (n = 7) studied in a placebo-controlled trial, lamotrigine had no effect on carbamazepine-epoxide plasma concentrations, but in a small, uncontrolled study (n = 9), carbamazepine-epoxide levels increased.

The addition of carbamazepine decreases lamotrigine steady-state concentrations by approximately 40%.

Felbamate: In a study of 21 healthy volunteers, coadministration of felbamate (1,200 mg twice daily) with lamotrigine (100 mg twice daily for 10 days) appeared to have no clinically relevant effects on the pharmacokinetics of lamotrigine.

Folate Inhibitors: Lamotrigine is a weak inhibitor of dihydrofolate reductase. Prescribers should be aware of this action when prescribing other medications that inhibit folate metabolism.

Gabapentin: Based on a retrospective analysis of plasma levels in 34 patients who received lamotrigine both with and without gabapentin, gabapentin does not appear to change the apparent clearance of lamotrigine.

Levetiracetam: Potential drug interactions between levetiracetam and lamotrigine were assessed by evaluating serum concentrations of both agents during placebo-controlled clinical trials. These data indicate that lamotrigine does not influence the pharmacokinetics of levetiracetam and that levetiracetam does not influence the pharmacokinetics of lamotrigine.

Lithium: The pharmacokinetics of lithium were not altered in healthy subjects (n = 20) by coadministration of lamotrigine (100 mg/day) for 6 days.

Olanzapine: The AUC and C_{max} of olanzapine were similar following the addition of olanzapine (15 mg once daily) to lamotrigine (200 mg once daily) in healthy male volunteers (n = 16) compared with the AUC and C_{max} in healthy male volunteers receiving olanzapine alone (n = 16).

In the same study, the AUC and C_{max} of lamotrigine were reduced on average by 24% and 20%, respectively, following the addition of olanzapine to lamotrigine in healthy male volunteers compared with those receiving lamotrigine alone. This reduction in lamotrigine plasma concentrations is not expected to be clinically relevant.

Oxcarbazepine: The AUC and C_{max} of oxcarbazepine and its active 10-monohydroxy oxcarbazepine metabolite were not significantly different following the addition of oxcarbazepine (600 mg twice daily) to lamtrigine (200 mg once daily) in healthy male volunteers (n = 13) compared with healthy male volunteers receiving oxcarbazepine alone (n = 13).

In the same study, the AUC and C_{max} of lamotrigine were similar following the addition of oxcarbazepine (600 mg twice daily) to LAMICTAL in healthy male volunteers compared with those receiving LAMICTAL alone. Limited clinical data suggest a higher incidence of headache, dizziness, nausea, and somnolence with coadministration of lamotrigine and oxcarbazepine compared with lamotrigine alone or oxcarbazepine alone.

Phenobarbital, Primidone: The addition of phenobarbital or primidone decreases lamotrigine steady-state concentrations by approximately 40%.

Phenytoin: Lamotrigine has no appreciable effect on steady-state phenytoin plasma concentrations in patients with epilepsy. The addition of phenytoin decreases lamotrigine steady-state concentrations by approximately 40%.

Pregabalin: Steady-state trough plasma concentrations of lamotrigine were not affected by concomitant pregabalin (200 mg 3 times daily) administration. There are no pharmacokinetic interactions between lamotrigine and pregabalin.

Rifampin: In 10 male volunteers, rifampin (600 mg/day for 5 days) significantly increased the apparent clearance of a single 25-mg dose of lamotrigine by approximately 2-fold (AUC decreased by approximately 40%).

Topiramate: Topiramate resulted in no change in plasma concentrations of lamotrigine. Administration of lamotrigine resulted in a 15% increase in topiramate concentrations.

Valproate: When lamotrigine was administered to healthy volunteers (n = 18) receiving valproate, the trough steady-state valproate plasma concentrations decreased by an average of 25% over a 3-week period, and then stabilized. However, adding lamotrigine to the existing therapy did not cause a change in valproate plasma concentrations in either adult or pediatric patients in controlled clinical trials.

The addition of valproate increased lamotrigine steady-state concentrations in normal volunteers by slightly more than 2-fold. In one study, maximal inhibition of lamotrigine clearance was reached at valproate doses between 250 mg/day and 500 mg/day and did not increase as the valproate dose was further increased.

Zonisamide: In a study of 18 patients with epilepsy, coadministration of zonisamide (200 to 400 mg/day) with lamotrigine (150 to 500 mg/day for 35 days) had no significant effect on the pharmacokinetics of lamotrigine.

Known Inducers or Inhibitors of Glucuronidation: Drugs other than those listed above have not been systematically evaluated in combination with lamotrigine. Since lamotrigine is metabolized predominately by glucuronic acid conjugation, drugs that are known to induce or inhibit glucuronidation may affect the apparent clearance of lamotrigine and doses of lamotrigine may require adjustment based on clinical response.

Other: Results of in vitro experiments suggest that clearance of lamotrigine is unlikely to be reduced by concomitant administration of amitriptyline, clonazepam, clozapine, fluoxetine, haloperidol, lorazepam, phenelzine, risperidone, sertraline, or trazodone).

Results of in vitro experiments suggest that lamotrigine does not reduce the clearance of drugs eliminated predominantly by CYP2D6.

Special Populations: *Patients With Renal Impairment:* Twelve volunteers with chronic renal failure (mean creatinine clearance: 13 mL/min; range: 6 to 23) and another 6 individuals undergoing hemodialysis were each given a single 100-mg dose of lamotrigine. The mean plasma half-lives determined in the study were 42.9 hours (chronic renal failure), 13.0 hours (during hemodialysis), and 57.4 hours (between hemodialysis) compared with 26.2 hours in healthy volunteers. On average, approximately 20% (range: 5.6 to 35.1) of the amount of lamotrigine present in the body was eliminated by hemodialysis during a 4-hour session *[see Dosage and Administration (2.1)]*.

Hepatic Disease: The pharmacokinetics of lamotrigine following a single 100-mg dose of lamotrigine were evaluated in 24 subjects with mild, moderate, and severe hepatic impairment (Child-Pugh Classification system) and compared with 12 subjects without hepatic impairment. The patients with severe hepatic impairment were without ascites (n = 2) or with ascites (n = 5). The mean apparent clearances of lamotrigine in patients with mild (n = 12), moderate (n = 5), severe without ascites (n = 2), and severe with ascites (n = 5) liver impairment were 0.30 ± 0.09, 0.24 ± 0.1, 0.21 ± 0.04, and 0.15 ± 0.09 mL/min/kg, respectively, as compared with 0.37 ± 0.1 mL/min/kg in the healthy controls. Mean half-lifes of lamotrigine in patients with mild, moderate, severe without ascites, and severe with ascites hepatic impairment were 46 ± 20, 72 ± 44, 67 ± 11, and 100 ± 48 hours, respectively, as compared with 33 ± 7 hours in healthy controls *[see Dosage and Administration (2.1)]*.

Age: Pediatric Patients: The pharmacokinetics of lamotrigine following a single 2-mg/kg dose were evaluated in 2 studies of pediatric patients (n = 29 for patients 10 months to 5.9 years of age and n = 26 for patients 5 to 11 years of age). Forty-three patients received concomitant therapy with other AEDs and 12 patients received lamotrigine as monotherapy. Lamotrigine pharmacokinetic parameters for pediatric patients are summarized in Table 16.

Population pharmacokinetic analyses involving patients 2 to 18 years of age demonstrated that lamotrigine clearance was influenced predominantly by total body weight and concurrent AED therapy. The oral clearance of lamotrigine was higher, on a body weight basis, in pediatric patients than in adults. Weight-normalized lamotrigine clearance was higher in those subjects weighing less than 30 kg, compared with those weighing greater than 30 kg. Accordingly, patients weighing less than 30 kg may need an increase of as much as 50% in maintenance doses, based on clinical response, as compared with subjects weighing more than 30 kg being administered the same AEDs *[see Dosage and Administration (2.2)]*. These analyses also revealed that, af-

ter accounting for body weight, lamotrigine clearance was not significantly influenced by age. Thus, the same weight-adjusted doses should be administered to children irrespective of differences in age. Concomitant AEDs which influence lamotrigine clearance in adults were found to have similar effects in children.
[See table 16 above]

Elderly: The pharmacokinetics of lamotrigine following a single 150-mg dose of LAMICTAL were evaluated in 12 elderly volunteers between the ages of 65 and 76 years (mean creatinine clearance = 61 mL/min, range: 33 to 108 mL/min). The mean half-life of lamotrigine in these subjects was 31.2 hours (range: 24.5 to 43.4 hours), and the mean clearance was 0.40 mL/min/kg (range: 0.26 to 0.48 mL/min/kg).

Gender: The clearance of lamotrigine is not affected by gender. However, during dose escalation of LAMICTAL in one clinical trial in patients with epilepsy on a stable dose of valproate (n = 77), mean trough lamotrigine concentrations, unadjusted for weight, were 24% to 45% higher (0.3 to 1.7 mcg/mL) in females than in males.

Race: The apparent oral clearance of lamotrigine was 25% lower in non-Caucasians than Caucasians.

Table 16. Mean Pharmacokinetic Parameters in Pediatric Patients With Epilepsy

Pediatric Study Population	Number of Subjects	T_{max} (hr)	$t_{1/2}$ (hr)	Cl/F (mL/min/kg)
Ages 10 months-5.3 years				
Patients taking carbamazepine, phenytoin, phenobarbital, or primidone*	10	3.0 (1.0-5.9)	7.7 (5.7-11.4)	3.62 (2.44-5.28)
Patients taking AEDs with no known effect on the apparent clearance of lamotrigine	7	5.2 (2.9-6.1)	19.0 (12.9-27.1)	1.2 (0.75-2.42)
Patients taking valproate only	8	2.9 (1.0-6.0)	44.9 (29.5-52.5)	0.47 (0.23-0.77)
Ages 5-11 years				
Patients taking carbamazepine, phenytoin, phenobarbital, or primidone*	7	1.6 (1.0-3.0)	7.0 (3.8-9.8)	2.54 (1.35-5.58)
Patients taking carbamazepine, phenytoin, phenobarbital, or primidone* plus valproate	8	3.3 (1.0-6.4)	19.1 (7.0-31.2)	0.89 (0.39-1.93)
Patients taking valproate only†	3	4.5 (3.0-6.0)	65.8 (50.7-73.7)	0.24 (0.21-0.26)
Ages 13-18 years				
Patients taking carbamazepine, phenytoin, phenobarbital, or primidone*	11	‡	‡	1.3
Patients taking carbamazepine, phenytoin, phenobarbital, or primidone* plus valproate	8	‡	‡	0.5
Patients taking valproate only	4	‡	‡	0.3

* Carbamazepine, phenobarbital, phenytoin, and primidone have been shown to increase the apparent clearance of lamotrigine. Estrogen-containing oral contraceptives and rifampin have also been shown to increase the apparent clearance of lamotrigine *[see Drug Interactions (7)]*.
† Two subjects were included in the calculation for mean T_{max}.
‡ Parameter not estimated.

13 NONCLINICAL TOXICOLOGY

13.1 Carcinogenesis, Mutagenesis, Impairment of Fertility

No evidence of carcinogenicity was seen in 1 mouse study or 2 rat studies following oral administration of lamotrigine for up to 2 years at maximum tolerated doses (30 mg/kg/day for mice and 10 to 15 mg/kg/day for rats, doses that are equivalent to 90 mg/m^2 and 60 to 90 mg/m^2, respectively). Steady-state plasma concentrations ranged from 1 to 4 mcg/mL in the mouse study and 1 to 10 mcg/mL in the rat study. Plasma concentrations associated with the recommended human doses of 300 to 500 mg/day are generally in the range of 2 to 5 mcg/mL, but concentrations as high as 19 mcg/mL have been recorded.

Lamotrigine was not mutagenic in the presence or absence of metabolic activation when tested in 2 gene mutation assays (the Ames test and the in vitro mammalian mouse lymphoma assay). In 2 cytogenetic assays (the in vitro human lymphocyte assay and the in vivo rat bone marrow assay), lamotrigine did not increase the incidence of structural or numerical chromosomal abnormalities.

No evidence of impairment of fertility was detected in rats given oral doses of lamotrigine up to 2.4 times the highest usual human maintenance dose of 8.33 mg/kg/day or 0.4 times the human dose on a mg/m^2 basis. The effect of lamotrigine on human fertility is unknown.

14 CLINICAL STUDIES

14.1 Epilepsy

Monotherapy With LAMICTAL in Adults With Partial Seizures Already Receiving Treatment With Carbamazepine, Phenytoin, Phenobarbital, or Primidone as the Single AED: The effectiveness of monotherapy with LAMICTAL was established in a multicenter, double-blind clinical trial enrolling 156 adult outpatients with partial seizures. The patients experienced at least 4 simple partial, complex partial, and/or secondarily generalized seizures during each of 2 consecutive 4-week periods while receiving carbamazepine or phenytoin monotherapy during baseline. LAMICTAL (target dose of 500 mg/day) or valproate (1,000 mg/day) was added to either carbamazepine or phenytoin monotherapy over a 4-week period. Patients were then converted to monotherapy with LAMICTAL or valproate during the next 4 weeks, then continued on monotherapy for an additional 12-week period.

Study endpoints were completion of all weeks of study treatment or meeting an escape criterion. Criteria for escape relative to baseline were: (1) doubling of average monthly seizure count, (2) doubling of highest consecutive 2-day seizure frequency, (3) emergence of a new seizure type (defined as a seizure that did not occur during the 8-week baseline) that is more severe than seizure types that occur during study treatment, or (4) clinically significant prolongation of generalized tonic-clonic (GTC) seizures. The primary efficacy variable was the proportion of patients in each treatment group who met escape criteria.

The percentages of patients who met escape criteria were 42% (32/76) in the group receiving LAMICTAL and 69% (55/80) in the valproate group. The difference in the percentage of patients meeting escape criteria was statistically significant (p= 0.0012) in favor of LAMICTAL. No differences in efficacy based on age, sex, or race were detected.

Patients in the control group were intentionally treated with a relatively low dose of valproate; as such, the sole objective of this study was to demonstrate the effectiveness and safety of monotherapy with LAMICTAL, and cannot be interpreted to imply the superiority of LAMICTAL to an adequate dose of valproate.

Adjunctive Therapy With LAMICTAL in Adults With Partial Seizures: The effectiveness of LAMICTAL as adjunctive therapy (added to other AEDs) was established in 3 multicenter, placebo-controlled, double-blind clinical trials in 355 adults with refractory partial seizures. The patients had a history of at least 4 partial seizures per month in spite of receiving one or more AEDs at therapeutic concentrations and, in 2 of the studies, were observed on their established AED regimen during baselines that varied between 8 to 12 weeks. In the third, patients were not observed in a prospective baseline. In patients continuing to have at least 4 seizures per month during the baseline, LAMICTAL or placebo was then added to the existing therapy. In all 3 studies, change from baseline in seizure frequency was the primary measure of effectiveness. The results given below are for all partial seizures in the intent-to-treat population (all patients who received at least one dose of treatment) in each study, unless otherwise indicated. The median seizure frequency at baseline was 3 per week while the mean at baseline was 6.6 per week for all patients enrolled in efficacy studies.

One study (n = 216) was a double-blind, placebo-controlled, parallel trial consisting of a 24-week treatment period. Patients could not be on more than 2 other anticonvulsants and valproate was not allowed. Patients were randomized to receive placebo, a target dose of 300 mg/day of LAMICTAL, or a target dose of 500 mg/day of LAMICTAL. The median reductions in the frequency of all partial seizures relative to baseline were 8% in patients receiving placebo, 20% in patients receiving 300 mg/day of LAMICTAL, and 36% in patients receiving 500 mg/day of LAMICTAL. The seizure frequency reduction was statistically significant in the 500-mg/day group compared with the placebo group, but not in the 300-mg/day group.

A second study (n = 98) was a double-blind, placebo-controlled, randomized, crossover trial consisting of two 14-week treatment periods (the last 2 weeks of which consisted of dose tapering) separated by a 4-week washout period. Patients could not be on more than 2 other anticonvulsants and valproate was not allowed. The target dose of LAMICTAL was 400 mg/day. When the first 12 weeks of the treatment periods were analyzed, the median change in seizure frequency was a 25% reduction on LAMICTAL compared with placebo (p<0.001).

The third study (n = 41) was a double-blind, placebo-controlled, crossover trial consisting of two 12-week treatment periods separated by a 4-week washout period. Patients could not be on more than 2 other anticonvulsants. Thirteen patients were on concomitant valproate; these patients received 150 mg/day of LAMICTAL. The 28 other patients had a target dose of 300 mg/day of LAMICTAL. The median change in seizure frequency was a 26% reduction on LAMICTAL compared with placebo (p<0.01).

No differences in efficacy based on age, sex, or race, as measured by change in seizure frequency, were detected.

Adjunctive Therapy With LAMICTAL in Pediatric Patients With Partial Seizures: The effectiveness of LAMICTAL as adjunctive therapy in pediatric patients with partial seizures was established in a multicenter, double-blind, placebo-controlled trial in 199 patients 2 to 16 years of age (n = 98 on LAMICTAL, n = 101 on placebo). Following an 8-week baseline phase, patients were randomized to 18 weeks of treatment with LAMICTAL or placebo added to their current AED regimen of up to 2 drugs. Patients were dosed based on body weight and valproate use. Target doses were designed to approximate 5 mg/kg/day for patients taking valproate (maximum dose: 250 mg/day) and 15 mg/kg/day for the patients not taking valproate (maximum dose: 750 mg/day). The primary efficacy endpoint was percentage change from baseline in all partial seizures. For the intent-to-treat population, the median reduction of all partial seizures was 36% in patients treated with LAMICTAL and 7% on placebo, a difference that was statistically significant (p<0.01).

Adjunctive Therapy With LAMICTAL in Pediatric and Adult Patients With Lennox-Gastaut Syndrome: The effectiveness of LAMICTAL as adjunctive therapy in patients with Lennox-Gastaut syndrome was established in a multicenter, double-blind, placebo-controlled trial in 169 patients 3 to 25 years of age (n = 79 on LAMICTAL, n = 90 on placebo). Following a 4-week single-blind, placebo phase, patients were randomized to 16 weeks of treatment with LAMICTAL or placebo added to their current AED regimen of up to 3 drugs. Patients were dosed on a fixed-dose regimen based on body weight and valproate use. Target doses were designed to approximate 5 mg/kg/day for patients taking valproate (maximum dose: 200 mg/day) and 15 mg/kg/day for patients not taking valproate (maximum dose: 400 mg/day). The primary efficacy endpoint was percentage change from baseline in major motor seizures (atonic, tonic, major myoclonic, and tonic-clonic seizures). For the intent-to-treat population, the median reduction of major motor seizures was 32% in patients treated with LAMICTAL and 9% on placebo, a difference that was statistically significant (p<0.05). Drop attacks were significantly reduced by LAMICTAL (34%) compared with placebo (9%), as were tonic-clonic seizures (36% reduction versus 10% increase for LAMICTAL and placebo, respectively).

Adjunctive Therapy With LAMICTAL in Pediatric and Adult Patients With Primary Generalized Tonic-Clonic Seizures: The effectiveness of LAMICTAL as adjunctive therapy in patients with primary generalized tonic-clonic seizures was established in a multicenter, double-blind, placebo-controlled trial in 117 pediatric and adult patients ≥2 years (n = 58 on LAMICTAL, n = 59 on placebo). Patients with at least 3 primary generalized tonic-clonic seizures during an 8-week baseline phase were randomized to 19 to 24 weeks of treatment with LAMICTAL or placebo added to their current AED regimen of up to 2 drugs. Patients were dosed on a fixed-dose regimen, with target doses ranging from 3 mg/kg/day to 12 mg/kg/day for pediatric patients and from 200 mg/day to 400 mg/day for adult patients based on concomitant AED.

The primary efficacy endpoint was percentage change from baseline in primary generalized tonic-clonic seizures. For the intent-to-treat population, the median percent reduction of primary generalized tonic-clonic seizures was 66% in patients treated with LAMICTAL and 34% on placebo, a difference that was statistically significant (p = 0.006).

14.2 Bipolar Disorder

The effectiveness of LAMICTAL in the maintenance treatment of Bipolar I Disorder was established in 2 multicenter,

Continued on next page

Product information on these pages is effective as of June 2009. Further information is available at 1-888-825-5249 or www.gsk.com.

Lamictal—Cont.

double-blind, placebo-controlled studies in adult patients who met DSM-IV criteria for Bipolar I Disorder. Study 1 enrolled patients with a current or recent (within 60 days) depressive episode as defined by DSM-IV and Study 2 included patients with a current or recent (within 60 days) episode of mania or hypomania as defined by DSM-IV. Both studies included a cohort of patients (30% of 404 patients in Study 1 and 28% of 171 patients in Study 2) with rapid cycling Bipolar Disorder (4 to 6 episodes per year).

In both studies, patients were titrated to a target dose of 200 mg of LAMICTAL, as add-on therapy or as monotherapy, with gradual withdrawal of any psychotropic medications during an 8- to 16-week open-label period. Overall 81% of 1,305 patients participating in the open-label period were receiving 1 or more other psychotropic medications, including benzodiazepines, selective serotonin reuptake inhibitors (SSRIs), atypical antipsychotics (including olanzapine), valproate, or lithium, during titration of LAMICTAL. Patients with a CGI-severity score of 3 or less maintained for at least 4 continuous weeks, including at least the final week on monotherapy with LAMICTAL, were randomized to a placebo-controlled, double-blind treatment period for up to 18 months. The primary endpoint was TIME (time to intervention for a mood episode or one that was emerging, time to discontinuation for either an adverse event that was judged to be related to Bipolar Disorder, or for lack of efficacy). The mood episode could be depression, mania, hypomania, or a mixed episode.

In Study 1, patients received double-blind monotherapy with LAMICTAL 50 mg/day (n = 50), LAMICTAL 200 mg/day (n = 124), LAMICTAL 400 mg/day (n = 47), or placebo (n = 121). LAMICTAL (200- and 400-mg/day treatment groups combined) was superior to placebo in delaying the time to occurrence of a mood episode. Separate analyses of the 200- and 400-mg/day dose groups revealed no added benefit from the higher dose.

In Study 2, patients received double-blind monotherapy with LAMICTAL (100 to 400 mg/day, n = 59), or placebo (n = 70). LAMICTAL was superior to placebo in delaying time to occurrence of a mood episode. The mean dose of LAMICTAL was about 211 mg/day.

Although these studies were not designed to separately evaluate time to the occurrence of depression or mania, a combined analysis for the 2 studies revealed a statistically significant benefit for LAMICTAL over placebo in delaying the time to occurrence of both depression and mania, although the finding was more robust for depression.

16 HOW SUPPLIED/STORAGE AND HANDLING

LAMICTAL (lamotrigine) Tablets

25 mg, white, scored, shield-shaped tablets debossed with "LAMICTAL" and "25", bottles of 100 (NDC 0173-0633-02).

Store at 25°C (77°F); excursions permitted to 15-30°C (59-86°F) [see USP Controlled Room Temperature] in a dry place.

100 mg, peach, scored, shield-shaped tablets debossed with "LAMICTAL" and "100", bottles of 100 (NDC 0173-0642-55).

150 mg, cream, scored, shield-shaped tablets debossed with "LAMICTAL" and "150", bottles of 60 (NDC 0173-0643-60).

200 mg, blue, scored, shield-shaped tablets debossed with "LAMICTAL" and "200", bottles of 60 (NDC 0173-0644-60).

Store at 25°C (77°F); excursions permitted to 15-30°C (59-86°F) [see USP Controlled Room Temperature] in a dry place and protect from light.

LAMICTAL Starter Kit for Patients Taking Valproate (Blue)

25 mg, white, scored, shield-shaped tablets debossed with "LAMICTAL" and "25", blisterpack of 35 tablets (NDC 0173-0633-10).

Store at 25°C (77°F); excursions permitted to 15-30°C (59-86°F) [see USP Controlled Room Temperature] in a dry place.

LAMICTAL Starter Kit for Patients Taking Carbamazepine, Phenytoin, Phenobarbital, Primidone, or Rifampin and Not Taking Valproate (Green)

25 mg, white, scored, shield-shaped tablets debossed with "LAMICTAL" and "25" and 100 mg, peach, scored, shield-shaped tablets debossed with "LAMICTAL" and "100", blisterpack of 98 tablets (84/25 mg tablets and 14/100 mg tablets) (NDC 0173-0594-01)

Store at 25°C (77°F); excursions permitted to 15-30°C (59-86°F) [see USP Controlled Room Temperature] in a dry place and protect from light.

LAMICTAL Starter Kit for Patients Not Taking Carbamazepine, Phenytoin, Phenobarbital, Primidone, Rifampin, or Valproate (Orange)

25 mg, white, scored, shield-shaped tablets debossed with "LAMICTAL" and "25" and 100 mg, peach, scored, shield-shaped tablets debossed with "LAMICTAL" and "100", blisterpack of 49 tablets (42/25mg tablets and 7/100 mg tablets) (NDC 0173-0594-02).

Store at 25°C (77°F); excursions permitted to 15-30°C (59-86°F) [see USP Controlled Room Temperature] in a dry place and protect from light.

LAMICTAL (lamotrigine) Chewable Dispersible Tablets

2 mg, white to off-white, round tablets debossed with "LTG" over "2", bottles of 30 (NDC 0173-0699-00). ORDER DIRECTLY FROM GlaxoSmithKline 1-800-334-4153.

5 mg, white to off-white, caplet-shaped tablets debossed with "GX CL2", bottles of 100 (NDC 0173-0526-00).

25 mg, white, super elliptical-shaped tablets debossed with "GX CL5", bottles of 100 (NDC 0173-0527-00).

Store at 25°C (77°F); excursions permitted to 15-30°C (59-86°F) [see USP Controlled Room Temperature] in a dry place.

LAMICTAL ODT (lamotrigine) Orally Disintegrating Tablets

25 mg, white to off-white, round, flat-faced, radius edge, tablets debossed with "LMT" on one side and "25" on the other, Maintenance Packs of 30 (NDC 0173-0772-02).

50 mg, white to off-white, round, flat-faced, radius edge, tablets debossed with "LMT" on one side and "50" on the other, Maintenance Packs of 30 (NDC 0173-0774-02).

100 mg, white to off-white, round, flat-faced, radius edge, tablets debossed with "LAMICTAL" on one side and "100" on the other, Maintenance Packs of 30 (NDC 0173-0776-02).

200 mg, white to off-white, round, flat-faced, radius edge, tablets debossed with "LAMICTAL" on one side and "200" on the other, Maintenance Packs of 30 (NDC 0173-0777-02).

Store between 20°C to 25°C (68°F to 77°F); with excursions permitted between 15°C and 30°C (59°F and 86°F).

LAMICTAL ODT Patient Titration Kit for Patients Taking Valproate (Blue)

25 mg, white to off-white, round, flat-faced, radius edge, tablets debossed with "LMT" on one side and "25" on the other, and 50 mg, white to off-white, round, flat-faced, radius edge, tablets debossed with "LMT" on one side and "50" on the other, blisterpack of 28 tablets (21/25-mg tablets and 7/50-mg tablets) (NDC 0173-0779-00).

LAMICTAL ODT Patient Titration Kit for Patients Taking Carbamazepine, Phenytoin, Phenobarbital, Primidone, or Rifampin and Not Taking Valproate (Green)

50 mg, white to off-white, round, flat-faced, radius edge, tablets debossed with "LMT" on one side and "50" on the other, and 100 mg, white to off-white, round, flat-faced, radius edge, tablets debossed with "LAMICTAL" on one side and "100" on the other, blisterpack of 56 tablets (42/50-mg tablets and 14/100-mg tablets) (NDC 0173-0780-00).

LAMICTAL ODT Patient Titration Kit for Patients Not Taking Carbamazepine, Phenytoin, Phenobarbital, Primidone, Rifampin, or Valproate (Orange)

25 mg, white to off-white, round, flat-faced, radius edge, tablets debossed with "LMT" on one side and "25" on the other, 50 mg, white to off-white, round, flat-faced, radius edge, tablets debossed with "LMT" on one side and "50" on the other, and 100 mg, white to off-white, round, flat-faced, radius edge, tablets debossed with "LAMICTAL" on one side and "100" on the other, blisterpack of 35 (14/25-mg tablets, 14/50-mg tablets, and 7/100-mg tablets) (NDC 0173-0778-00).

Store between 20°C to 25°C (68°F to 77°F); with excursions permitted between 15°C and 30°C (59°F and 86°F).

Blisterpacks: If the product is dispensed in a blisterpack, the patient should be advised to examine the blisterpack before use and not use if blisters are torn, broken, or missing.

17 PATIENT COUNSELING INFORMATION

See Medication Guide that accompanies the product.

17.1 Rash

Prior to initiation of treatment with LAMICTAL, the patient should be instructed that a rash or other signs or symptoms of hypersensitivity (e.g., fever, lymphadenopathy) may herald a serious medical event and that the patient should report any such occurrence to a physician immediately.

17.2 Suicidal Thinking and Behavior

Patients, their caregivers, and families should be counseled that AEDs, including LAMICTAL, may increase the risk of suicidal thoughts and behavior and should be advised of the need to be alert for the emergence or worsening of symptoms of depression, any unusual changes in mood or behavior, or the emergence of suicidal thoughts, behavior, or thoughts about self-harm. Behaviors of concern should be reported immediately to healthcare providers.

17.3 Worsening of Seizures

Patients should be advised to notify their physician if worsening of seizure control occurs.

17.4 CNS Adverse Effects

Patients should be advised that LAMICTAL may cause dizziness, somnolence, and other symptoms and signs of central nervous system (CNS) depression. Accordingly, they should be advised neither to drive a car nor to operate other complex machinery until they have gained sufficient experience on LAMICTAL to gauge whether or not it adversely affects their mental and/or motor performance.

17.5 Blood Dyscrasias and/or Acute Multiorgan Failure

Patients should be advised of the possibility of blood dyscrasias and/or acute multiorgan failure and to contact their physician immediately if they experience any signs or symptoms of these conditions *[see Warnings and Precautions (5.3, 5.4)]*.

17.6 Pregnancy

Patients should be advised to notify their physicians if they become pregnant or intend to become pregnant during therapy. Patients should be advised to notify their physicians if they intend to breastfeed or are breastfeeding an infant.

Patients should also be encouraged to enroll in the North American Antiepileptic Drug (NAAED) Pregnancy Registry if they become pregnant. This registry is collecting information about the safety of antiepileptic drugs during pregnancy. To enroll, patients can call the toll-free number 1-888-233-2334 *[see Use in Specific Populations (8.1)]*.

17.7 Oral Contraceptive Use

Women should be advised to notify their physician if they plan to start or stop use of oral contraceptives or other female hormonal preparations. Starting estrogen-containing oral contraceptives may significantly decrease lamotrigine plasma levels and stopping estrogen-containing oral contraceptives (including the "pill-free" week) may significantly increase lamotrigine plasma levels *[see Warnings and Precautions (5.8), Clinical Pharmacology (12.3)]*. Women should also be advised to promptly notify their physician if they experience adverse events or changes in menstrual pattern (e.g., break-through bleeding) while receiving LAMICTAL in combination with these medications.

17.8 Discontinuing LAMICTAL

Patients should be advised to notify their physician if they stop taking LAMICTAL for any reason and not to resume LAMICTAL without consulting their physician.

17.9 Potential Medication Errors

Medication errors involving LAMICTAL have occurred. In particular the name LAMICTAL or lamotrigine can be confused with the names of other commonly used medications. Medication errors may also occur between the different formulations of LAMICTAL. To reduce the potential of medication errors, write and say LAMICTAL clearly. Depictions of the LAMICTAL Tablets, Chewable Dispersible Tablets, and Orally Disintegrating Tablets can be found in the Medication Guide that accompanies the product to highlight the distinctive markings, colors, and shapes that serve to identify the different presentations of the drug and thus may help reduce the risk of medication errors. **To avoid a medication error of using the wrong drug or formulation, patients should be strongly advised to visually inspect their tablets to verify that they are LAMICTAL, as well as the correct formulation of LAMICTAL, each time they fill their prescription** *[see Dosage Forms and Strengths (3.1, 3.2, 3.3), How Supplied/Storage and Handling (16)]*.

GlaxoSmithKline
Research Triangle Park, NC 27709

LAMICTAL Tablets and Chewable Dispersible Tablets are manufactured by
DSM Pharmaceuticals, Inc., Greenville, NC 27834 or
GlaxoSmithKline, Research Triangle Park, NC 27709

LAMICTAL Orally Disintegrating Tablets are manufactured by
Eurand, Inc., Vandalia, OH 45377

LAMICTAL is a registered trademark of GlaxoSmithKline.
*Microcaps and AdvaTab are registered trademarks of Eurand, Inc.

MEDICATION GUIDE

LAMICTAL® (la-MIK-tal) (lamotrigine) Tablets and Chewable Dispersible Tablets

LAMICTAL® ODT™ (lamotrigine) Orally Disintegrating Tablets

Read this Medication Guide before you start taking LAMICTAL and each time you get a refill. There may be new information. This information does not take the place of talking with your healthcare provider about your medical condition or treatment. If you have questions about LAMICTAL, ask your healthcare provider or pharmacist.

What is the most important information I should know about LAMICTAL?

1. **LAMICTAL may cause a serious skin rash that may cause you to be hospitalized or to stop LAMICTAL; it may rarely cause death.**

   There is no way to tell if a mild rash will develop into a more serious reaction. These serious skin reactions are more likely to happen when you begin taking LAMICTAL, within the first 2 to 8 weeks of treatment. But it can happen in people who have taken LAMICTAL for any period of time. Children between 2 to 16 years of age have a higher chance of getting this serious skin reaction while taking LAMICTAL.

The risk of getting a rash is higher if you:

- take LAMICTAL while taking valproate (DEPAKENE (valproic acid) or DEPAKOTE (divalproex sodium))
- take a higher starting dose of LAMICTAL than your healthcare provider prescribed

- increase your dose of LAMICTAL faster than prescribed.

LAMICTAL can also cause other types of allergic reactions or serious problems which may affect organs and other parts of your body like the liver or blood cells. You may or may not have a rash with these types of reactions.

Call your healthcare provider right away if you have any of the following:

- **a skin rash**
- **hives**
- **fever**
- **swollen lymph glands**
- **painful sores in the mouth or around your eyes**
- **swelling of your lips or tongue**
- **yellowing of your skin or eyes**
- **unusual bruising or bleeding**
- **severe fatigue or weakness**
- **severe muscle pain**
- **frequent infections**

These symptoms may be the first signs of a serious reaction. A healthcare provider should examine you to decide if you should continue taking LAMICTAL.

2. Like other antiepileptic drugs, LAMICTAL may cause suicidal thoughts or actions in a very small number of people, about 1 in 500.

Call a healthcare provider right away if you have any of these symptoms, especially if they are new, worse, or worry you:

- thoughts about suicide or dying
- attempt to commit suicide
- new or worse depression
- new or worse anxiety
- feeling agitated or restless
- panic attacks
- trouble sleeping (insomnia)
- new or worse irritability
- acting aggressive, being angry, or violent
- acting on dangerous impulses
- an extreme increase in activity and talking (mania)
- other unusual changes in behavior or mood

Do not stop LAMICTAL without first talking to a healthcare provider.

- Stopping LAMICTAL suddenly can cause serious problems.
- Suicidal thoughts or actions can be caused by things other than medicines. If you have suicidal thoughts or actions, your healthcare provider may check for other causes.

How can I watch for early symptoms of suicidal thoughts and actions?

- Pay attention to any changes, especially sudden changes, in mood, behaviors, thoughts, or feelings.
- Keep all follow-up visits with your healthcare provider as scheduled.
- Call your healthcare provider between visits as needed, especially if you are worried about symptoms.

LAMICTAL can have other serious side effects. For more information ask your healthcare provider or pharmacist. Tell your healthcare provider if you have any side effect that bothers you. Be sure to read the section below entitled "What are the possible side effects of LAMICTAL?"

3. Patients prescribed LAMICTAL have sometimes been given the wrong medicine because many medicines have names similar to LAMICTAL, so always check that you receive LAMICTAL.

Taking the wrong medication can cause serious health problems. When your healthcare provider gives you a prescription for LAMICTAL:

- Make sure you can read it clearly.
- Talk to your pharmacist to check that you are given the correct medicine.
- Each time you fill your prescription, check the tablets you receive against the pictures of the tablets below.

These pictures show the distinct wording, colors, and shapes of the tablets that help to identify the right strength of LAMICTAL Tablets, Chewable Dispersible Tablets, and Orally Disintegrating Tablets. Immediately call your pharmacist if you receive a LAMICTAL tablet that does not look like one of the tablets shown below, as you may have received the wrong medication.

[See figure above]

What is LAMICTAL?

LAMICTAL is a prescription medicine used:

1. together with other medicines to treat certain types of seizures (partial seizures, primary generalized tonic-clonic seizures, generalized seizures of Lennox-Gastaut syndrome) in people 2 years or older.
2. alone when changing from other medicines to treat partial seizures in people 16 years or older.
3. for the long-term treatment of Bipolar I Disorder to lengthen the time between mood episodes in people 18 years or older who have been treated for mood episodes with other medicine.

LAMICTAL (lamotrigine) Tablets

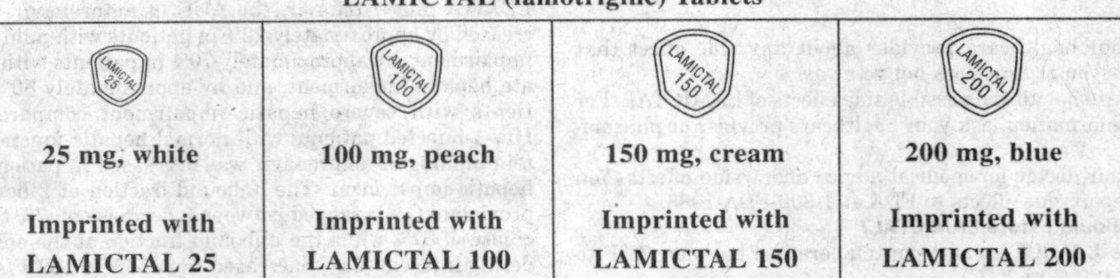

25 mg, white	100 mg, peach	150 mg, cream	200 mg, blue
Imprinted with LAMICTAL 25	Imprinted with LAMICTAL 100	Imprinted with LAMICTAL 150	Imprinted with LAMICTAL 200

LAMICTAL (lamotrigine) Chewable Dispersible Tablets

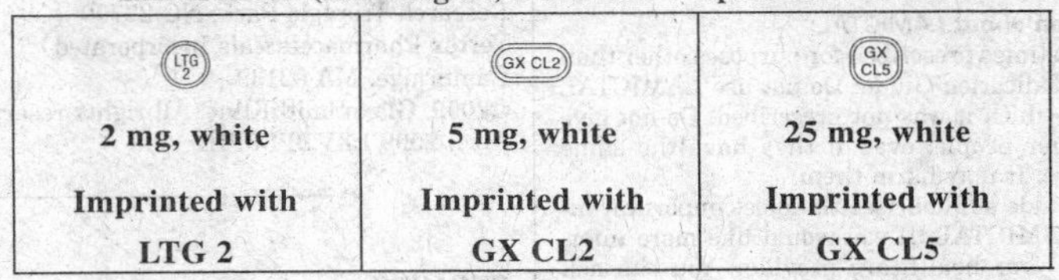

2 mg, white	5 mg, white	25 mg, white
Imprinted with LTG 2	Imprinted with GX CL2	Imprinted with GX CL5

LAMICTAL ODT (lamotrigine) Orally Disintegrating Tablets

25 mg, white to off-white	50 mg, white to off-white	100 mg, white to off-white	200 mg, white to off-white
Imprinted with LMT on one side 25 on the other	Imprinted with LMT on one side 50 on the other	Imprinted with LAMICTAL on one side 100 on the other	Imprinted with LAMICTAL on one side 200 on the other

It is not known if LAMICTAL is safe or effective in children or teenagers under the age of 18 with mood disorders such as bipolar disorder or depression.

It is not known if LAMICTAL is safe or effective when used alone as the first treatment of seizures in adults.

Who should not take LAMICTAL?

You should not take LAMICTAL if you have had an allergic reaction to lamotrigine or to any of the inactive ingredients in LAMICTAL. See the end of this leaflet for a complete list of ingredients in LAMICTAL.

What should I tell my healthcare provider before taking LAMICTAL?

Before taking LAMICTAL, tell your healthcare provider about all of your medical conditions, including if you:

- have had a rash or allergic reaction to another antiseizure medicine.
- have or have had depression, mood problems or suicidal thoughts or behavior.
- are taking oral contraceptives (birth control pills) or other female hormonal medicines. Do not start or stop taking birth control pills or other female hormonal medicine until you have talked with your healthcare provider. Tell your healthcare provider if you have any changes in your menstrual pattern such as breakthrough bleeding. Stopping or starting these medicines may cause side effects (such as dizziness, lack of coordination, or double vision) or lessen how well LAMICTAL works.
- are pregnant or plan to become pregnant. It is not known if LAMICTAL will harm your unborn baby. If you become pregnant while taking LAMICTAL, talk to your healthcare provider about registering with the North American Antiepileptic Drug Pregnancy Registry. You can enroll in this registry by calling 1-888-233-2334. The purpose of this registry is to collect information about the safety of antiepileptic drugs during pregnancy.
- are breastfeeding. LAMICTAL can pass into your breast milk. You and your healthcare provider should decide if you should take LAMICTAL or breastfeed. Breastfeeding while taking LAMICTAL is not recommended.

Tell your healthcare provider about all the medicines you take or if you are planning to take a new medicine, including prescription and non-prescription medicines, vitamins, and herbal supplements. Using LAMICTAL with certain other medicines can affect each other, causing side effects.

How should I take LAMICTAL?

- Take LAMICTAL exactly as prescribed.
- Your healthcare provider may change your dose. Do not change your dose without talking to your healthcare provider.
- Do not stop taking LAMICTAL without talking to your healthcare provider. Stopping LAMICTAL suddenly may cause serious problems. For example, if you have epilepsy and you stop taking LAMICTAL suddenly, you may get seizures that do not stop. Talk with your healthcare provider about how to stop LAMICTAL slowly.
- If you miss a dose of LAMICTAL, take it as soon as you remember. If it is almost time for your next dose, just skip the missed dose. Take the next dose at your regular time. **Do not take two doses at the same time.**
- You may not feel the full effect of LAMICTAL for several weeks.
- If you have epilepsy, tell your healthcare provider if your seizures get worse or if you have any new types of seizures.
- Swallow LAMICTAL tablets whole.
- If you have trouble swallowing LAMICTAL Tablets, there may be another form of LAMICTAL you can take.
- LAMICTAL ODT should be placed on the tongue and moved around the mouth. The tablet will rapidly disintegrate, can be swallowed with or without water, and can be taken with or without food.
- LAMICTAL Chewable Dispersible tablets may be swallowed whole, chewed, or mixed in water or diluted fruit juice. If the tablets are chewed, drink a small amount of water or diluted fruit juice to help in swallowing. To break up LAMICTAL Chewable Dispersible tablets, add the tablets to a small amount of liquid (1 teaspoon, or enough to cover the medicine) in a glass or spoon. Wait at least 1 minute or until the tablets are completely broken up, mix the solution together and take the whole amount right away.
- If you receive LAMICTAL in a blisterpack, examine the blisterpack before use. Do not use if blisters are torn, broken, or missing.

What should I avoid while taking LAMICTAL?

- Do not drive a car or operate complex, hazardous machinery until you know how LAMICTAL affects you.

What are possible side effects of LAMICTAL?

- See "What is the most important information I should know about LAMICTAL?"

Common side effects of LAMICTAL include:

• dizziness	• tremor
• headache	• rash
• blurred or double vision	• fever
• lack of coordination	• abdominal pain
• sleepiness	• back pain
• nausea, vomiting	• tiredness
• insomnia	• dry mouth

Continued on next page

Product information on these pages is effective as of June 2009. Further information is available at 1-888-825-5249 or www.gsk.com.

Lamictal—Cont.

Tell your healthcare provider about any side effect that bothers you or that does not go away.
These are not all the possible side effects of LAMICTAL. For more information, ask your healthcare provider or pharmacist.
Call your doctor for medical advice about side effects. You may report side effects to FDA at 1-800-FDA-1088.

How should I store LAMICTAL?

- Store LAMICTAL at room temperature between 68°F to 77°F (20°C to 25°C).
- **Keep LAMICTAL and all medicines out of the reach of children.**

General information about LAMICTAL

Medicines are sometimes prescribed for purposes other than those listed in a Medication Guide. Do not use LAMICTAL for a condition for which it was not prescribed. Do not give LAMICTAL to other people, even if they have the same symptoms you have. It may harm them.
This Medication Guide summarizes the most important information about LAMICTAL. If you would like more information, talk with your healthcare provider. You can ask your healthcare provider or pharmacist for information about LAMICTAL that is written for healthcare professionals.
For more information, go to www.lamictal.com or call 1-888-825-5249.

What are the ingredients in LAMICTAL?

LAMICTAL Tablets

Active ingredient: lamotrigine.
Inactive ingredients: lactose; magnesium stearate, microcrystalline cellulose, povidone, sodium starch glycolate, FD&C Yellow No. 6 Lake (100-mg tablet only), ferric oxide, yellow (150-mg tablet only), and FD&C Blue No. 2 Lake (200-mg tablet only).

LAMICTAL Chewable Dispersible Tablets

Active ingredient: lamotrigine.
Inactive ingredients: blackcurrant flavor, calcium carbonate, low-substituted hydroxypropylcellulose, magnesium aluminum silicate, magnesium stearate, povidone, saccharin sodium, and sodium starch glycolate.

LAMICTAL ODT Orally Disintegrating Tablets

Active ingredient: lamotrigine.
Inactive ingredients: artificial cherry flavor, crospovidone, ethylcellulose, magnesium stearate, mannitol, polyethylene, and sucralose.

This Medication Guide has been approved by the U.S. Food and Drug Administration.

GlaxoSmithKline
Research Triangle Park, NC 27709
LAMICTAL Tablets and Chewable Dispersible Tablets are manufactured by
DSM Pharmaceuticals, Inc.,
Greenville, NC 27834 or
GlaxoSmithKline
Research Triangle Park, NC 27709
LAMICTAL Orally Disintegrating Tablets are manufactured by
Eurand, Inc., Vandalia, OH 45377
LAMICTAL is a registered trademark of GlaxoSmithKline.
DEPAKENE and DEPAKOTE are registered trademarks of Abbott Laboratories.

May 2009 LMT:2MG

LEXIVA® ℞
[*lex-ē' va*]
(fosamprenavir calcium)
Tablets

LEXIVA® ℞
(fosamprenavir calcium)
Oral Suspension

Prescribing information for this product, which appears on pages 1510–1520 of the 2009 PDR, has been revised as follows. Please write "See Supplement B" next to the product heading.

The following paragraph was revised under **2 DOSAGE AND ADMINISTRATION: 2.3 Patients With Hepatic Impairment:**

Severe Hepatic Impairment (Child-Pugh score ranging from 10 to 15): LEXIVA should be used with caution at a reduced dosage of 350 mg twice daily without ritonavir (therapy-naive) or 300 mg twice daily plus ritonavir 100 mg once daily (therapy-naive or protease inhibitor-experienced).

The following paragraph was revised under **12 CLINICAL PHARMACOLOGY: 12.3 Pharmacokinetics:**

Special Populations: ***Hepatic Impairment:*** The pharmacokinetics of amprenavir have been studied after the administration of LEXIVA in combination with ritonavir to adult HIV-1-infected patients with mild, moderate, and severe hepatic impairment. Following 2 weeks of dosing with LEXIVA plus ritonavir, the AUC of amprenavir was increased by approximately 22% in patients with mild hepatic impairment, by approximately 70% in patients with moderate hepatic impairment, and by approximately 80% in patients with severe hepatic impairment compared with HIV-1-infected patients with normal hepatic function. Protein binding of amprenavir was decreased in patients with hepatic impairment. The unbound fraction at 2 hours (approximate C_{max}) ranged between a decrease of -7% to an increase of 57% while the unbound fraction at the end of the dosing interval (C_{min}) increased from 50% to 102% *[see Dosage and Administration (2.3)]*.

GlaxoSmithKline
Research Triangle Park, NC 27709
Vertex Pharmaceuticals Incorporated
Cambridge, MA 02139

April 2009 LXV:9PI-6PIL

REQUIP® ℞
[*rē' kwip*]
(ropinirole tablets)

Patient Information Included

Prescribing information for this product, which appears on pages 1561–1568 of the 2009 PDR, has been revised as follows. Please write "See Supplement B" next to the product heading.

The following sentence was revised in the **DESCRIPTION** *section:*

Each pentagonal film-coated TILTAB® tablet with beveled edges contains 0.29 mg, 0.57 mg, 1.14 mg, 2.28 mg, 3.42 mg, 4.56 mg, or 5.70 mg ropinirole hydrochloride equivalent to ropinirole, 0.25 mg, 0.5 mg, 1 mg, 2 mg, 3 mg, 4 mg, or 5 mg.

The following sentence was revised in the **CLINICAL PHARMACOLOGY: Population Subgroups:** ***Age*** *section:*

Oral clearance of ropinirole is reduced by 15% in patients above 65 years of age compared to younger patients.

The **CONTRAINDICATIONS** *section was revised to:*

REQUIP is contraindicated for patients known to have hypersensitivity reaction (including urticaria, angioedema, rash, pruritus) to ropinirole or to any of the excipients.

The **PRECAUTIONS: General:** ***Melanoma*** *section was revised to:*

Epidemiologic studies have shown that patients with Parkinson's disease have a higher risk (2- to approximately 6-fold higher) of developing melanoma than the general population. Whether the increased risk observed was due to Parkinson's disease or other factors, such as drugs used to treat Parkinson's disease, is unclear.
For the reasons stated above, patients and providers are advised to monitor for melanomas frequently and on a regular basis when using REQUIP for *any* indication. Ideally, periodic skin examinations should be performed by appropriately qualified individuals (e.g., dermatologists).

The **ADVERSE REACTIONS: Postmarketing Reports** *section was revised to:*

The following adverse events (not listed above in clinical trials or other sections of the prescribing information) have been identified during postapproval use of ropinirole. Because these events are reported voluntarily from a population of uncertain size, it is not always possible to reliably estimate their frequency or establish a causal relationship to drug exposure.
Immune Systems Disorders: Hypersensitivity reactions (including urticaria, angioedema, rash, and pruritus).
Psychiatric Disorders: Impulse control symptoms, pathological gambling, increased libido including hypersexuality.

The **HOW SUPPLIED** *section was revised to delete the 2-Week Starter Kit for Treatment of Moderate-to-Severe Primary Restless Legs Syndrome.*

The following was added to the **PATIENT INFORMATION: What is the most important information I should know about REQUIP?** *section:*

Unusual urges: Some patients taking REQUIP or REQUIP XL get urges to behave in a way unusual for them. Examples of this are an unusual urge to gamble or increased sexual urges and behaviors. If you notice or your family notices that you are developing any unusual behaviors, talk to your healthcare provider.

The **How should I take REQUIP for RLS?** *section was revised to delete the bullet point discussing the starting kit with doses marked by day.*

GlaxoSmithKline
Research Triangle Park, NC 27709

May 2009 REP:3PI April 2009 REP:2PIL

TIMENTIN® ℞
[*tī-měn' tin*]
(sterile ticarcillin disodium and clavulanate potassium) for Intravenous Administration
ADD-VANTAGE® ANTIBIOTIC VIAL

Prescribing information for this product, which appears on pages 1606-1609 of the 2009 PDR, has been revised as follows. Please write "See Supplement B" next to the product heading.

The following was added to the PRECAUTIONS *section:*

Geriatric Use: An analysis of clinical studies of TIMENTIN was conducted to determine whether subjects aged 65 and over respond differently from younger subjects. Of the 1,078 subjects treated with at least one dose of TIMENTIN, 67.5% were <65 years old, and 32.5% were ≥65 years old. No overall differences in safety or efficacy were observed between these subjects and younger subjects, and other reported clinical experience have not identified differences in responses between the elderly and younger patients, but a greater sensitivity of some older individuals cannot be ruled out.
This drug is known to be substantially excreted by the kidney, and the risk of toxic reactions to this drug may be greater in patients with impaired renal function. Because elderly patients are more likely to have decreased renal function, care should be taken in dose selection, and it may be useful to monitor renal function (see DOSAGE and ADMINISTRATION).
TIMENTIN contains 103.6 mg (4.51 mEq) of sodium per gram of TIMENTIN. At the usual recommended doses, patients would receive between 1,285 and 1,927 mg/day (56 and 84 mEq) of sodium. The geriatric population may respond with a blunted natriuresis to salt loading. This may be clinically important with regard to such diseases as congestive heart failure.

GlaxoSmithKline
Research Triangle Park, NC 27709

February 2009 TMV:17PI

TWINRIX® ℞
[*twin' rix*]
[Hepatitis A Inactivated & Hepatitis B (Recombinant) Vaccine]

Prescribing information for this product, which appears on pages 1628-1631 of the 2009 PDR, has been revised as follows. Please write "See Supplement B" next to the product heading.

The following subsections were revised in the ADVERSE REACTIONS *section:*

Incidence 1% to 10% of Injections, Seen in Clinical Trials With TWINRIX:

Infections and Infestations: Upper respiratory tract infections.
General Disorders and Administration Site Conditions: Injection site induration.

Incidence <1% of Injections, Seen in Clinical Trials With TWINRIX:

Infections and Infestations: Respiratory tract illnesses.
Metabolism and Nutrition Disorders: Anorexia.
Psychiatric Disorders: Agitation, insomnia.
Nervous System Disorders: Dizziness, migraine, paresthesia, somnolence, syncope.
Ear and Labyrinth Disorders: Vertigo.
Vascular Disorders: Flushing.
Gastrointestinal System: Abdominal pain, vomiting.
Skin and Subcutaneous Tissue Disorders: Erythema, petechiae, rash, sweating, urticaria.
Musculoskeletal and Connective Tissue Disorders: Arthralgia, back pain, myalgia.
General Disorders and Administration Site Conditions: Injection site ecchymosis, injection site pruritus, influenza-like symptoms, irritability, weakness.

Incidence <1% of Injections, Seen in Clinical Trials With HAVRIX[a] and/or ENGERIX-B[b]:

Blood and Lymphatic System Disorders: Lymphadenopathy.[a+b]
Nervous System: Dysgeusia,[a] hypertonic episode,[a] tingling.[b]
Eye Disorders: Photophobia.[a]
Vascular Disorders: Hypotension.[b]
Gastrointestinal Disorders: Constipation.[b]
Investigations: Elevation of creatine phosphokinase.[a]

Postmarketing Reports: Worldwide voluntary reports of adverse events received for TWINRIX, HAVRIX, and/or ENGERIX-B since market introduction of these vaccines are listed below. These lists include serious events or events which have suspected causal connections to components of these or other vaccines or drugs. Because these events are reported voluntarily from a population of uncertain size, it is not possible to reliably estimate their frequency or establish a causal relationship to vaccine exposure.

Postmarketing Reports With TWINRIX:
Infections and Infestations: Herpes zoster, meningitis.
Blood and Lymphatic System Disorders: Thrombocytopenia, thrombocytopenic purpura.
Immune System Disorders: Allergic reaction, anaphylactoid reaction, anaphylaxis, serum sickness–like syndrome days to weeks after vaccination including arthralgia/arthritis (usually transient), fever, urticaria, erythema multiforme, ecchymoses, and erythema nodosum.
Nervous System Disorders: Bell's palsy, convulsions, encephalitis, encephalopathy, Guillain-Barré syndrome, hypoesthesia, myelitis, multiple sclerosis, neuritis, neuropathy, optic neuritis, paralysis, paresis, transverse myelitis.
Eye Disorders: Conjunctivitis, visual disturbances.
Ear and Labyrinth Disorders: Earache, tinnitus.
Cardiac Disorders: Palpitations, tachycardia.
Vascular Disorders: Vasculitis.
Respiratory, Thoracic and Mediastinal Disorders: Bronchospasm including asthma-like symptoms, dyspnea.
Gastrointestinal Disorders: Dyspepsia.
Hepatobiliary disorders: Hepatitis, jaundice.
Skin and Subcutaneous Tissue Disorders: Alopecia, angioedema, eczema, erythema multiforme, erythema nodosum, hyperhydrosis, lichen planus.
Musculoskeletal and Connective Tissue Disorders: Arthritis, muscular weakness.
General Disorders and Administration Site Conditions: Chills, injection site reaction, malaise.
Investigations: Abnormal liver function tests.
Postmarketing Reports With HAVRIX and/or ENGERIX-B: Worldwide voluntary reports of adverse events received for HAVRIX and/or ENGERIX-B but not already reported for TWINRIX are listed below.
Eye Disorders: Keratitis.[b]
Skin and Subcutaneous Tissue Disorders: Stevens-Johnson syndrome.[b]
Congenital, Familial and Genetic Disorders: Congenital abnormality.[a]
[a]Following HAVRIX.
[b]Following ENGERIX-B.
[a+b]Following either HAVRIX or ENGERIX-B.

Manufactured by **GlaxoSmithKline Biologicals**
Rixensart, Belgium, US License No. 1617
Distributed by **GlaxoSmithKline**
Research Triangle Park, NC 27709

May 2009 TWR:11PI

WELLBUTRIN® ℞
[wel' byü-trin]
(bupropion hydrochloride)
Tablets

Prescribing information for this product, which appears on pages 1648-1653 of the 2009 PDR, has been revised as follows. Please write "See Supplement B" next to the product heading.
The Black Box Warning *was revised to:*

WARNING
Suicidality and Antidepressant Drugs
Use in Treating Psychiatric Disorders: Antidepressants increased the risk compared to placebo of suicidal thinking and behavior (suicidality) in children, adolescents, and young adults in short-term studies of major depressive disorder (MDD) and other psychiatric disorders. Anyone considering the use of WELLBUTRIN or any other antidepressant in a child, adolescent, or young adult must balance this risk with the clinical need. Short-term studies did not show an increase in the risk of suicidality with antidepressants compared to placebo in adults beyond age 24; there was a reduction in risk with antidepressants compared to placebo in adults aged 65 and older. Depression and certain other psychiatric disorders are themselves associated with increases in the risk of suicide. Patients of all ages who are started on antidepressant therapy should be monitored appropriately and observed closely for clinical worsening, suicidality, or unusual changes in behavior. Families and caregivers should be advised of the need for close observation and communication with the prescriber. WELLBUTRIN is not approved for use in pediatric patients. (See WARNINGS: Clinical Worsening and Suicide Risk in Treating Psychiatric Disorders, PRECAUTIONS: Information for Patients, and PRECAUTIONS: Pediatric Use.)
Use in Smoking Cessation Treatment: WELLBUTRIN®, WELLBUTRIN SR®, and WELLBUTRIN XL® are not approved for smoking cessation treatment, but bupropion under the name ZYBAN® is approved for this use. Serious neuropsychiatric events, including but not limited to depression, suicidal ideation, suicide attempt, and completed suicide have been reported in patients taking bupropion for smoking cessation. Some cases may have been complicated by the symptoms of nicotine withdrawal in patients who stopped smoking. Depressed mood may be a symptom of nicotine withdrawal. Depression, rarely including suicidal ideation, has been reported in smokers undergoing a smoking cessation attempt without medication. However, some of these symptoms have occurred in patients taking bupropion who continued to smoke. All patients being treated with bupropion for smoking cessation treatment should be observed for neuropsychiatric symptoms including changes in behavior, hostility, agitation, depressed mood, and suicide-related events, including ideation, behavior, and attempted suicide. These symptoms, as well as worsening of pre-existing psychiatric illness and completed suicide have been reported in some patients attempting to quit smoking while taking ZYBAN in the postmarketing experience. When symptoms were reported, most were during treatment with ZYBAN, but some were following discontinuation of treatment with ZYBAN. These events have occurred in patients with and without pre-existing psychiatric disease; some have experienced worsening of their psychiatric illnesses. Patients with serious psychiatric illness such as schizophrenia, bipolar disorder, and major depressive disorder did not participate in the premarketing studies of ZYBAN.
Advise patients and caregivers that the patient using bupropion for smoking cessation should stop taking bupropion and contact a healthcare provider immediately if agitation, hostility, depressed mood, or changes in thinking or behavior that are not typical for the patient are observed, or if the patient develops suicidal ideation or suicidal behavior. In many postmarketing cases, resolution of symptoms after discontinuation of ZYBAN was reported, although in some cases the symptoms persisted; therefore, ongoing monitoring and supportive care should be provided until symptoms resolve. The risks of using bupropion for smoking cessation should be weighed against the benefits of its use. ZYBAN has been demonstrated to increase the likelihood of abstinence from smoking for as long as 6 months compared to treatment with placebo. The health benefits of quitting smoking are immediate and substantial. (See WARNINGS: Neuropsychiatric Symptoms and Suicide Risk in Smoking Cessation Treatment and PRECAUTIONS: Information for Patients.)

The following subsection was added to the WARNINGS *section:*
WELLBUTRIN, WELLBUTRIN SR, and WELLBUTRIN XL are not approved for smoking cessation treatment, but bupropion under the name ZYBAN is approved for this use. Serious neuropsychiatric symptoms have been reported in patients taking bupropion for smoking cessation **(see BOXED WARNING, ADVERSE REACTIONS). These have included changes in mood (including depression and mania), psychosis, hallucinations, paranoia, delusions, homicidal ideation, hostility, agitation, aggression, anxiety, and panic, as well as suicidal ideation, suicide attempt, and completed suicide.** Some reported cases may have been complicated by the symptoms of nicotine withdrawal in patients who stopped smoking. Depressed mood may be a symptom of nicotine withdrawal. Depression, rarely including suicidal ideation, has been reported in smokers undergoing a smoking cessation attempt without medication. However, some of these symptoms have occurred in patients taking bupropion who continued to smoke. When symptoms were reported, most were during bupropion treatment, but some were following discontinuation of bupropion therapy. These events have occurred in patients with and without pre-existing psychiatric disease; some have experienced worsening of their psychiatric illnesses. All patients being treated with bupropion as part of smoking cessation treatment should be observed for neuropsychiatric symptoms or worsening of pre-existing psychiatric illness.
Patients with serious psychiatric illness such as schizophrenia, bipolar disorder, and major depressive disorder did not participate in the pre-marketing studies of ZYBAN.
Advise patients and caregivers that the patient using bupropion for smoking cessation should stop taking bupropion and contact a healthcare provider immediately if agitation, depressed mood, or changes in behavior or thinking that are not typical for the patient are observed, or if the patient develops suicidal ideation or suicidal behavior. In many postmarketing cases, resolution of symptoms after discontinuation of ZYBAN was reported, although in some cases the symptoms persisted, therefore, ongoing monitoring and supportive care should be provided until symptoms resolve.
The risks of using bupropion for smoking cessation should be weighed against the benefits of its use. ZYBAN has been demonstrated to increase the likelihood of abstinence from smoking for as long as six months compared to treatment with placebo. The health benefits of quitting smoking are immediate and substantial.

The following subsection was added to the PRECAUTIONS: Information for Patients *section:*
Neuropsychiatric Symptoms and Suicide Risk in Smoking Cessation Treatment: Although WELLBUTRIN is not indicated for smoking cessation treatment, it contains the same active ingredient as ZYBAN which is approved for this use. Patients should be informed that quitting smoking, with or without ZYBAN, may be associated with nicotine withdrawal symptoms (including depression or agitation), or exacerbation of pre-existing psychiatric illness. Furthermore, some patients have experienced changes in mood (including depression and mania), psychosis, hallucinations, paranoia, delusions, homicidal ideation aggression, anxiety, and panic, as well as suicidal ideation, suicide attempt, and completed suicide when attempting to quit smoking while taking ZYBAN. If patients develop agitation, hostility, depressed mood, or changes in thinking or behavior that are not typical for them, or if patients develop suicidal ideation or behavior, they should be urged to report these symptoms to their healthcare provider immediately.
The ADVERSE REACTIONS: **Postintroduction Reports:** *Nervous System section was revised to:*
aggression, coma, completed suicide, delirium, dream abnormalities, paranoid ideation, paresthesia, restlessness, suicide attempt, unmasking of tardive dyskinesia.
The MEDICATION GUIDE *was revised to:*

MEDICATION GUIDE
WELLBUTRIN® (WELL byu-trin)
(bupropion hydrochloride) Tablets
Read this Medication Guide carefully before you start using WELLBUTRIN and each time you get a refill. There may be new information. This information does not take the place of talking with your doctor about your medical condition or your treatment. If you have any questions about WELLBUTRIN, ask your doctor or pharmacist.
IMPORTANT: Be sure to read the three sections of this Medication Guide. The first section is about the risk of suicidal thoughts and actions with antidepressant medicines; the second section is about the risk of changes in thinking and behavior, depression and suicidal thoughts or actions with medicines used to quit smoking; and the third section is entitled "What Other Important Information Should I Know About WELLBUTRIN?"
Antidepressant Medicines, Depression and Other Serious Mental Illnesses, and Suicidal Thoughts or Actions
This section of the Medication Guide is only about the risk of suicidal thoughts and actions with antidepressant medicines. **Talk to your, or your family member's, healthcare provider about:**
- all risks and benefits of treatment with antidepressant medicines
- all treatment choices for depression or other serious mental illness

What is the most important information I should know about antidepressant medicines, depression and other serious mental illnesses, and suicidal thoughts or actions?
1. **Antidepressant medicines may increase suicidal thoughts or actions in some children, teenagers, and young adults within the first few months of treatment.**
2. **Depression and other serious mental illnesses are the most important causes of suicidal thoughts and actions. Some people may have a particularly high risk of having suicidal thoughts or actions.** These include people who have (or have a family history of) bipolar illness (also called manic-depressive illness) or suicidal thoughts or actions.
3. **How can I watch for and try to prevent suicidal thoughts and actions in myself or a family member?**

- Pay close attention to any changes, especially sudden changes, in mood, behaviors, thoughts, or feelings. This is very important when an antidepressant medicine is started or when the dose is changed.
- Call the healthcare provider right away to report new or sudden changes in mood, behavior, thoughts, or feelings.
- Keep all follow-up visits with the healthcare provider as scheduled. Call the healthcare provider between visits as needed, especially if you have concerns about symptoms.

Call a healthcare provider right away if you or your family member has any of the following symptoms, especially if they are new, worse, or worry you:
- thoughts about suicide or dying
- attempts to commit suicide
- new or worse depression
- new or worse anxiety
- feeling very agitated or restless
- panic attacks
- trouble sleeping (insomnia)

Continued on next page

Product information on these pages is effective as of June 2009. Further information is available at 1-888-825-5249 or www.gsk.com.

Wellbutrin—Cont.

- new or worse irritability
- acting aggressive, being angry, or violent
- acting on dangerous impulses
- an extreme increase in activity and talking (mania)
- other unusual changes in behavior or mood

What else do I need to know about antidepressant medicines?

- **Never stop an antidepressant medicine without first talking to a healthcare provider.** Stopping an antidepressant medicine suddenly can cause other symptoms.
- **Antidepressants are medicines used to treat depression and other illnesses.** It is important to discuss all the risks of treating depression and also the risks of not treating it. Patients and their families or other caregivers should discuss all treatment choices with the healthcare provider, not just the use of antidepressants.
- **Antidepressant medicines have other side effects.** Talk to the healthcare provider about the side effects of the medicine prescribed for you or your family member.
- **Antidepressant medicines can interact with other medicines.** Know all of the medicines that you or your family member takes. Keep a list of all medicines to show the healthcare provider. Do not start new medicines without first checking with your healthcare provider.
- **Not all antidepressant medicines prescribed for children are FDA approved for use in children.** Talk to your child's healthcare provider for more information.

WELLBUTRIN has not been studied in children under the age of 18 and is not approved for use in children and teenagers.

Quitting Smoking, Quit-Smoking Medications, Changes in Thinking and Behavior, Depression, and Suicidal Thoughts or Actions

This section of the Medication Guide is only about the risk of changes in thinking and behavior, depression and suicidal thoughts or actions with drugs used to quit smoking. Although WELLBUTRIN is not a treatment for quitting smoking, it contains the same active ingredient (bupropion hydrochloride) as ZYBAN® which is used to help patients quit smoking.

Some people have had changes in behavior, hostility, agitation, depression, suicidal thoughts or actions while taking bupropion to help them quit smoking. These symptoms can develop during treatment with bupropion or after stopping treatment with bupropion.

If you, your family member, or your caregiver notice agitation, hostility, depression, or changes in thinking or behavior that are not typical for you, or you have any of the following symptoms, stop taking bupropion and call your healthcare provider right away:

- thoughts about suicide or dying
- attempts to commit suicide
- new or worse depression
- new or worse anxiety
- panic attacks
- feeling very agitated or restless
- acting aggressive, being angry, or violent
- acting on dangerous impulses
- an extreme increase in activity and talking (mania)
- abnormal thoughts or sensations
- seeing or hearing things that are not there (hallucinations)
- feeling people are against you (paranoia)
- feeling confused
- other unusual changes in behavior or mood

When you try to quit smoking, with or without bupropion, you may have symptoms that may be due to nicotine withdrawal, including urge to smoke, depressed mood, trouble sleeping, irritability, frustration, anger, feeling anxious, difficulty concentrating, restlessness, decreased heart rate, and increased appetite or weight gain. Some people have even experienced suicidal thoughts when trying to quit smoking without medication. Sometimes quitting smoking can lead to worsening of mental health problems that you already have, such as depression.

Before taking bupropion, tell your healthcare provider if you have ever had depression or other mental illnesses. You should also tell your doctor about any symptoms you had during other times you tried to quit smoking, with or without bupropion.

What Other Important Information Should I Know About WELLBUTRIN?

- **Seizures: There is a chance of having a seizure (convulsion, fit) with WELLBUTRIN, especially in people:**
  - with certain medical problems.
  - who take certain medicines.

The chance of having seizures increases with higher doses of WELLBUTRIN. For more information, see the sections "Who should not take WELLBUTRIN?" and "What should I tell my doctor before using WELLBUTRIN?" Tell your doctor about all of your medical conditions and all the medicines you take. **Do not take any other medicines while you are using WELLBUTRIN unless your doctor has said it is okay to take them.**

If you have a seizure while taking WELLBUTRIN, stop taking the tablets and call your doctor right away. Do not take WELLBUTRIN again if you have a seizure.

- **High blood pressure (hypertension). Some people get high blood pressure, that can be severe, while taking WELLBUTRIN.** The chance of high blood pressure may be higher if you also use nicotine replacement therapy (such as a nicotine patch) to help you stop smoking.
- **Severe allergic reactions. Some people have severe allergic reaction to WELLBUTRIN. Stop taking WELLBUTRIN and call your doctor right away** if you get a rash, itching, hives, fever, swollen lymph glands, painful sores in the mouth or around the eyes, swelling of the lips or tongue, chest pain, or have trouble breathing. These could be signs of a serious allergic reaction.
- **Unusual thoughts or behaviors.** Some patients have unusual thoughts or behaviors while taking WELLBUTRIN, including delusions (believe you are someone else), hallucinations (seeing or hearing things that are not there), paranoia (feeling that people are against you), or feeling confused. If this happens to you, call your doctor.

What is WELLBUTRIN?

WELLBUTRIN is a prescription medicine used to treat adults with a certain type of depression called major depressive disorder.

Who should not take WELLBUTRIN?

Do not take WELLBUTRIN if you

- have or had a seizure disorder or epilepsy.
- **are taking ZYBAN (used to help people stop smoking) or any other medicines that contain bupropion hydrochloride, such as WELLBUTRIN SR Sustained-Release Tablets or WELLBUTRIN XL Extended-Release Tablets.** Bupropion is the same ingredient that is in WELLBUTRIN.
- drink a lot of alcohol and abruptly stop drinking, or use medicines called sedatives (these make you sleepy) or benzodiazepines and you stop using them all of a sudden.
- have taken within the last 14 days medicine for depression called a monoamine oxidase inhibitor (MAOI), such as NARDIL®* (phenelzine sulfate), PARNATE® (tranylcypromine sulfate), or MARPLAN®* (isocarboxazid).
- have or had an eating disorder such as anorexia nervosa or bulimia.
- are allergic to the active ingredient in WELLBUTRIN, bupropion, or to any of the inactive ingredients. See the end of this leaflet for a complete list of ingredients in WELLBUTRIN.

What should I tell my doctor before using WELLBUTRIN?

Tell your doctor if you have ever had depression, suicidal thoughts or actions, or other mental health problems. See "Antidepressant Medicines, Depression and Other Serious Mental Illnesses, and Suicidal Thoughts or Actions."

- **Tell your doctor about your other medical conditions including if you:**
  - **are pregnant or plan to become pregnant.** It is not known if WELLBUTRIN can harm your unborn baby.
  - **are breastfeeding.** WELLBUTRIN passes through your milk. It is not known if WELLBUTRIN can harm your baby.
  - **have liver problems,** especially cirrhosis of the liver.
  - have kidney problems.
  - have an eating disorder, such as anorexia nervosa or bulimia.
  - have had a head injury.
  - have had a seizure (convulsion, fit).
  - have a tumor in your nervous system (brain or spine).
  - have had a heart attack, heart problems, or high blood pressure.
  - are a diabetic taking insulin or other medicines to control your blood sugar.
  - drink a lot of alcohol.
  - abuse prescription medicines or street drugs.
- **Tell your doctor about all the medicines you take,** including prescription and nonprescription medicines, vitamins, and herbal supplements. Many medicines increase your chances of having seizures or other serious side effects if you take them while you are using WELLBUTRIN.

How should I take WELLBUTRIN?

- Take WELLBUTRIN exactly as prescribed by your doctor.
- Take WELLBUTRIN at the same time each day.
- Take your doses of WELLBUTRIN at least 6 hours apart.
- You may take WELLBUTRIN with or without food.
- If you miss a dose, do not take an extra tablet to make up for the dose you forgot. Wait and take your next tablet at the regular time. **This is very important.** Too much WELLBUTRIN can increase your chance of having a seizure.
- If you take too much WELLBUTRIN, or overdose, call your local emergency room or poison control center right away.
- **Do not take any other medicines while using WELLBUTRIN unless your doctor has told you it is okay.**
- It may take several weeks for you to feel that WELLBUTRIN is working. Once you feel better, it is important to keep taking WELLBUTRIN exactly as directed by your doctor. Call your doctor if you do not feel WELLBUTRIN is working for you.
- Do not change your dose or stop taking WELLBUTRIN without talking with your doctor first.

What should I avoid while taking WELLBUTRIN?

- Do not drink a lot of alcohol while taking WELLBUTRIN. If you usually drink a lot of alcohol, talk with your doctor before suddenly stopping. If you suddenly stop drinking alcohol, you may increase your risk of having seizures.
- Do not drive a car or use heavy machinery until you know how WELLBUTRIN affects you. WELLBUTRIN can impair your ability to perform these tasks.

What are possible side effects of WELLBUTRIN?

WELLBUTRIN can cause serious side effects. Read this entire Medication Guide for more information about these serious side effects.

The most common side effects of WELLBUTRIN are nervousness, constipation, trouble sleeping, dry mouth, headache, nausea, vomiting, and shakiness (tremor).

If you have nausea, take your medicine with food. If you have trouble sleeping, do not take your medicine too close to bedtime.

These are not all the side effects of WELLBUTRIN. For a complete list, ask your doctor or pharmacist.

Call your doctor for medical advice about side effects. You may report side effects to FDA at 1-800-FDA-1088.

How should I store WELLBUTRIN?

- Store WELLBUTRIN at room temperature. Store out of direct sunlight. Keep WELLBUTRIN in its tightly closed bottle.

General Information about WELLBUTRIN.

- Medicines are sometimes prescribed for purposes other than those listed in a Medication Guide. Do not use WELLBUTRIN for a condition for which it was not prescribed. Do not give WELLBUTRIN to other people, even if they have the same symptoms you have. It may harm them. Keep WELLBUTRIN out of the reach of children.

This Medication Guide summarizes important information about WELLBUTRIN. For more information, talk to your doctor. You can ask your doctor or pharmacist for information about WELLBUTRIN that is written for health professionals.

What are the ingredients in WELLBUTRIN?

Active ingredient: bupropion hydrochloride.

Inactive ingredients: 75-mg tablet – D&C Yellow No. 10 Lake, FD&C Yellow No. 6 Lake, hydroxypropyl cellulose, hypromellose, microcrystalline cellulose, polyethylene glycol, talc, and titanium dioxide; 100-mg tablet – FD&C Red No. 40 Lake, FD&C Yellow No. 6 Lake, hydroxypropyl cellulose, hypromellose, microcrystalline cellulose, polyethylene glycol, talc, and titanium dioxide.

WELLBUTRIN, WELLBUTRIN SR, WELLBUTRIN XL, and PARNATE are registered trademarks of GlaxoSmithKline.

*The following are registered trademarks of their respective manufacturers: NARDIL® /Warner Lambert Company; MARPLAN® /Oxford Pharmaceutical Services, Inc.

Rx only

This Medication Guide has been approved by the U.S. Food and Drug Administration.

July 2009 WLT:5MG

Distributed by:
GlaxoSmithKline
Research Triangle Park, NC 27709
Manufactured by:
DSM Pharmaceuticals, Inc.
Greenville, NC 27834 for
GlaxoSmithKline
Research Triangle Park, NC 27709

July 2009 WLT:4PI

WELLBUTRIN SR® ℞

[wel'byü-trin]
(bupropion hydrochloride)
Sustained-Release Tablets

Prescribing information for this product, which appears on pages 1653-1659 of the 2009 PDR, has been revised as follows. Please write "See Supplement B" next to the product heading.

The Black Box Warning *was revised to:*

WARNING

Suicidality and Antidepressant Drugs

Use in Treating Psychiatric Disorders: Antidepressants increased the risk compared to placebo of suicidal thinking and behavior (suicidality) in children, adolescents, and young adults in short-term studies of major depressive disorder (MDD) and other psychiatric disorders. Anyone considering the use of WELLBUTRIN SR or any other antidepressant in a child, adolescent, or young adult must balance this risk with the clinical need. Short-term studies did not show an increase in the risk of suicidality with antidepressants compared to placebo in adults beyond age 24; there was a reduction in risk with antidepressants compared to placebo in adults aged 65 and older. Depression and certain other psychiatric disorders are themselves associated with increases in the risk of suicide. Patients of all ages who are started on antidepressant therapy should be monitored appropriately and observed closely for clinical worsening, suicidality, or unusual changes in behavior. Families and caregivers should be advised of the need for close observation and communication with the prescriber. WELLBUTRIN SR is not approved for use in pediatric patients. (See WARNINGS: Clinical Worsening and Suicide Risk in Treating Psychiatric Disorders, PRECAUTIONS: Information for Patients, and PRECAUTIONS: Pediatric Use.)

Use in Smoking Cessation Treatment: WELLBUTRIN®, WELLBUTRIN SR®, and WELLBUTRIN XL® are not approved for smoking cessation treatment, but bupropion under the name ZYBAN® is approved for this use. Serious neuropsychiatric events, including but not limited to depression, suicidal ideation, suicide attempt, and completed suicide have been reported in patients taking bupropion for smoking cessation. Some cases may have been complicated by the symptoms of nicotine withdrawal in patients who stopped smoking. Depressed mood may be a symptom of nicotine withdrawal. Depression, rarely including suicidal ideation, has been reported in smokers undergoing a smoking cessation attempt without medication. However, some of these symptoms have occurred in patients taking bupropion who continued to smoke.

All patients being treated with bupropion for smoking cessation treatment should be observed for neuropsychiatric symptoms including changes in behavior, hostility, agitation, depressed mood, and suicide-related events, including ideation, behavior, and attempted suicide. These symptoms, as well as worsening of pre-existing psychiatric illness and completed suicide have been reported in some patients attempting to quit smoking while taking ZYBAN in the postmarketing experience. When symptoms were reported, most were during treatment with ZYBAN, but some were following discontinuation of treatment with ZYBAN. These events have occurred in patients with and without pre-existing psychiatric disease; some have experienced worsening of their psychiatric illnesses. Patients with serious psychiatric illness such as schizophrenia, bipolar disorder, and major depressive disorder did not participate in the premarketing studies of ZYBAN.

Advise patients and caregivers that the patient using bupropion for smoking cessation should stop taking bupropion and contact a healthcare provider immediately if agitation, hostility, depressed mood, or changes in thinking or behavior that are not typical for the patient are observed, or if the patient develops suicidal ideation or suicidal behavior. In many postmarketing cases, resolution of symptoms after discontinuation of ZYBAN was reported, although in some cases the symptoms persisted; therefore, ongoing monitoring and supportive care should be provided until symptoms resolve.

The risks of using bupropion for smoking cessation should be weighed against the benefits of its use. ZYBAN has been demonstrated to increase the likelihood of abstinence from smoking for as long as 6 months compared to treatment with placebo. The health benefits of quitting smoking are immediate and substantial. (See WARNINGS: Neuropsychiatric Symptoms and Suicide Risk in Smoking Cessation Treatment and PRECAUTIONS: Information for Patients.)

The following subsection was added to the WARNINGS *section:*

Neuropsychiatric Symptoms and Suicide Risk in Smoking Cessation Treatment:

WELLBUTRIN, WELLBUTRIN SR, and WELLBUTRIN XL are not approved for smoking cessation treatment, but bupropion under the name ZYBAN is approved for this use. Serious neuropsychiatric symptoms have been reported in patients taking bupropion for smoking cessation **(see BOXED WARNING, ADVERSE REACTIONS). These have included changes in mood (including depression and mania), psychosis, hallucinations, paranoia, delusions, homicidal ideation, hostility, agitation, aggression, anxiety, and panic, as well as suicidal ideation, suicide attempt, and completed suicide.** Some reported cases may have been complicated by the symptoms of nicotine withdrawal in patients who stopped smoking. Depressed mood may be a symptom of nicotine withdrawal. Depression, rarely including suicidal ideation, has been reported in smokers undergoing a smoking cessation attempt without medication. However, some of these symptoms have occurred in patients taking bupropion who continued to smoke. When symptoms were reported, most were during bupropion treatment, but some were following discontinuation of bupropion therapy.

These events have occurred in patients with and without pre-existing psychiatric disease; some have experienced worsening of their psychiatric illnesses. All patients being treated with bupropion as part of smoking cessation treatment should be observed for neuropsychiatric symptoms or worsening of pre-existing psychiatric illness.

Patients with serious psychiatric illness such as schizophrenia, bipolar disorder, and major depressive disorder did not participate in the pre-marketing studies of ZYBAN.

Advise patients and caregivers that the patient using bupropion for smoking cessation should stop taking bupropion and contact a healthcare provider immediately if agitation, depressed mood, or changes in behavior or thinking that are not typical for the patient are observed, or if the patient develops suicidal ideation or suicidal behavior. In many postmarketing cases, resolution of symptoms after discontinuation of ZYBAN was reported, although in some cases the symptoms persisted, therefore, ongoing monitoring and supportive care should be provided until symptoms resolve.

The risks of using bupropion for smoking cessation should be weighed against the benefits of its use. ZYBAN has been demonstrated to increase the likelihood of abstinence from smoking for as long as six months compared to treatment with placebo. The health benefits of quitting smoking are immediate and substantial.

The following subsection was added to the PRECAUTIONS: Information for Patients *section:*

Neuropsychiatric Symptoms and Suicide Risk in Smoking Cessation Treatment: Although WELLBUTRIN SR is not indicated for smoking cessation treatment, it contains the same active ingredient as ZYBAN which is approved for this use. Patients should be informed that quitting smoking, with or without ZYBAN, may be associated with nicotine withdrawal symptoms (including depression or agitation), or exacerbation of pre-existing psychiatric illness. Furthermore, some patients have experienced changes in mood (including depression and mania), psychosis, hallucinations, paranoia, delusions, homicidal ideation, aggression, anxiety, and panic, as well as suicidal ideation, suicide attempt, and completed suicide when attempting to quit smoking while taking ZYBAN. If patients develop agitation, hostility, depressed mood, or changes in thinking or behavior that are not typical for them, or if patients develop suicidal ideation or behavior, they should be urged to report these symptoms to their healthcare provider immediately.

The ADVERSE REACTIONS: Other Events Observed During the Clinical Development and Postmarketing Experience of Bupropion: *Nervous System section was revised to:* Infrequent were abnormal coordination, decreased libido, depersonalization, dysphoria, emotional lability, hostility, hyperkinesia, hypertonia, hypesthesia, suicidal ideation, and vertigo. Rare were amnesia, ataxia, derealization, and hypomania. Also observed were abnormal electroencephalogram (EEG), akinesia, aggression, aphasia, coma, completed suicide, delirium, delusions, dysarthria, dyskinesia, dystonia, euphoria, extrapyramidal syndrome, hallucinations, hypokinesia, increased libido, manic reaction, neuralgia, neuropathy, paranoid ideation, restlessness, suicide attempt, and unmasking tardive dyskinesia.

The MEDICATION GUIDE *was revised to:*

MEDICATION GUIDE

WELLBUTRIN SR® (WELL byu-trin)

(bupropion hydrochloride) Sustained-Release Tablets

Read this Medication Guide carefully before you start using WELLBUTRIN SR and each time you get a refill. There may be new information. This information does not take the place of talking with your doctor about your medical condition or your treatment. If you have any questions about WELLBUTRIN SR, ask your doctor or pharmacist.

IMPORTANT: Be sure to read the three sections of this Medication Guide. The first section is about the risk of suicidal thoughts and actions with antidepressant medicines; the second section is about the risk of changes in thinking and behavior, depression and suicidal thoughts or actions with medicines used to quit smoking; and the third section is entitled "What Other Important Information Should I Know About WELLBUTRIN SR?"

Antidepressant Medicines, Depression and Other Serious Mental Illnesses, and Suicidal Thoughts or Actions

This section of the Medication Guide is only about the risk of suicidal thoughts and actions with antidepressant medicines. **Talk to your, or your family member's, healthcare provider about:**

- all risks and benefits of treatment with antidepressant medicines
- all treatment choices for depression or other serious mental illness

What is the most important information I should know about antidepressant medicines, depression and other serious mental illnesses, and suicidal thoughts or actions?

1. **Antidepressant medicines may increase suicidal thoughts or actions in some children, teenagers, and young adults within the first few months of treatment.**
2. **Depression and other serious mental illnesses are the most important causes of suicidal thoughts and actions. Some people may have a particularly high risk of having suicidal thoughts or actions.** These include people who have (or have a family history of) bipolar illness (also called manic-depressive illness) or suicidal thoughts or actions.
3. **How can I watch for and try to prevent suicidal thoughts and actions in myself or a family member?**
   - Pay close attention to any changes, especially sudden changes, in mood, behaviors, thoughts, or feelings. This is very important when an antidepressant medicine is started or when the dose is changed.
   - Call the healthcare provider right away to report new or sudden changes in mood, behavior, thoughts, or feelings.
   - Keep all follow-up visits with the healthcare provider as scheduled. Call the healthcare provider between visits as needed, especially if you have concerns about symptoms.

Call a healthcare provider right away if you or your family member has any of the following symptoms, especially if they are new, worse, or worry you:

- thoughts about suicide or dying
- attempts to commit suicide
- new or worse depression
- new or worse anxiety
- feeling very agitated or restless
- panic attacks
- trouble sleeping (insomnia)
- new or worse irritability
- acting aggressive, being angry, or violent
- acting on dangerous impulses
- an extreme increase in activity and talking (mania)
- other unusual changes in behavior or mood

What else do I need to know about antidepressant medicines?

- **Never stop an antidepressant medicine without first talking to a healthcare provider.** Stopping an antidepressant medicine suddenly can cause other symptoms.
- **Antidepressants are medicines used to treat depression and other illnesses.** It is important to discuss all the risks of treating depression and also the risks of not treating it. Patients and their families or other caregivers should discuss all treatment choices with the healthcare provider, not just the use of antidepressants.
- **Antidepressant medicines have other side effects.** Talk to the healthcare provider about the side effects of the medicine prescribed for you or your family member.
- **Antidepressant medicines can interact with other medicines.** Know all of the medicines that you or your family member takes. Keep a list of all medicines to show the healthcare provider. Do not start new medicines without first checking with your healthcare provider.
- **Not all antidepressant medicines prescribed for children are FDA approved for use in children.** Talk to your child's healthcare provider for more information.

WELLBUTRIN SR has not been studied in children under the age of 18 and is not approved for use in children and teenagers.

Quitting Smoking, Quit-Smoking Medications, Changes in Thinking and Behavior, Depression, and Suicidal Thoughts or Actions

This section of the Medication Guide is only about the risk of changes in thinking and behavior, depression and suicidal thoughts or actions with drugs used to quit smoking. Although WELLBUTRIN SR is not a treatment for quitting smoking, it contains the same active ingredient (bupropion hydrochloride) as ZYBAN® which is used to help patients quit smoking.

Some people have had changes in behavior, hostility, agitation, depression, suicidal thoughts or actions while taking bupropion to help them quit smoking. These symptoms can develop during treatment with bupropion or after stopping treatment with bupropion.

If you, your family member, or your caregiver notice agitation, hostility, depression, or changes in thinking or behavior that are not typical for you, or you have any of the following symptoms, stop taking bupropion and call your healthcare provider right away:

- thoughts about suicide or dying

Continued on next page

Product information on these pages is effective as of June 2009. Further information is available at 1-888-825-5249 or www.gsk.com.

Wellbutrin SR—Cont.

- attempts to commit suicide
- new or worse depression
- new or worse anxiety
- panic attacks
- feeling very agitated or restless
- acting aggressive, being angry, or violent
- acting on dangerous impulses
- an extreme increase in activity and talking (mania)
- abnormal thoughts or sensations
- seeing or hearing things that are not there (hallucinations)
- feeling people are against you (paranoia)
- feeling confused
- other unusual changes in behavior or mood

When you try to quit smoking, with or without bupropion, you may have symptoms that may be due to nicotine withdrawal, including urge to smoke, depressed mood, trouble sleeping, irritability, frustration, anger, feeling anxious, difficulty concentrating, restlessness, decreased heart rate, and increased appetite or weight gain. Some people have even experienced suicidal thoughts when trying to quit smoking without medication. Sometimes quitting smoking can lead to worsening of mental health problems that you already have, such as depression.

Before taking bupropion, tell your healthcare provider if you have ever had depression or other mental illnesses. You should also tell your doctor about any symptoms you had during other times you tried to quit smoking, with or without bupropion.

What Other Important Information Should I Know About WELLBUTRIN SR?

- **Seizures: There is a chance of having a seizure (convulsion, fit) with WELLBUTRIN SR, especially in people:**
  - with certain medical problems.
  - who take certain medicines.

  The chance of having seizures increases with higher doses of WELLBUTRIN SR. For more information, see the sections "Who should not take WELLBUTRIN SR?" and "What should I tell my doctor before using WELLBUTRIN SR?" Tell your doctor about all of your medical conditions and all the medicines you take. **Do not take any other medicines while you are using WELLBUTRIN SR unless your doctor has said it is okay to take them.**

  If you have a seizure while taking WELLBUTRIN SR, stop taking the tablets and call your doctor right away. Do not take WELLBUTRIN SR again if you have a seizure.
- **High blood pressure (hypertension). Some people get high blood pressure, that can be severe, while taking WELLBUTRIN SR.** The chance of high blood pressure may be higher if you also use nicotine replacement therapy (such as a nicotine patch) to help you stop smoking.
- **Severe allergic reactions. Some people have severe allergic reaction to WELLBUTRIN SR. Stop taking WELLBUTRIN SR and call your doctor right away** if you get a rash, itching, hives, fever, swollen lymph glands, painful sores in the mouth or around the eyes, swelling of the lips or tongue, chest pain, or have trouble breathing. These could be signs of a serious allergic reaction.
- **Unusual thoughts or behaviors.** Some patients have unusual thoughts or behaviors while taking WELLBUTRIN SR, including delusions (believe you are someone else), hallucinations (seeing or hearing things that are not there), paranoia (feeling that people are against you), or feeling confused. If this happens to you, call your doctor.

What is WELLBUTRIN SR?

WELLBUTRIN SR is a prescription medicine used to treat adults with a certain type of depression called major depressive disorder.

Who should not take WELLBUTRIN SR?

Do not take WELLBUTRIN SR if you

- have or had a seizure disorder or epilepsy.
- **are taking ZYBAN® (used to help people stop smoking) or any other medicines that contain bupropion hydrochloride, such as WELLBUTRIN® Tablets or WELLBUTRIN XL® Extended-Release Tablets.** Bupropion is the same active ingredient that is in WELLBUTRIN SR.
- drink a lot of alcohol and abruptly stop drinking, or use medicines called sedatives (these make you sleepy) or benzodiazepines and you stop using them all of a sudden.
- have taken within the last 14 days medicine for depression called a monoamine oxidase inhibitor (MAOI), such as NARDIL®* (phenelzine sulfate), PARNATE® (tranylcypromine sulfate), or MARPLAN®* (isocarboxazid).
- have or had an eating disorder such as anorexia nervosa or bulimia.
- are allergic to the active ingredient in WELLBUTRIN SR, bupropion, or to any of the inactive ingredients. See the end of this leaflet for a complete list of ingredients in WELLBUTRIN SR.

What should I tell my doctor before using WELLBUTRIN SR?

Tell your doctor if you have ever had depression, suicidal thoughts or actions, or other mental health problems. See "Antidepressant Medicines, Depression and Other Serious Mental Illnesses, and Suicidal Thoughts or Actions."

- **Tell your doctor about your other medical conditions including if you:**
  - **are pregnant or plan to become pregnant.** It is not known if WELLBUTRIN SR can harm your unborn baby.
  - **are breastfeeding.** WELLBUTRIN SR passes through your milk. It is not known if WELLBUTRIN SR can harm your baby.
  - **have liver problems,** especially cirrhosis of the liver.
  - have kidney problems.
  - have an eating disorder such as anorexia nervosa or bulimia.
  - have had a head injury.
  - have had a seizure (convulsion, fit).
  - have a tumor in your nervous system (brain or spine).
  - have had a heart attack, heart problems, or high blood pressure.
  - are a diabetic taking insulin or other medicines to control your blood sugar.
  - drink a lot of alcohol.
  - abuse prescription medicines or street drugs.
- **Tell your doctor about all the medicines you take,** including prescription and non-prescription medicines, vitamins, and herbal supplements. Many medicines increase your chances of having seizures or other serious side effects if you take them while you are using WELLBUTRIN SR.

How should I take WELLBUTRIN SR?

- Take WELLBUTRIN SR exactly as prescribed by your doctor.
- **Do not chew, cut, or crush WELLBUTRIN SR Tablets.** You must swallow the tablets whole. **Tell your doctor if you cannot swallow medicine tablets.**
- Take WELLBUTRIN SR at the same time each day.
- Take your doses of WELLBUTRIN SR at least 8 hours apart.
- You may take WELLBUTRIN SR with or without food.
- If you miss a dose, do not take an extra tablet to make up for the dose you forgot. Wait and take your next tablet at the regular time. **This is very important.** Too much WELLBUTRIN SR can increase your chance of having a seizure.
- If you take too much WELLBUTRIN SR, or overdose, call your local emergency room or poison control center right away.
- **Do not take any other medicines while using WELLBUTRIN SR unless your doctor has told you it is okay.**
- It may take several weeks for you to feel that WELLBUTRIN SR is working. Once you feel better, it is important to keep taking WELLBUTRIN SR exactly as directed by your doctor. Call your doctor if you do not feel WELLBUTRIN SR is working for you.
- Do not change your dose or stop taking WELLBUTRIN SR without talking with your doctor first.

What should I avoid while taking WELLBUTRIN SR?

- Do not drink a lot of alcohol while taking WELLBUTRIN SR. If you usually drink a lot of alcohol, talk with your doctor before suddenly stopping. If you suddenly stop drinking alcohol, you may increase your chance of having seizures.
- Do not drive a car or use heavy machinery until you know how WELLBUTRIN SR affects you. WELLBUTRIN SR can impair your ability to perform these tasks.

What are possible side effects of WELLBUTRIN SR?

WELLBUTRIN SR can cause serious side effects. Read this entire Medication Guide for more information about these serious side effects.

The most common side effects of WELLBUTRIN SR are loss of appetite, dry mouth, skin rash, sweating, ringing in the ears, shakiness, stomach pain, agitation, anxiety, dizziness, trouble sleeping, muscle pain, nausea, fast heartbeat, sore throat, and urinating more often.

If you have nausea, take your medicine with food. If you have trouble sleeping, do not take your medicine too close to bedtime.

These are not all the side effects of WELLBUTRIN SR. For a complete list, ask your doctor or pharmacist.

Call your doctor for medical advice about side effects. You may report side effects to FDA at 1-800-FDA-1088.

How should I store WELLBUTRIN SR?

- Store WELLBUTRIN SR at room temperature. Store out of direct sunlight. Keep WELLBUTRIN SR in its tightly closed bottle.
- WELLBUTRIN SR tablets may have an odor.

General Information about WELLBUTRIN SR.

Medicines are sometimes prescribed for purposes other than those listed in a Medication Guide. Do not use WELLBUTRIN SR for a condition for which it was not prescribed. Do not give WELLBUTRIN SR to other people, even if they have the same symptoms you have. It may harm them. Keep WELLBUTRIN SR out of the reach of children.

This Medication Guide summarizes important information about WELLBUTRIN SR. For more information, talk with your doctor. You can ask your doctor or pharmacist for information about WELLBUTRIN SR that is written for health professionals.

What are the ingredients in WELLBUTRIN SR?

Active ingredient: bupropion hydrochloride.

Inactive ingredients: carnauba wax, cysteine hydrochloride, hypromellose, magnesium stearate, microcrystalline cellulose, polyethylene glycol, polysorbate 80, and titanium dioxide. In addition, the 100-mg tablet contains FD&C Blue No. 1 Lake, the 150-mg tablet contains FD&C Blue No. 2 Lake and FD&C Red No. 40 Lake, and the 200-mg tablet contains FD&C Red No. 40 Lake. The tablets are printed with edible black ink.

WELLBUTRIN, WELLBUTRIN SR, WELLBUTRIN XL, ZYBAN, and PARNATE are registered trademarks of GlaxoSmithKline.

*The following are registered trademarks of their respective manufacturers: NARDIL® /Warner Lambert Company; MARPLAN® /Oxford Pharmaceutical Services, Inc.

Rx only

This Medication Guide has been approved by the U.S. Food and Drug Administration.

July 2009 WLS:5MG

Distributed by:
GlaxoSmithKline
Research Triangle Park, NC 27709
Manufactured by:
GlaxoSmithKline
Research Triangle Park, NC 27709
or DSM Pharmaceuticals, Inc.
Greenville, NC 27834

July 2009 WLS:4PI

ZANTAC® ℞
[*zan' tak*]
(ranitidine hydrochloride)
Injection

ZANTAC® ℞
(ranitidine hydrochloride)
Injection Premixed

Prescribing information for this product, which appears on pages 1666-1668 of the 2009 PDR, has been revised as follows. Please write "See Supplement B" next to the product heading.

The ADVERSE REACTIONS: Endocrine *subsection was revised to:*

Controlled studies in animals and humans have shown no stimulation of any pituitary hormone by ZANTAC and no antiandrogenic activity, and cimetidine-induced gynecomastia and impotence in hypersecretory patients have resolved when ZANTAC has been substituted. However, occasional cases of impotence and loss of libido have been reported in male patients receiving ZANTAC, but the incidence did not differ from that in the general population. Rare cases of breast symptoms and conditions, including galactorrhea and gynecomastia, have been reported in both males and females.

GlaxoSmithKline
Research Triangle Park, NC 27709
ZANTAC® Injection:
GlaxoSmithKline
Research Triangle Park, NC 27709
ZANTAC® Injection Premixed:
Manufactured for GlaxoSmithKline
Research Triangle Park, NC 27709
by Hospira, Inc., Lake Forest, IL 60045

April 2009 ZNJ:5PI

ZANTAC® ℞
[*zan'tak*]
(ranitidine hydrochloride)
Injection

Pharmacy Bulk Package—Not for Direct Infusion

Prescribing information for this product, which appears on pages 1668-1670 of the 2009 PDR, has been revised as follows. Please write "See Supplement B" next to the product heading.

The ADVERSE REACTIONS: Endocrine *subsection was revised to:*

Controlled studies in animals and humans have shown no stimulation of any pituitary hormone by ZANTAC and no antiandrogenic activity, and cimetidine-induced gynecomastia and impotence in hypersecretory patients have resolved when ZANTAC has been substituted. However, occasional cases of impotence, and loss of libido have been reported in male patients receiving ZANTAC, but the incidence did not differ from that in the general population. Rare cases of breast symptoms and conditions, including galactorrhea and gynecomastia, have been reported in both males and females.
GlaxoSmithKline
Research Triangle Park, NC 27709

April 2009 ZNP:4PI

ZANTAC® 150 ℞
[zan'tak]
(ranitidine hydrochloride)
Tablets, USP

ZANTAC® 300 ℞
(ranitidine hydrochloride)
Tablets, USP

ZANTAC® 25 ℞
(ranitidine hydrochloride effervescent)
EFFERdose® Tablets

ZANTAC® ℞
(ranitidine hydrochloride)
Syrup, USP

Prescribing information for this product, which appears on pages 1670-1673 of the 2009 PDR, has been revised as follows. Please write "See Supplement B" next to the product heading.
The ADVERSE REACTIONS: Endocrine *subsection was revised to:*
Controlled studies in animals and man have shown no stimulation of any pituitary hormone by ZANTAC and no antiandrogenic activity, and cimetidine-induced gynecomastia and impotence in hypersecretory patients have resolved when ZANTAC has been substituted. However, occasional cases of impotence and loss of libido have been reported in male patients receiving ZANTAC, but the incidence did not differ from that in the general population. Rare cases of breast symptoms and conditions, including galactorrhea and gynecomastia, have been reported in both males and females.
GlaxoSmithKline
Research Triangle Park, NC 27709

April 2009 ZNT:5PI

ZIAGEN® ℞
[zī' ə-jin]
(abacavir sulfate)
Tablets and Oral Solution

Prescribing information for this product, which appears on pages 1673-1678 of the 2009 PDR, has been revised as follows. Please write "See Supplement B" next to the product heading.
The DOSAGE AND ADMINISTRATION: Pediatric Patients *subsection was revised to:*
The recommended oral dose of ZIAGEN Oral Solution in HIV-1-infected pediatric patients aged 3 months and older is 8 mg/kg twice daily (up to a maximum of 300 mg twice daily) in combination with other antiretroviral agents.
ZIAGEN is also available as a scored tablet for HIV-1-infected pediatric patients weighing greater than or equal to 14 kg for whom a solid dosage form is appropriate. Before prescribing ZIAGEN Tablets, children should be assessed for the ability to swallow tablets. If a child is unable to reliably swallow ZIAGEN Tablets, the oral solution formulation should be prescribed. The recommended oral dosage of ZIAGEN Tablets for HIV-1-infected pediatric patients is presented in Table 1.

Table 1. Dosing Recommendations for ZIAGEN Tablets in Pediatric Patients

Weight (kg)	Dosage Regimen Using Scored Tablet		Total Daily Dose
	AM Dose	PM Dose	
14 to 21	½ tablet (150 mg)	½ tablet (150 mg)	300 mg
>21 to <30	½ tablet (150 mg)	1 tablet (300 mg)	450 mg
≥30	1 tablet (300 mg)	1 tablet (300 mg)	600 mg

The following paragraph was revised in the DOSAGE FORMS AND STRENGTHS *section:*
ZIAGEN Tablets, containing abacavir sulfate equivalent to 300 mg abacavir, are yellow, biconvex, scored, capsule-shaped, film-coated, and imprinted with "GX 623" on both sides.
The CLINICAL PHARMACOLOGY: Pharmacokinetics: Special Populations: *Pediatric Patients subsection was revised to:*
The pharmacokinetics of abacavir have been studied after either single or repeat doses of ZIAGEN in 68 pediatric patients. Following multiple-dose administration of ZIAGEN 8 mg/kg twice daily, steady-state $AUC_{(0-12\ hr)}$ and C_{max} were 9.8 ± 4.56 mcg•hr/mL and 3.71 ± 1.36 mcg/mL (mean ± SD), respectively *[see Use in Specific Populations (8.4)]*. In addition, to support dosing of ZIAGEN scored tablet (300 mg) for pediatric patients 14 to greater than 30 kg, analysis of actual and simulated pharmacokinetic data indicated comparable exposures are expected following administration of 300 mg scored tablet and the 8 mg/kg dosing regimen using oral solution.
The first paragraph of the HOW SUPPLIED/STORAGE AND HANDLING *section was revised to:*
ZIAGEN Tablets, containing abacavir sulfate equivalent to 300 mg abacavir are yellow, biconvex, scored, capsule-shaped, film-coated, and imprinted with "GX 623" on both sides.
The following bullet point was added to section 17 PATIENT COUNSELING INFORMATION:
17.2 FDA Approved Patient Labeling: MEDICATION GUIDE: **How should I take ZIAGEN?**:

- Children aged 3 months and older can also take ZIAGEN. The child's healthcare professional will decide the right dose and formulation based on the child's weight. The dose should not exceed the recommended adult dose.

GlaxoSmithKline
Research Triangle Park, NC 27709

January 2009 ZGN:2MG
December 2008 ZGN:2PI

ZYBAN® ℞
[zī'ban]
(bupropion hydrochloride)
Sustained-Release Tablets

Prescribing information for this product, which appears on pages 1692–1698 of the 2009 PDR, has been revised as follows. Please write "See Supplement B" next to the product heading.
The Black Box Warning *was revised to:*

WARNING
Serious neuropsychiatric events, including but not limited to depression, suicidal ideation, suicide attempt, and completed suicide have been reported in patients taking ZYBAN for smoking cessation. Some cases may have been complicated by the symptoms of nicotine withdrawal in patients who stopped smoking. Depressed mood may be a symptom of nicotine withdrawal. Depression, rarely including suicidal ideation, has been reported in smokers undergoing a smoking cessation attempt without medication. However, some of these symptoms have occurred in patients taking ZYBAN who continued to smoke.
All patients being treated with ZYBAN should be observed for neuropsychiatric symptoms including changes in behavior, hostility, agitation, depressed mood, and suicide-related events, including ideation, behavior, and attempted suicide. These symptoms, as well as worsening of pre-existing psychiatric illness and completed suicide have been reported in some patients attempting to quit smoking while taking ZYBAN in the postmarketing experience. When symptoms were reported, most were during treatment with ZYBAN, but some were following discontinuation of treatment with ZYBAN. These events have occurred in patients with and without pre-existing psychiatric disease; some have experienced worsening of their psychiatric illnesses. Patients with serious psychiatric illness such as schizophrenia, bipolar disorder, and major depressive disorder did not participate in the premarketing studies of ZYBAN.
Advise patients and caregivers that the patient should stop taking ZYBAN and contact a healthcare provider immediately if agitation, hostility, depressed mood, or changes in thinking or behavior that are not typical for the patient are observed, or if the patient develops suicidal ideation or suicidal behavior. In many postmarketing cases, resolution of symptoms after discontinuation of ZYBAN was reported, although in some cases the symptoms persisted; therefore, ongoing monitoring and supportive care should be provided until symptoms resolve.
The risks of ZYBAN should be weighed against the benefits of its use. ZYBAN has been demonstrated to increase the likelihood of abstinence from smoking for as long as 6 months compared to treatment with placebo. The health benefits of quitting smoking are immediate and substantial. (See WARNINGS: Neuropsychiatric Symptoms and Suicide Risk in Smoking Cessation Treatment and PRECAUTIONS: Information for Patients.)

Use in Treating Psychiatric Disorders: Although ZYBAN is not indicated for treatment of depression, it contains the same active ingredient as the antidepressant medications WELLBUTRIN®, WELLBUTRIN SR®, and WELLBUTRIN XL®. Antidepressants increased the risk compared to placebo of suicidal thinking and behavior (suicidality) in children, adolescents, and young adults in short-term studies of major depressive disorder (MDD) and other psychiatric disorders. Anyone considering the use of ZYBAN or any other antidepressant in a child, adolescent, or young adult must balance this risk with the clinical need. Short-term studies did not show an increase in the risk of suicidality with antidepressants compared to placebo in adults beyond age 24; there was a reduction in risk with antidepressants compared to placebo in adults aged 65 and older. Depression and certain other psychiatric disorders are themselves associated with increases in the risk of suicide. Patients of all ages who are started on antidepressant therapy should be monitored appropriately and observed closely for clinical worsening, suicidality, or unusual changes in behavior. Families and caregivers should be advised of the need for close observation and communication with the prescriber. ZYBAN is not approved for use in pediatric patients. (See WARNINGS: Clinical Worsening and Suicide Risk in Treating Psychiatric Disorders, PRECAUTIONS: Information for Patients, and PRECAUTIONS: Pediatric Use.)

The following subsection was added to the **WARNINGS** *section:*
Neuropsychiatric Symptoms and Suicide Risk in Smoking Cessation Treatment: Serious neuropsychiatric symptoms have been reported in patients taking ZYBAN for smoking cessation **(see BOXED WARNING, ADVERSE REACTIONS). These have included changes in mood (including depression and mania), psychosis, hallucinations, paranoia, delusions, homicidal ideation, hostility, agitation, aggression, anxiety, and panic, as well as suicidal ideation, suicide attempt, and completed suicide.** Some reported cases may have been complicated by the symptoms of nicotine withdrawal in patients who stopped smoking. Depressed mood may be a symptom of nicotine withdrawal. Depression, rarely including suicidal ideation, has been reported in smokers undergoing a smoking cessation attempt without medication. However, some of these symptoms have occurred in patients taking ZYBAN who continued to smoke. When symptoms were reported, most were during treatment with ZYBAN, but some were following discontinuation of treatment with ZYBAN.
These events have occurred in patients with and without pre-existing psychiatric disease; some patients have experienced worsening of their psychiatric illnesses. All patients being treated with ZYBAN should be observed for neuropsychiatric symptoms or worsening of pre-existing psychiatric illness.
Patients with serious psychiatric illness such as schizophrenia, bipolar disorder, and major depressive disorder did not participate in the premarketing studies of ZYBAN.
Advise patients and caregivers that the patient should stop taking ZYBAN and contact a healthcare provider immediately if agitation, depressed mood, or changes in behavior or thinking that are not typical for the patient are observed, or if the patient develops suicidal ideation or suicidal behavior. In many post-marketing cases, resolution of symptoms after discontinuation of ZYBAN was reported, although in some cases the symptoms persisted, therefore, ongoing monitoring and supportive care should be provided until symptoms resolve.

Continued on next page

Product information on these pages is effective as of June 2009. Further information is available at 1-888-825-5249 or www.gsk.com.

Zyban—Cont.

The risks of ZYBAN should be weighed against the benefits of its use. ZYBAN has been demonstrated to increase the likelihood of abstinence from smoking for as long as six months compared to treatment with placebo. The health benefits of quitting smoking are immediate and substantial.
The PRECAUTIONS: General: *Psychosis, Confusion, and Other Neuropsychiatric Phenomena subsection was revised to:*
Depressed patients treated with bupropion in depression trials have been reported to show a variety of neuropsychiatric signs and symptoms including delusions, hallucinations, psychosis, concentration disturbance, paranoia, and confusion. In some cases, these symptoms abated upon dose reduction and/or withdrawal of treatment. In clinical trials with ZYBAN conducted in nondepressed smokers, the incidence of neuropsychiatric side effects was generally comparable to placebo. However, in the postmarketing experience, patients taking ZYBAN to quit smoking have reported similar types of neuropsychiatric symptoms to those reported by patients in the clinical trials of bupropion for depression.
The following subsection was added to the **PRECAUTIONS:** Information for Patients *section:*
Neuropsychiatric Symptoms and Suicide Risk in Smoking Cessation Treatment: Patients should be informed that quitting smoking, with or without ZYBAN, may be associated with nicotine withdrawal symptoms (including depression or agitation), or exacerbation of pre-existing psychiatric illness. Furthermore, some patients have experienced changes in mood (including depression and mania), psychosis, hallucinations, paranoia, delusions, homicidal ideation, aggression, anxiety, and panic, as well as suicidal ideation, suicide attempt, and completed suicide when attempting to quit smoking while taking ZYBAN. If patients develop agitation, hostility, depressed mood, or changes in thinking or behavior that are not typical for them, or if patients develop suicidal ideation or behavior, they should be urged to report these symptoms to their healthcare provider immediately.
The **ADVERSE REACTIONS:** Other Events Observed During the Clinical Development and Postmarketing Experience of Bupropion: *Nervous System section was revised to:*
Frequent were agitation, depression, and irritability. Infrequent were abnormal coordination, CNS stimulation, confusion, decreased libido, decreased memory, depersonalization, emotional lability, hostility, hyperkinesia, hypertonia, hypesthesia, paresthesia, suicidal ideation, and vertigo. Rare were amnesia, ataxia, derealization, and hypomania. Also observed were abnormal electroencephalogram (EEG), aggression, akinesia, aphasia, coma, completed suicide, delirium, delusions, dysarthria, dyskinesia, dystonia, euphoria, extrapyramidal syndrome, hallucinations, hypokinesia, increased libido, manic reaction, neuralgia, neuropathy, paranoid ideation, restlessness, suicide attempt, and unmasking tardive dyskinesia.
The **MEDICATION GUIDE** *was revised to:*

MEDICATION GUIDE
ZYBAN® (zi ban)
(bupropion hydrochloride) Sustained-Release Tablets

Read this Medication Guide carefully before you start using ZYBAN and each time you get a refill. There may be new information. This information does not take the place of talking with your doctor about your medical condition or your treatment. If you have any questions about ZYBAN, ask your doctor or pharmacist.
IMPORTANT: Be sure to read the three sections of this Medication Guide. The first section is about the risk of changes in thinking and behavior, depression and suicidal thoughts or actions with medicines used to quit smoking; the second section is about the risk of suicidal thoughts and actions with antidepressant medicines; and the third section is entitled "What Other Important Information Should I Know About ZYBAN?"

Quitting Smoking, Quit-Smoking Medications, Changes in Thinking and Behavior, Depression, and Suicidal Thoughts or Actions
This section of the Medication Guide is only about the risk of changes in thinking and behavior depression and suicidal thoughts or actions with drugs used to quit smoking.
Some people have had changes in behavior, hostility, agitation, depression, suicidal thoughts or actions while taking ZYBAN to help them quit smoking. These symptoms can develop during treatment with ZYBAN or after stopping treatment with ZYBAN.
If you, your family member, or your caregiver notice agitation, hostility, depression, or changes in thinking or behavior that are not typical for you, or you have any of the following symptoms, stop taking ZYBAN and call your healthcare provider right away:
- thoughts about suicide or dying
- attempts to commit suicide
- new or worse depression
- new or worse anxiety
- panic attacks
- feeling very agitated or restless
- acting aggressive, being angry, or violent
- acting on dangerous impulses
- an extreme increase in activity and talking (mania)
- abnormal thoughts or sensations
- seeing or hearing things that are not there (hallucinations)
- feeling people are against you (paranoia)
- feeling confused
- other unusual changes in behavior or mood

When you try to quit smoking, with or without ZYBAN, you may have symptoms that may be due to nicotine withdrawal, including urge to smoke, depressed mood, trouble sleeping, irritability, frustration, anger, feeling anxious, difficulty concentrating, restlessness, decreased heart rate, and increased appetite or weight gain. Some people have even experienced suicidal thoughts when trying to quit smoking without medication. Sometimes quitting smoking can lead to worsening of mental health problems that you already have, such as depression.
Before taking ZYBAN, tell your healthcare provider if you have ever had depression or other mental health problems. You should also tell your doctor about any symptoms you had during other times you tried to quit smoking, with or without ZYBAN.

Antidepressant Medicines, Depression and Other Serious Mental Illnesses, and Suicidal Thoughts or Actions
Although ZYBAN is not a treatment for depression, it contains bupropion, the same active ingredient as the antidepressant medications WELLBUTRIN®, WELLBUTRIN SR®, and WELLBUTRIN XL®.
This section of the Medication Guide is only about the risk of suicidal thoughts and actions with antidepressant medicines. **Talk to your doctor, or your family member's healthcare provider about:**
- all risks and benefits of treatment with antidepressant medicines
- all treatment choices for depression or other serious mental illness

What is the most important information I should know about antidepressant medicines, depression and other serious mental illnesses, and suicidal thoughts or actions?
1. **Antidepressant medicines may increase suicidal thoughts or actions in some children, teenagers, and young adults within the first few months of treatment.**
2. **Depression and other serious mental illnesses are the most important causes of suicidal thoughts and actions. Some people may have a particularly high risk of having suicidal thoughts or actions.**These include people who have (or have a family history of) bipolar illness (also called manic-depressive illness) or suicidal thoughts or actions.
3. **How can I watch for and try to prevent suicidal thoughts and actions in myself or a family member?**
   - Pay close attention to any changes, especially sudden changes, in mood, behaviors, thoughts, or feelings. This is very important when an antidepressant medicine is started or when the dose is changed.
   - Call the healthcare provider right away to report new or sudden changes in mood, behavior, thoughts, or feelings.
   - Keep all follow-up visits with the healthcare provider as scheduled. Call the healthcare provider between visits as needed, especially if you have concerns about symptoms.

Call a healthcare provider right away if you or your family member has any of the following symptoms, especially if they are new, worse, or worry you:
- thoughts about suicide or dying
- attempts to commit suicide
- new or worse depression
- new or worse anxiety
- feeling very agitated or restless
- panic attacks
- trouble sleeping (insomnia)
- new or worse irritability
- acting aggressive, being angry, or violent
- acting on dangerous impulses
- an extreme increase in activity and talking (mania)
- other unusual changes in behavior or mood

What else do I need to know about antidepressant medicines?
- **Never stop an antidepressant medicine without first talking to a healthcare provider.** Stopping an antidepressant medicine suddenly can cause other symptoms.
- **Antidepressants are medicines used to treat depression and other illnesses.** It is important to discuss all the risks of treating depression and also the risks of not treating it. Patients and their families or other caregivers should discuss all treatment choices with the healthcare provider, not just the use of antidepressants.
- **Antidepressant medicines have other side effects.** Talk to the healthcare provider about the side effects of the medicine prescribed for you or your family member.
- **Antidepressant medicines can interact with other medicines.** Know all of the medicines that you or your family member takes. Keep a list of all medicines to show the healthcare provider. Do not start new medicines without first checking with your healthcare provider.
- **Not all antidepressant medicines prescribed for children are FDA approved for use in children.** Talk to your child's healthcare provider for more information.

ZYBAN has not been studied in children under the age of 18 and is not approved for use in children and teenagers.

What Other Important Information Should I Know About ZYBAN?
- **Seizures: There is a chance of having a seizure (convulsion, fit) with ZYBAN, especially in people:**
  - with certain medical problems.
  - who take certain medicines.

  The chance of having seizures increases with higher doses of ZYBAN. For more information, see the sections "Who should not take ZYBAN?" and "What should I tell my doctor before using ZYBAN?" Tell your doctor about all of your medical conditions and all the medicines you take. **Do not take any other medicines while you are using ZYBAN unless your doctor has said it is okay to take them.**
  If you have a seizure while taking ZYBAN, stop taking the tablets and call your doctor right away. Do not take ZYBAN again if you have a seizure.
- **High blood pressure (hypertension): Some people get high blood pressure that can be severe, while taking ZYBAN.** The chance of high blood pressure may be higher if you also use nicotine replacement therapy (such as a nicotine patch) to help you stop smoking (see "Can ZYBAN be used at the same time as nicotine patches?").
- **Severe allergic reactions: Some people have severe allergic reactions to ZYBAN. Stop taking ZYBAN and call your doctor right away** if you get a rash, itching, hives, fever, swollen lymph glands, painful sores in your mouth or around your eyes, swelling of your lips or tongue, chest pain, or have trouble breathing. These could be signs of a serious allergic reaction.

What is ZYBAN?
ZYBAN is a prescription medicine to help people quit smoking. Studies have shown that more than one third of people quit smoking for at least 1 month while taking ZYBAN and participating in a patient support program. For many patients, ZYBAN reduces withdrawal symptoms and the urge to smoke. ZYBAN should be used with a patient support program. It is important to participate in the behavioral program, counseling, or other support program your healthcare professional recommends.

Who should not take ZYBAN?
Do not take ZYBAN if you:
- have or had a seizure disorder or epilepsy.
- **are taking WELLBUTRIN, WELLBUTRIN SR, WELLBUTRIN XL, or any other medicines that contain bupropion hydrochloride.** Bupropion is the same active ingredient that is in ZYBAN.
- drink a lot of alcohol and abruptly stop drinking, or use medicines called sedatives (these make you sleepy) or benzodiazepines and you stop using them all of a sudden.
- have taken within the last 14 days medicine for depression called a monoamine oxidase inhibitor (MAOI), such as NARDIL®* (phenelzine sulfate), PARNATE® (tranylcypromine sulfate), or MARPLAN®* (isocarboxazid).
- have or had an eating disorder such as anorexia nervosa or bulimia.
- are allergic to the active ingredient in ZYBAN, bupropion, or to any of the inactive ingredients. See the end of this leaflet for a complete list of ingredients in ZYBAN.

What should I tell my doctor before using ZYBAN?
Tell your doctor if you have ever had depression, suicidal thoughts or actions, or other mental health problems. You should also tell your doctor about any symptoms you had during other times you tried to quit smoking, with or without ZYBAN. See "Quitting Smoking, Quit-Smoking Medications, Changes in Thinking and Behavior, Depression, and Suicidal Thoughts or Actions."
- **Tell your doctor about your other medical conditions, including if you:**
  - **are pregnant or plan to become pregnant.** It is not known if ZYBAN can harm your unborn baby.
  - **are breastfeeding.** ZYBAN passes through your milk. It is not known if ZYBAN can harm your baby.
  - **have liver problems,** especially cirrhosis of the liver.
  - have kidney problems.
  - have an eating disorder such as anorexia nervosa or bulimia.
  - have had a head injury.
  - have had a seizure (convulsion, fit).
  - have a tumor in your nervous system (brain or spine).
  - have had a heart attack, heart problems, or high blood pressure.
  - are a diabetic taking insulin or other medicines to control your blood sugar.
  - drink a lot of alcohol.
  - abuse prescription medicines or street drugs.

- **Tell your doctor about all the medicines you take,** including prescription and non-prescription medicines, vitamins, and herbal supplements. Many medicines increase your chances of getting seizures or other serious side effects if you take them while you are using ZYBAN.

How should I take ZYBAN?

- Take ZYBAN exactly as prescribed by your doctor.
- **Do not chew, cut, or crush ZYBAN Tablets.** You must swallow the tablets whole. **Tell your doctor if you cannot swallow medicine tablets.**
- Take ZYBAN at the same time each day.
- Take your doses of ZYBAN at least 8 hours apart.
- If you miss a dose, do not take an extra tablet to make up for the dose you forgot. Wait and take your next tablet at the regular time. **This is very important.** Too much ZYBAN can increase your chance of having a seizure.
- If you take too much ZYBAN, or overdose, call your local emergency room or poison control center right away.
- **Do not take any other medicines while using ZYBAN unless your doctor has told you it is okay.**
- Do not change your dose or stop taking ZYBAN without talking with your doctor first.

How long should I take ZYBAN?

Most people should take ZYBAN for at least 7 to 12 weeks. Some people may need to take ZYBAN for a longer period of time to assist in their smoking cessation efforts. Follow your doctor's instructions.

When should I stop smoking?

It takes about 1 week for ZYBAN to start working. For your best chance of quitting, you should not stop smoking until you have been taking ZYBAN for 1 week. You should set a date to stop smoking during the second week you're taking ZYBAN.

Can I smoke while taking ZYBAN?

It is not physically dangerous to smoke and use ZYBAN at the same time. But you will seriously lower your chance of breaking your smoking habit if you smoke after the date you set to stop smoking.

Can ZYBAN be used at the same time as nicotine patches?

Yes, ZYBAN and nicotine patches can be used at the same time but should only be used together under the supervision of your doctor. Using ZYBAN and nicotine patches together may raise your blood pressure, sometimes severely. Tell your doctor if you are planning to use nicotine replacement therapy because your doctor should check your blood pressure regularly.

Do not smoke at any time if you are using a nicotine patch or any other nicotine product along with ZYBAN. It is possible to get too much nicotine and have serious side effects.

What should I avoid while taking ZYBAN?

- Do not drink a lot of alcohol while taking ZYBAN. If you usually drink a lot of alcohol, talk with your doctor before suddenly stopping. If you suddenly stop drinking alcohol, you may increase your chance of having seizures.
- Do not drive a car or use heavy machinery until you know how ZYBAN affects you. ZYBAN can affect your ability to do these things safely.

What are possible side effects of ZYBAN?

ZYBAN can cause serious side effects. Read this entire Medication Guide for more information about these serious side effects.

The most common side effects of ZYBAN are dry mouth and trouble sleeping. These side effects are generally mild and often disappear after a few weeks. If you have trouble sleeping, do not take ZYBAN too close to bedtime.

These are not all the side effects of ZYBAN. For a complete list, ask your doctor or pharmacist.

Call your doctor for medical advice about side effects. You may report side effects to FDA at 1-800-FDA-1088.

How should I store ZYBAN?

- Store ZYBAN at room temperature.
- Store out of direct sunlight.
- Keep ZYBAN in its tightly closed bottle.
- ZYBAN may have an odor.

General Information about ZYBAN.

Medicines are sometimes prescribed for purposes other than those listed in a Medication Guide. Do not use ZYBAN for a condition for which it was not prescribed. Do not give ZYBAN to other people, even if they have the same symptoms you have. It may harm them. Keep ZYBAN out of the reach of children.

This Medication Guide summarizes important information about ZYBAN. For more information, talk with your doctor. You can ask your doctor or pharmacist for information about ZYBAN that is written for health professionals.

What are the ingredients in ZYBAN?

Active ingredient: bupropion hydrochloride.

Inactive ingredients: carnauba wax, cysteine hydrochloride, hypromellose, magnesium stearate, microcrystalline cellulose, polyethylene glycol, polysorbate 80 and titanium dioxide. The tablets are printed with edible black ink. In addition, the 150-mg tablet contains FD&C Blue No. 2 Lake and FD&C Red No. 40 Lake.

WELLBUTRIN, WELLBUTRIN SR, WELLBUTRIN XL, and PARNATE are registered trademarks of GlaxoSmithKline.

*The following are registered trademarks of their respective manufacturers: NARDIL®/Warner Lambert Company; MARPLAN® /Oxford Pharmaceutical Services, Inc.

Rx only

This Medication Guide has been approved by the U.S. Food and Drug Administration.

July 2009 ZYB:5MG

Distributed by:
GlaxoSmithKline
Research Triangle Park, NC 27709
Manufactured by:
GlaxoSmithKline
Research Triangle Park, NC 27709
or DSM Pharmaceuticals, Inc.
Greenville, NC 27834

July 2009 ZYB:4PI

Janssen®

Division of Ortho-McNeil-Janssen Pharmaceuticals, Inc.
1125 TRENTON-HARBOURTON ROAD
P.O. BOX 200
TITUSVILLE, NJ 08560-0200
www.janssen.com

For Medical Information
(800) 526-7736
FAX: (609) 730-3138

RISPERDAL® CONSTA® ℞
(risperidone)
Long-Lasting Injection

Prescribing information for this product, which appears on pages 1753–1759 of the 2009 PDR, has been revised as follows. Please write "See Supplement B" next to the product heading.

In **1 INDICATIONS AND USAGE, 1.2 Biploar Disorder** has been added:

1.2 Bipolar Disorder

RISPERDAL® CONSTA® is indicated as monotherapy or as adjunctive therapy to lithium or valproate for the maintenance treatment of Bipolar I Disorder *[see Clinical Studies (14.2, 14.3)]*.

Section **2 DOSAGE AND ADMINISTRATION** has been revised after the first 2 paragraphs (rest of copy deleted), a new section **2.1 Schizophrenia** was added:

The recommended dose is 25 mg IM every 2 weeks. Although dose response for effectiveness has not been established for RISPERDAL® CONSTA®, some patients not responding to 25 mg may benefit from a higher dose of 37.5 mg or 50 mg. The maximum dose should not exceed 50 mg RISPERDAL® CONSTA® every 2 weeks. No additional benefit was observed with dosages greater than 50 mg RISPERDAL® CONSTA®, however, a higher incidence of adverse effects was observed.

The efficacy of RISPERDAL® CONSTA® in the treatment of schizophrenia has not been evaluated in controlled clinical trials for longer than 12 weeks. Although controlled studies have not been conducted to answer the question of how long patients with schizophrenia should be treated with RISPERDAL® CONSTA®, oral risperidone has been shown to be effective in delaying time to relapse in longer-term use. It is recommended that responding patients be continued on treatment with RISPERDAL® CONSTA® at the lowest dose needed. The physician who elects to use RISPERDAL® CONSTA® for extended periods should periodically re-evaluate the long-term risks and benefits of the drug for the individual patient.

In **2 DOSAGE AND ADMINISTRATION** another new section, **2.2 Bipolar Disorder,** was added:

Bipolar Disorder

The recommended dose for monotherapy or adjunctive therapy to lithium or valproate for the maintenance treatment of Bipolar I Disorder is 25 mg IM every 2 weeks. Some patients may benefit from a higher dose of 37.5 mg or 50 mg. Dosages above 50 mg have not been studied in this population. The physician who elects to use RISPERDAL® CONSTA® for extended periods should periodically re-evaluate the long-term risks and benefits of the drug for the individual patient.

In **2 DOSAGE AND ADMINISTRATION,** section **2.3 General Dosing Information** has been added as follows:

2.3 General Dosing Information

A lower initial dose of 12.5 mg may be appropriate when clinical factors warrant dose adjustment, such as in patients with hepatic or renal impairment, for certain drug interactions that increase risperidone plasma concentrations *[see Drug Interactions (7.11)]* or in patients who have a history of poor tolerability to psychotropic medications. The efficacy of the 12.5 mg dose has not been investigated in clinical trials.

Oral RISPERDAL® (or another antipsychotic medication) should be given with the first injection of RISPERDAL® CONSTA® and continued for 3 weeks (and then discontinued) to ensure that adequate therapeutic plasma concentrations are maintained prior to the main release phase of risperidone from the injection site *[see Clinical Pharmacology (12.3)]*.

Upward dose adjustment should not be made more frequently than every 4 weeks. The clinical effects of this dose adjustment should not be anticipated earlier than 3 weeks after the first injection with the higher dose.

In patients with clinical factors such as hepatic or renal impairment or certain drug interactions that increase risperidone plasma concentrations *[see Drug Interactions (7.11)]*, dose reduction as low as 12.5 mg may be appropriate. The efficacy of the 12.5 mg dose has not been investigated in clinical trials.

Do not combine two different dose strengths of RISPERDAL® CONSTA® in a single administration.

Under **5 WARNINGS AND PRECAUTIONS,** a new section has been added:

5.8 Leukopenia, Neutropenia, and Agranulocytosis

Class Effect: In clinical trial and postmarketing experience, events of leukopenia/neutropenia have been reported temporally related to antipsychotic agents, including risperidone. Agranulocytosis (including fatal cases) has also been reported.

Possible risk factors for leukopenia/neutropenia include preexisting low white blood cell count (WBC) and a history of drug induced leukopenia/neutropenia. Patients with a preexisting low WBC or a history of drug induced leukopenia/neutropenia should have their complete blood count (CBC) monitored frequently during the first few months of therapy and discontinuation of RISPERDAL® CONSTA® should be considered at the first sign of a decline in WBC in the absence of other causative factors.

Patients with neutropenia should be carefully monitored for fever or other symptoms or signs of infection and treated promptly if such symptoms or signs occur. Patients with severe neutropenia (absolute neutrophil count $<1000/mm^3$) should discontinue RISPERDAL® CONSTA® and have their WBC followed until recovery.

Under **5 WARNINGS AND PRECAUTIONS,** section **5.17 Suicide** has been updated as follows:

5.17 Suicide

There is an increased risk of suicide attempt in patients with schizophrenia or bipolar disorder, and close supervision of high-risk patients should accompany drug therapy. RISPERDAL® CONSTA® is to be administered by a health care professional *[see Dosage and Administration (2)]*; therefore, suicide due to an overdose is unlikely.

Under **6 ADVERSE REACTIONS,** a new bullet was added after the 7th bullet:

- Leukopenia/Neutropenia and Agranulocytosis *[see Warnings and Precautions (5.8)]*

Under **6 ADVERSE REACTIONS,** the first 2 paragraphs after the bulleted list were revised as follows:

The most common adverse reactions in clinical trials in patients with schizophrenia (≥5%) were headache, parkinsonism, dizziness, akathisia, fatigue, constipation, dyspepsia, sedation, weight increased, pain in extremity, and dry mouth. The most common adverse reactions in the double-blind, placebo-controlled periods of the bipolar disorder trials were weight increased (5% in the monotherapy trial) and tremor and parkinsonism (≥ 10% in the adjunctive treatment trial).

The most common adverse reactions that were associated with discontinuation from the 12-week double-blind, placebo-controlled trial in patients with schizophrenia (causing discontinuation in ≥1% of patients) were agitation, depression, anxiety, and akathisia. Adverse reactions that were associated with discontinuation from the double-blind, placebo-controlled periods of the bipolar disorder trials were hyperglycemia (one patient in the monotherapy trial) and hypokinesia and tardive dyskinesia (one patient each in the adjunctive treatment trial).

Under **6 ADVERSE REACTIONS,** the 4th & 5th paragraphs after the bulleted list were added:

In addition to the studies in patients with schizophrenia, safety data are presented from a trial assessing the efficacy and safety of RISPERDAL® CONSTA® when administered as monotherapy for maintenance treatment in patients with bipolar I disorder. The subjects in this multi-center, double-blind, placebo-controlled study were adult patients who met

Continued on next page

Risperdal Consta—Cont.

DSM-IV criteria for Bipolar Disorder Type I and who were stable on risperidone (oral or long-acting injection), were stable on other antipsychotics or mood stabilizers, or were experiencing an acute episode. After a 3-week period of treatment with open-label oral risperidone (n=440), subjects who demonstrated an initial response to oral risperidone in this period and those who were stable on risperidone (oral or long-acting injection) at study entry entered into a 26-week stabilization period of open-label RISPERDAL® CONSTA® (n=501). Subjects who demonstrated a maintained response during this period were then randomized into a 24-month double-blind, placebo-controlled period in which they received RISPERDAL® CONSTA® (n=154) or placebo (n=149) as monotherapy. Subjects who relapsed or who completed the double-blind period could choose to enter an 8-week open-label RISPERDAL® CONSTA® extension period (n=160).

Safety data are also presented from a trial assessing the efficacy and safety of RISPERDAL® CONSTA® when administered as adjunctive maintenance treatment in patients with bipolar disorder. The subjects in this multi-center, double-blind, placebo-controlled study were adult patients who met DSM-IV criteria for Bipolar Disorder Type I or Type II and who experienced at least 4 episodes of mood disorder requiring psychiatric/clinical intervention in the previous 12 months, including at least 2 episodes in the 6 months prior to the start of the study. At the start of this study, all patients (n = 275) entered into a 16-week open-label treatment phase in which they received RISPERDAL® CONSTA® in addition to continuing their treatment as usual, which consisted of various mood stabilizers (primarily lithium and valproate), antidepressants, and/or anxiolytics. Patients who reached remission at the end of this 16-week open-label treatment phase (n = 139) were then randomized into a 52-week double-blind, placebo-controlled phase in which they received RISPERDAL® CONSTA® (n = 72) or placebo (n = 67) as adjunctive treatment in addition to continuing their treatment as usual. Patients who did not reach remission at the end of the 16-week open-label treatment phase could choose to continue to receive RISPERDAL® CONSTA® as adjunctive therapy in an open-label manner, in addition to continuing their treatment as usual, for up to an additional 36 weeks as clinically indicated for a total period of up to 52 weeks; these patients (n = 70) were also included in the evaluation of safety.

Under **6 ADVERSE REACTIONS,** section **6.2 Commonly-Observed Adverse Reactions in Double-Blind, Placebo-Controlled Clinical Trials – Bipolar Disorder** has been added:

6.2 Commonly-Observed Adverse Reactions in Double-Blind, Placebo-Controlled Clinical Trials – Bipolar Disorder

Table 2 lists the treatment-emergent adverse reactions reported in 2% or more of RISPERDAL® CONSTA®-treated patients in the 24-month double-blind, placebo-controlled treatment period of the trial assessing the efficacy and safety of RISPERDAL® CONSTA® when administered as monotherapy for maintenance treatment in patients with Bipolar I Disorder.

Table 2. Adverse Reactions in ≥2% of Patients with Bipolar I Disorder Treated with RISPERDAL® CONSTA® as Monotherapy in a 24-Month Double-Blind, Placebo-Controlled Trial

System/Organ Class Adverse Reaction	Percentage of Patients Reporting Event	
	RISPERDAL® CONSTA® (N=154)	**Placebo (N=149)**
Investigations		
Weight increased	5	1
Nervous system disorders		
Dizziness	3	1
Vascular disorders		
Hypertension	3	1

Table 3 lists the treatment-emergent adverse reactions reported in 4% or more of patients in the 52-week double-blind, placebo-controlled treatment phase of a trial assessing the efficacy and safety of RISPERDAL® CONSTA® when administered as adjunctive maintenance treatment in patients with bipolar disorder.

[See table 3 above]

Table 3. Adverse Reactions in ≥4% of Patients with Bipolar Disorder Treated with RISPERDAL® CONSTA® as Adjunctive Therapy in a 52-Week Double-Blind, Placebo-Controlled Trial

System/Organ Class Adverse Reaction	Percentage of Patients Reporting Event	
	RISPERDAL® CONSTA® + Treatment as Usual[a] (N=72)	**Placebo + Treatment as Usual[a] (N=67)**
General disorders and administration site conditions		
Gait abnormal	4	0
Infections and infestations		
Upper respiratory tract infection	6	3
Investigations		
Weight increased	7	1
Metabolism and nutrition disorders		
Decreased appetite	6	1
Increased appetite	4	0
Musculoskeletal and connective tissue disorders		
Arthralgia	4	3
Nervous system disorders		
Tremor	24	16
Parkinsonism[b]	15	6
Dyskinesia[b]	6	3
Sedation[c]	7	1
Disturbance in attention	4	0
Reproductive system and breast disorders		
Amenorrhea	4	1
Respiratory, thoracic and mediastinal disorders		
Cough	4	1

[a] Patients received double-blind RISPERDAL® CONSTA® or placebo in addition to continuing their treatment as usual, which included mood stabilizers, antidepressants, and/or anxiolytics.

[b] Parkinsonism includes muscle rigidity, hypokinesia, cogwheel rigidity, and bradykinesia. Dyskinesia includes muscle twitching and dyskinesia.

[c] Sedation includes sedation and somnolence.

Under **6 ADVERSE REACTIONS, 6.3 Other Adverse Reactions Observed During the Premarketing Evaluation of RISPERDAL® CONSTA®**, the 1st paragraph and list of disorders was revised as follows:

6.3 Other Adverse Reactions Observed During the Premarketing Evaluation of RISPERDAL® CONSTA®

The following additional adverse reactions occurred in <2% of the RISPERDAL® CONSTA®-treated patients in the above schizophrenia double-blind, placebo-controlled trial dataset, in <2% of the RISPERDAL® CONSTA®-treated patients in the above double-blind, placebo-controlled period of the monotherapy bipolar disorder trial dataset, or in < 4% of the RISPERDAL® CONSTA®-treated patients in the above double-blind, placebo-controlled period of the adjunctive treatment bipolar disorder trial dataset. The following also includes additional adverse reactions reported at any frequency in RISPERDAL® CONSTA®-treated patients who participated in other studies, including double-blind, active-controlled and open-label studies in schizophrenia, and in the open-label phases of the bipolar disorder studies.

Blood and lymphatic system disorders: anemia, neutropenia

Cardiac disorders: tachycardia, atrioventricular block first degree, palpitations, sinus bradycardia, bundle branch block left, bradycardia, sinus tachycardia, bundle branch block right

Ear and labyrinth disorders: ear pain, vertigo

Endocrine disorders: hyperprolactinemia

Eye disorders: conjunctivitis, visual acuity reduced

Gastrointestinal disorders: diarrhea, vomiting, abdominal pain, stomach discomfort, gastritis

General disorders and administration site conditions: injection site pain, chest discomfort, chest pain, influenza like illness, sluggishness, malaise, induration, injection site induration, injection site swelling, injection site reaction, face edema

Immune system disorders: hypersensitivity

Infections and infestations: nasopharyngitis, influenza, bronchitis, urinary tract infection, rhinitis, ear infection, pneumonia, lower respiratory tract infection, pharyngitis, sinusitis, viral infection, infection, localized infection, cystitis, gastroenteritis, subcutaneous abscess

Injury and poisoning: fall, procedural pain

Investigations: blood prolactin increased, alanine aminotransferase increased, electrocardiogram abnormal, gamma-glutamyl transferase increased, blood glucose increased, hepatic enzyme increased, aspartate aminotransferase increased, electrocardiogram QT prolonged

Metabolism and nutritional disorders: anorexia, hyperglycemia

Musculoskeletal, connective tissue and bone disorders: posture abnormal, myalgia, back pain, buttock pain, muscular weakness, neck pain, musculoskeletal chest pain

Nervous system disorders: coordination abnormal, dystonia, tardive dyskinesia, drooling, paresthesia, dizziness postural, convulsion, akinesia, hypokinesia, dysarthria

Psychiatric disorders: insomnia, agitation, anxiety, sleep disorder, depression, libido decreased, nervousness

Renal and urinary disorders: urinary incontinence

Reproductive system and breast disorders: oligomenorrhea, erectile dysfunction, galactorrhea, sexual dysfunction, ejaculation disorder, gynecomastia, breast discomfort, menstruation irregular, menstruation delayed, menstrual disorder

Respiratory, thoracic and mediastinal disorders: nasal congestion, pharyngolaryngeal pain, dyspnea, rhinorrhea

Skin and subcutaneous tissue disorders: rash, eczema, pruritus

Vascular disorders: hypotension, orthostatic hypotension

Under **6 ADVERSE REACTIONS, 6.4 Discontinuations Due to Adverse Reactions**, a new section on Biploar Disorder was added:

Bipolar Disorder

In the 24-month double-blind, placebo-controlled treatment period of the trial assessing the efficacy and safety of RISPERDAL® CONSTA® when administered as monotherapy for maintenance treatment in patients with bipolar I disorder, 1 (0.6%) of 154 RISPERDAL® CONSTA®-treated patients discontinued due to an adverse reaction (hyperglycemia).

In the 52-week double-blind phase of the placebo-controlled trial in which RISPERDAL® CONSTA® was administered as adjunctive therapy to patients with bipolar disorder in addition to continuing with their treatment as usual, approximately 4% (3/72) of RISPERDAL® CONSTA®-treated patients discontinued treatment due to an adverse event, compared with 1.5% (1/67) of placebo-treated patients. Adverse reactions associated with discontinuation in RISPERDAL® CONSTA®-treated patients were: hypokinesia (one patient) and tardive dyskinesia (one patient).

Under **6 ADVERSE REACTIONS, 6.6 Changes in Body Weight** was revised as follows:

6.6 Changes in Body Weight

In the 12-week double-blind, placebo-controlled trial in patients with schizophrenia, 9% of patients treated with RISPERDAL® CONSTA®, compared with 6% of patients treated with placebo, experienced a weight gain of >7% of body weight at endpoint.

In the 24-month double-blind, placebo-controlled treatment period of a trial assessing the efficacy and safety of RISPERDAL® CONSTA® when administered as monotherapy for maintenance treatment in patients with bipolar I disorder, 11.6% of patients treated with RISPERDAL® CONSTA® compared with 2.8% of patients treated with placebo experienced a weight gain of >7% of body weight at endpoint.

In the 52-week double-blind, placebo-controlled trial in patients with bipolar disorder, 26.8% of patients treated with RISPERDAL® CONSTA® as adjunctive treatment in addition to continuing their treatment as usual, compared with 27.3% of patients treated with placebo in addition to continuing their treatment as usual, experienced a weight gain of >7% of body weight at endpoint.

Under **6 ADVERSE REACTIONS, 6.7 Changes in ECG** was revised as follows:

6.7 Changes in ECG

The electrocardiograms of 202 schizophrenic patients treated with 25 mg or 50 mg RISPERDAL® CONSTA® and 98 schizophrenic patients treated with placebo in the 12-week double-blind, placebo-controlled trial were evaluated. Compared with placebo, there were no statistically significant differences in QTc intervals (using Fridericia's and linear correction factors) during treatment with RISPERDAL® CONSTA®.

The electrocardiograms of 227 patients with Bipolar I Disorder were evaluated in the 24-month double-blind, placebo-controlled period. There were no clinically relevant differences in QTc intervals (using Fridericia's and linear correction factors) during treatment with RISPERDAL® CONSTA® compared to placebo.

The electrocardiograms of 85 patients with bipolar disorder were evaluated in the 52-week double-blind, placebo-controlled trial. There were no statistically significant differences in QTc intervals (using Fridericia's and linear correction factors) during treatment with RISPERDAL® CONSTA® 25 mg, 37.5 mg, or 50 mg when administered as adjunctive treatment in addition to continuing treatment as usual compared to placebo.

Under **14 CLINICAL STUDIES,** sections **14.2 Bipolar Disorder - Monotherapy** and **14.3 Bipolar Disorder – Adjunctive Therapy** were added:

14.2 Bipolar Disorder - Monotherapy

The effectiveness of RISPERDAL® CONSTA® for the maintenance treatment of Bipolar I Disorder was established in a multicenter, double-blind, placebo-controlled study of adult patients who met DSM-IV criteria for Bipolar Disorder Type I, who were stable on medications or experiencing an acute manic or mixed episode.

A total of 501 patients were treated during a 26-week open-label period with RISPERDAL® CONSTA® (starting dose of 25 mg, and titrated, if deemed clinically desirable, to 37.5 mg or 50 mg; in patients not tolerating the 25 mg dose, the dose could be reduced to 12.5 mg). In the open-label phase, 303 (60%) patients were judged to be stable and were randomized to double-blind treatment with either the same dose of RISPERDAL® CONSTA® or placebo and monitored for relapse. The primary endpoint was time to relapse to any mood episode (depression, mania, hypomania, or mixed).

Time to relapse was delayed in patients receiving RISPERDAL® CONSTA® monotherapy as compared to placebo. The majority of relapses were due to manic rather than depressive symptoms. Based on their bipolar disorder history, subjects entering this study had had, on average, more manic episodes than depressive episodes.

14.3 Bipolar Disorder - Adjunctive Therapy

The effectiveness of RISPERDAL® CONSTA® as an adjunct to treatment with lithium or valproate for the maintenance treatment of Bipolar Disorder was established in a multicenter, randomized, double-blind, placebo-controlled study of adult patients who met DSM-IV criteria for Bipolar Disorder Type I and who experienced at least 4 episodes of mood disorder requiring psychiatric/clinical intervention in the previous 12 months, including at least 2 episodes in the 6 months prior to the start of the study.

A total of 240 patients were treated during a 16-week open-label period with RISPERDAL® CONSTA® (starting dose of 25 mg, and titrated, if deemed clinically desirable, to 37.5 mg or 50 mg), as adjunctive therapy in addition to continuing their treatment as usual for their bipolar disorder, which consisted of mood stabilizers (primarily lithium and valproate), antidepressants, and/or anxiolytics. All oral antipsychotics were discontinued after the first three weeks of the initial RISPERDAL® CONSTA® injection. In the open-label phase, 124 (51.7%) were judged to be stable for at least the last 4 weeks and were randomized to double-blind treatment with either the same dose of RISPERDAL® CONSTA® or placebo in addition to continuing their treatment as usual and monitored for relapse during a 52-week period. The primary endpoint was time to relapse to any new mood episode (depression, mania, hypomania, or mixed).

Time to relapse was delayed in patients receiving adjunctive therapy with RISPERDAL® CONSTA® as compared to placebo. The relapse types were about half depressive and half manic or mixed episodes.

10130502 **Revised May 2009**

To keep your **PDR** up to date throughout the year, note these revisions on the corresponding pages of the annual volume. Simply write **"See Supplement B"** next to the product heading.

McNeil Pediatrics™

Division of Ortho-McNeil-Janssen Pharmaceuticals, Inc
1125 TRENTON-HARBOURTON ROAD
TITUSVILLE, NJ 08560
www.mcneilpediatrics.net

For Medical Information:
(800) 526-7736

CONCERTA® Ⓒ℞
(methylphenidate HCl)
Extended-release Tablets

Prescribing information for this product, which appears on pages 1925–1931 of the 2009 PDR, has been revised as follows. Please write "See Supplement B" next to the product heading.

In the **Medication Guide**, the "What are possible side effects of CONCERTA®?" section, the following copy has been added at the end of that section:

Call your doctor for medical advice about side effects. You may report side effects to FDA at 1-800-FDA-1088.

You may also report side effects to McNeil Pediatrics at 1-888-440-7903.

10180601
Revised June 2009

Merck & Co., Inc.

PO BOX 4 ZB-714
WEST POINT, PA 19486-0004

For Medical Information and Product and Service Information Contact:
Call the Merck National Service Center, 8:00 AM to 7:00 PM (ET), Monday through Friday:
(800) NSC-MERCK
(800) 672-6372
FAX: (800) MERCK-68
FAX: (800) 637-2568
Adverse Drug Experiences:
Call the Merck National Service Center, 8:00 AM to 7:00 PM (ET), Monday through Friday:
(800) NSC-MERCK
(800) 672-6372
Pregnancy Registries
(800) 986-8999
In Emergencies:
24-hour emergency information for healthcare professionals:
(800) NSC-MERCK
(800) 672-6372
Sales and Ordering:
For product orders and direct account inquiries only, call the Order Management Center,
8:00 AM to 6:00 PM (ET), Monday through Friday:
(800) MERCK RX
(800) 637-2579

ISENTRESS® ℞
(raltegravir)
Tablets

Prescribing information for this product, which appears on pages 2036–2041 of the 2009 PDR and was updated in Supplement A page A237, has been completely revised as follows. Please write "See Supplement B" next to the product heading.

HIGHLIGHTS OF PRESCRIBING INFORMATION

These highlights do not include all the information needed to use ISENTRESS safely and effectively. See full prescribing information for ISENTRESS.

ISENTRESS (raltegravir) Tablets
Initial U.S. Approval: 2007

RECENT MAJOR CHANGES

Indications And Usage (1)	07/2009
Dosage And Administration (2)	01/2009
Warnings And Precautions (5.2) - removal	07/2009

INDICATIONS AND USAGE

ISENTRESS® is a human immunodeficiency virus integrase strand transfer inhibitor (HIV-1 INSTI) indicated:

- In combination with other antiretroviral agents for the treatment of HIV-1 infection in adult patients (1).

The safety and efficacy of ISENTRESS have not been established in pediatric patients (1).

DOSAGE AND ADMINISTRATION

- 400 mg administered orally, twice daily with or without food (2).
- During coadministration with rifampin, 800 mg twice daily (2).

DOSAGE FORMS AND STRENGTHS

Tablets: 400 mg (3).

CONTRAINDICATIONS

None (4)

WARNINGS AND PRECAUTIONS

Monitor for Immune Reconstitution Syndrome (5.1)

ADVERSE REACTIONS

- The most common adverse reactions of moderate to severe intensity (≥2%) which occurred at a higher rate than the comparator are insomnia, headache, nausea, asthenia and fatigue (6.1).
- Creatine kinase elevations were observed in subjects who received ISENTRESS. Myopathy and rhabdomyolysis have been reported; however, the relationship of ISENTRESS to these events is not known. Use with caution in patients at increased risk of myopathy or rhabdomyolysis, such as patients receiving concomitant medications known to cause these conditions (6.1).

To report SUSPECTED ADVERSE REACTIONS, contact Merck & Co., Inc. at 1-877-888-4231 or FDA at 1-800-FDA-1088 or www.fda.gov/medwatch.

DRUG INTERACTIONS

- Coadministration of ISENTRESS with drugs that are strong inducers of UGT1A1 may result in reduced plasma concentrations of raltegravir (7.2).

USE IN SPECIFIC POPULATIONS

Pregnancy:

- ISENTRESS should be used during pregnancy only if the potential benefit justifies the potential risk to the fetus. Physicians are encouraged to register pregnant women exposed to ISENTRESS by calling 1-800-258-4263 so that Merck can monitor maternal and fetal outcomes (8.1).

Nursing Mothers:

- Breast-feeding is not recommended while taking ISENTRESS (8.3).

See 17 for PATIENT COUNSELING INFORMATION and FDA-approved patient labeling.

Revised: 07/2009

FULL PRESCRIBING INFORMATION: CONTENTS*

*Sections or subsections omitted from the Full Prescribing Information are not listed.

FULL PRESCRIBING INFORMATION

1 INDICATIONS AND USAGE

ISENTRESS[1] is indicated in combination with other antiretroviral agents for the treatment of human immunodeficiency virus (HIV-1) infection in adult patients.

This indication is based on analyses of plasma HIV-1 RNA levels up through 48 weeks in three double-blind controlled

Continued on next page

Information on the Merck & Co., Inc., products listed on these pages is from the full prescribing information in use March 1, 2009. For information, please call 1-800-NSC-MERCK [1-800-672-6372].

Table 2: Selected Grade 2 to 4 Laboratory Abnormalities Reported in Treatment-Naïve Subjects (48 Week Analysis)

Laboratory Parameter Preferred Term (Unit)	Limit	Randomized Study Protocol 021 ISENTRESS 400 mg Twice Daily + Emtricitabine (+) Tenofovir (N = 281)	Efavirenz 600 mg At Bedtime + Emtricitabine (+) Tenofovir (N = 282)
Hematology			
Absolute neutrophil count (10^3/µL)			
Grade 2	0.75–0.999	3%	3%
Grade 3	0.50–0.749	1%	<1%
Grade 4	<0.50	<1%	0%
Hemoglobin (gm/dL)			
Grade 2	7.5–8.4	<1%	<1%
Grade 3	6.5–7.4	<1%	<1%
Grade 4	<6.5	0%	0%
Platelet count (10^3/µL)			
Grade 2	50–99.999	2%	0%
Grade 3	25–49.999	0%	<1%
Grade 4	<25	0%	.0%
Blood chemistry			
Fasting (non-random) serum glucose test (mg/dL)			
Grade 2	126–250	2%	3%
Grade 3	251–500	<1%	0%
Grade 4	>500	0%	0%
Total serum bilirubin			
Grade 2	1.6–2.5 × ULN	4%	0%
Grade 3	2.6–5.0 × ULN	<1%	0%
Grade 4	>5.0 × ULN	0%	0%
Serum aspartate aminotransferase			
Grade 2	2.6–5.0 × ULN	3%	4%
Grade 3	5.1–10.0 × ULN	1%	1%
Grade 4	>10.0 × ULN	<1%	<1%
Serum alanine aminotransferase			
Grade 2	2.6–5.0 × ULN	4%	6%
Grade 3	5.1–10.0 × ULN	<1%	2%
Grade 4	>10.0 × ULN	<1%	<1%
Serum alkaline phosphatase			
Grade 2	2.6–5.0 × ULN	<1%	2%
Grade 3	5.1–10.0 × ULN	0%	<1%
Grade 4	>10.0 × ULN	0%	0%

ULN = Upper limit of normal range

Table 3: Lipid Values, Mean Change from Baseline, Protocol 021

Laboratory Parameter Preferred Term	ISENTRESS 400 mg Twice Daily + Emtricitabine (+) Tenofovir N = 281			Efavirenz 600 mg At Bedtime + Emtricitabine (+) Tenofovir N = 282		
			Change from Baseline at Week 48			Change from Baseline at Week 48
	Baseline Mean (mg/dL)	Week 48 Mean (mg/dL)	Mean Change (mg/dL)	Baseline Mean (mg/dL)	Week 48 Mean (mg/dL)	Mean Change (mg/dL)
LDL-Cholesterol†	97	103	6	92	108	16
HDL-Cholesterol†	38	42	4	38	48	10
Total Cholesterol†	159	169	10	156	188	33
Triglyceride†	125	122	-3	136	174	37

†Fasting (non-random) laboratory tests.

Notes:

N = Number of subjects in the treatment group. The analysis is based on all available data.

If subjects initiated or increased serum lipid-reducing agents, the last available lipid values prior to the change in therapy were used in the analysis. If the missing data was due to other reasons, subjects were censored thereafter for the analysis.

At baseline, serum lipid-reducing agents were used in 5% of subjects in the group receiving ISENTRESS and 3% in the efavirenz group. Through Week 48, serum lipid-reducing agents were used in 6% of subjects in the group receiving ISENTRESS and 6% in the efavirenz group.

2 DOSAGE AND ADMINISTRATION

For the treatment of patients with HIV-1 infection, the dosage of ISENTRESS is 400 mg administered orally, twice daily with or without food. During coadministration with rifampin, the recommended dosage of ISENTRESS is 800 mg twice daily with or without food.

3 DOSAGE FORMS AND STRENGTHS

400 mg pink, oval-shaped, film-coated tablets with "227" on one side.

4 CONTRAINDICATIONS

None

5 WARNINGS AND PRECAUTIONS

5.1 Immune Reconstitution Syndrome

During the initial phase of treatment, patients responding to antiretroviral therapy may develop an inflammatory response to indolent or residual opportunistic infections (such as *Mycobacterium avium* complex, cytomegalovirus, *Pneumocystis jiroveci* pneumonia, *Mycobacterium* tuberculosis, or reactivation of varicella zoster virus), which may necessitate further evaluation and treatment.

6 ADVERSE REACTIONS

6.1 Clinical Trials Experience

Because clinical trials are conducted under widely varying conditions, adverse reaction rates observed in the clinical trials of a drug cannot be directly compared to rates in the clinical trials of another drug and may not reflect the rates observed in practice.

Treatment-Naïve Studies

The following safety assessment of ISENTRESS in treatment-naïve subjects is based on the randomized double-blind active controlled study of treatment-naïve subjects, STARTMRK (Protocol 021) with ISENTRESS 400 mg twice daily in combination with a fixed dose of emtricitabine 200 mg (+) tenofovir 300 mg, (N=281) versus efavirenz (EFV) 600 mg at bedtime in combination with emtricitabine (+) tenofovir, (N=282). During double-blind treatment, the total follow-up for subjects receiving ISENTRESS 400 mg twice daily + emtricitabine (+) tenofovir was 247 patient-years and 241 patient-years for subjects receiving efavirenz 600 mg at bedtime + emtricitabine (+) tenofovir.

In Protocol 021, the rate of discontinuation of therapy due to adverse reactions was 3% in subjects receiving ISENTRESS + emtricitabine (+) tenofovir and 6% in subjects receiving efavirenz + emtricitabine (+) tenofovir.

The clinical adverse drug reactions (ADRs) listed below were considered by investigators to be causally related to ISENTRESS + emtricitabine (+) tenofovir or efavirenz + emtricitabine (+) tenofovir. Clinical ADRs of moderate to severe intensity occurring in ≥2% of treatment-naïve subjects treated with ISENTRESS and occurring at a higher rate than efavirenz are presented in Table 1.

Table 1: Adverse Reactions* of Moderate to Severe Intensity† Occurring in ≥2% of Treatment-Naïve Adult Subjects Receiving ISENTRESS and at a Higher Rate Compared to Efavirenz (48 Week Analysis)

System Organ Class, Preferred Term	Randomized Study Protocol 021 ISENTRESS 400 mg Twice Daily + Emtricitabine (+) Tenofovir (n = 281)‡ %	Efavirenz 600 mg At Bedtime + Emtricitabine (+) Tenofovir (n = 282)‡ %
Psychiatric Disorders		
Insomnia	4	3

*Includes adverse experiences considered by investigators to be at least possibly, probably, or definitely related to the drug

†Intensities are defined as follows: Moderate (discomfort enough to cause interference with usual activity); Severe (incapacitating with inability to work or do usual activity).

‡n = total number of subjects per treatment group

Less Common Adverse Reactions

The following ADRs occurred in <2% of subjects receiving ISENTRESS + emtricitabine (+) tenofovir. These events have been included because of their seriousness, increased frequency on ISENTRESS compared with efavirenz or investigator's assessment of potential causal relationship.

General Disorders and Administration Site Conditions: fatigue

Psychiatric Disorders: abnormal dreams

Laboratory Abnormalities

The percentages of adult subjects treated with ISENTRESS 400 mg twice daily or efavirenz in Protocol 021 with se-

Isentress—Cont.

studies of ISENTRESS. Two of these studies were conducted in clinically advanced, 3-class antiretroviral (NNRTI, NRTI, PI) treatment-experienced adults and one was conducted in treatment-naïve adults.

The use of other active agents with ISENTRESS is associated with a greater likelihood of treatment response *[see Clinical Studies (14)]*.

The safety and efficacy of ISENTRESS have not been established in pediatric patients.

[1] Registered trademark of MERCK & CO., Inc.

lected Grades 2 to 4 laboratory abnormalities that represent a worsening from baseline are presented in Table 2.

[See table 2 at top of previous page]

Lipids, Change from Baseline

Changes from baseline in fasting lipids are shown in Table 3.

[See table 3 at top of previous page]

Treatment-Experienced Studies

The safety assessment of ISENTRESS in treatment-experienced subjects is based on the pooled safety data from the randomized, double-blind, placebo-controlled trials, BENCHMRK 1 and BENCHMRK 2 (Protocols 018 and 019) in antiretroviral treatment-experienced HIV-1 infected adult subjects. A total of 462 subjects received the recommended dose of ISENTRESS 400 mg twice daily in combination with optimized background therapy (OBT) compared to 237 subjects taking placebo in combination with OBT. The median duration of therapy in these trials was 48 weeks for subjects receiving ISENTRESS and 38 weeks for subjects receiving placebo. The total exposure to ISENTRESS was 387 patient-years versus 156 patient-years on placebo. The rates of discontinuation due to adverse events were 2% in subjects receiving ISENTRESS and 3% in subjects receiving placebo.

Clinical ADRs were considered by investigators to be causally related to ISENTRESS + OBT or placebo + OBT. Clinical ADRs of moderate to severe intensity occurring in ≥2% of subjects treated with ISENTRESS and occurring at a higher exposure adjusted rate compared to placebo are presented in Table 4.

Table 4: Adverse Drug Reactions* of Moderate to Severe Intensity† Occurring in ≥2% of Treatment-Experienced Adult Subjects Receiving ISENTRESS and at a Higher Exposure Adjusted Rate Compared to Placebo (48 Week Analysis, Exposure Adjusted Incidence Rates)

System Organ Class, Adverse Reactions	Randomized Studies Protocol 018 and 019	
	ISENTRESS 400 mg Twice Daily + OBT (n = 462)‡	Placebo + OBT (n = 237)‡
	Rate per 100 Patient-Years	Rate per 100 Patient-Years
Nervous System Disorders		
Headache	3	1
Gastrointestinal Disorders		
Nausea	2	1
General Disorders and Administration Site Conditions		
Asthenia	2	1
Fatigue	2	1

*Includes adverse reactions at least possibly, probably, or definitely related to the drug.

†Intensities are defined as follows: Moderate (discomfort enough to cause interference with usual activity); Severe (incapacitating with inability to work or do usual activity).

‡ n=total number of subjects per treatment group.

Table 5: Selected Grade 2 to 4 Laboratory Abnormalities Reported in Treatment-Experienced Subjects (48 Week Analysis)

Laboratory Parameter Preferred Term (Unit)	Limit	Randomized Studies Protocol 018 and 019: ISENTRESS 400 mg Twice Daily + OBT (N = 462)	Placebo + OBT (N = 237)
Hematology			
Absolute neutrophil count ($10^3/\mu L$)			
Grade 2	0.75–0.999	3%	5%
Grade 3	0.50–0.749	3%	3%
Grade 4	<0.50	1%	<1%
Hemoglobin (gm/dL)			
Grade 2	7.5–8.4	1%	3%
Grade 3	6.5–7.4	1%	<1%
Grade 4	<6.5	<1%	0%
Platelet count ($10^3/\mu L$)			
Grade 2	50–99.999	3%	5%
Grade 3	25–49.999	1%	<1%
Grade 4	<25	1%	<1%
Blood chemistry			
Fasting (non-random) serum glucose test (mg/dL)			
Grade 2	126–250	8%	5%
Grade 3	251–500	2%	1%
Grade 4	>500	0%	0%
Total serum bilirubin			
Grade 2	1.6–2.5 × ULN	5%	3%
Grade 3	2.6–5.0 × ULN	2%	2%
Grade 4	>5.0 × ULN	1%	0%
Serum aspartate aminotransferase			
Grade 2	2.6–5.0 × ULN	8%	6%
Grade 3	5.1–10.0 × ULN	3%	3%
Grade 4	>10.0 × ULN	<1%	1%
Serum alanine aminotransferase			
Grade 2	2.6–5.0 × ULN	7%	8%
Grade 3	5.1–10.0 × ULN	3%	2%
Grade 4	>10.0 × ULN	1%	2%
Serum alkaline phosphatase			
Grade 2	2.6–5.0 × ULN	2%	<1%
Grade 3	5.1–10.0 × ULN	<1%	1%
Grade 4	>10.0 × ULN	1%	<1%
Serum pancreatic amylase test			
Grade 2	1.6–2.0 × ULN	2%	1%
Grade 3	2.1–5.0 × ULN	3%	3%
Grade 4	>5.0 × ULN	<1%	0%
Serum lipase test			
Grade 2	1.6–3.0 × ULN	4%	3%
Grade 3	3.1–5.0 × ULN	1%	<1%
Grade 4	>5.0 × ULN	0%	0%
Serum creatine kinase			
Grade 2	6.0–9.9 × ULN	2%	2%
Grade 3	10.0–19.9 × ULN	3%	3%
Grade 4	≥20.0 × ULN	2%	1%

ULN = Upper limit of normal range

Less Common Adverse Reactions

The following ADRs occurred in <2% of subjects receiving ISENTRESS + OBT. These events have been included because of either their seriousness, increased frequency on ISENTRESS compared with placebo or investigator's assessment of potential causal relationship.

Gastrointestinal Disorders: abdominal pain, gastritis

Hepatobiliary Disorders: hepatitis

Immune System Disorders: hypersensitivity

Infections and Infestations: genital herpes, herpes zoster

Nervous System Disorders: dizziness

Renal and Urinary Disorders: renal failure

Laboratory Abnormalities

The percentages of adult subjects treated with ISENTRESS 400 mg twice daily or placebo in Protocols 018 and 019 with selected Grade 2 to 4 laboratory abnormalities representing a worsening from baseline are presented in Table 5.

[See table 5 above]

Selected Adverse Events

Regardless of Drug Relationship

Cancers were reported in treatment-experienced subjects who initiated ISENTRESS or placebo, both with OBT, and in treatment-naïve subjects who initiated ISENTRESS or efavirenz, both with emtricitabine (+) tenofovir; several were recurrent. The types and rates of specific cancers were those expected in a highly immunodeficient population (many had CD4+ counts below 50 cells/mm^3 and most had prior AIDS diagnoses). The risk of developing cancer in these studies was similar in the group receiving ISENTRESS and the group receiving the comparator.

Grade 2-4 creatine kinase laboratory abnormalities were observed in subjects treated with ISENTRESS (see Table 5). Myopathy and rhabdomyolysis have been reported; however, the relationship of ISENTRESS to these events is not known. Use with caution in patients at increased risk of myopathy or rhabdomyolysis, such as patients receiving concomitant medications known to cause these conditions.

Patients with Co-existing Conditions

Patients Co-infected with Hepatitis B and/or Hepatitis C Virus

In the randomized, double-blind, placebo-controlled trials, treatment-experienced subjects (N = 114/699 or 16%) and treatment-naïve subjects (N = 34/563 or 6%) with chronic (but not acute) active hepatitis B and/or hepatitis C virus co-infection were permitted to enroll provided that baseline liver function tests did not exceed 5 times the upper limit of normal (ULN). In general the safety profile of ISENTRESS in subjects with hepatitis B and/or hepatitis C virus co-infection was similar to that in subjects without hepatitis B and/or hepatitis C virus co-infection, although the rates of AST and ALT abnormalities were higher in the subgroup with hepatitis B and/or hepatitis C virus co-infection for all treatment groups. In treatment-experienced subjects, Grade 2 or higher laboratory abnormalities that represent a worsening Grade from baseline of AST, ALT or total bilirubin occurred in 25%, 31% and 12%, respectively, of co-infected subjects treated with ISENTRESS as compared to 8%, 7% and 8% of all other subjects treated with ISENTRESS. In treatment-naïve subjects, Grade 2 or higher laboratory abnormalities that represent a worsening Grade from baseline of AST, ALT or total bilirubin occurred in 17%, 22% and 11%, respectively, of co-infected subjects treated with ISENTRESS as compared to 4%, 4% and 3% of all other subjects treated with ISENTRESS.

Continued on next page

Information on the Merck & Co., Inc., products listed on these pages is from the full prescribing information in use March 1, 2009. For information, please call 1-800-NSC-MERCK [1-800-672-6372].

Isentress—Cont.

6.2 Postmarketing Experience

The following adverse reactions have been identified during postapproval use of ISENTRESS. Because these reactions are reported voluntarily from a population of uncertain size, it is not always possible to reliably estimate their frequency or establish a causal relationship to drug exposure.

Psychiatric Disorders: anxiety, depression (particularly in patients with a pre-existing history of psychiatric illness), including suicidal ideation and behaviors, paranoia

Skin and Subcutaneous Tissue Disorders: rash, Stevens-Johnson syndrome

7 DRUG INTERACTIONS

7.1 Effect of Raltegravir on the Pharmacokinetics of Other Agents

Raltegravir does not inhibit (IC_{50}>100 μM) CYP1A2, CYP2B6, CYP2C8, CYP2C9, CYP2C19, CYP2D6 or CYP3A *in vitro*. Moreover, *in vitro*, raltegravir did not induce CYP1A2, CYP2B6 or CYP3A4. A midazolam drug interaction study confirmed the low propensity of raltegravir to alter the pharmacokinetics of agents metabolized by CYP3A4 *in vivo* by demonstrating a lack of effect of raltegravir on the pharmacokinetics of midazolam, a sensitive CYP3A4 substrate. Similarly, raltegravir is not an inhibitor (IC_{50}>50 μM) of the UDP-glucuronosyltransferases (UGT) tested (UGT1A1, UGT2B7), and raltegravir does not inhibit P-glycoprotein-mediated transport. Based on these data, ISENTRESS is not expected to affect the pharmacokinetics of drugs that are substrates of these enzymes or P-glycoprotein (e.g., protease inhibitors, NNRTIs, opioid analgesics, statins, azole antifungals, proton pump inhibitors and anti-erectile dysfunction agents).

In drug interaction studies, raltegravir did not have a clinically meaningful effect on the pharmacokinetics of the following: hormonal contraceptives, methadone, lamivudine, tenofovir, etravirine.

7.2 Effect of Other Agents on the Pharmacokinetics of Raltegravir

Raltegravir is not a substrate of cytochrome P450 (CYP) enzymes. Based on *in vivo* and *in vitro* studies, raltegravir is eliminated mainly by metabolism via a UGT1A1-mediated glucuronidation pathway.

Rifampin, a strong inducer of UGT1A1, reduces plasma concentrations of ISENTRESS. Therefore, the dose of ISENTRESS should be increased during coadministration with rifampin *[see Dosage and Administration (2)]*. The impact of other inducers of drug metabolizing enzymes, such as phenytoin and phenobarbital, on UGT1A1 is unknown. Coadministration of ISENTRESS with drugs that inhibit UGT1A1 may increase plasma levels of raltegravir.

Selected drug interactions are presented in Table 6 *[see Clinical Pharmacology (12.3)]*.

[See table 6 below]

8 USE IN SPECIFIC POPULATIONS

8.1 Pregnancy

Pregnancy Category C

ISENTRESS should be used during pregnancy only if the potential benefit justifies the potential risk to the fetus. There are no adequate and well-controlled studies in pregnant women. In addition, there have been no pharmacokinetic studies conducted in pregnant patients.

Developmental toxicity studies were performed in rabbits (at oral doses up to 1000 mg/kg/day) and rats (at oral doses up to 600 mg/kg/day). The reproductive toxicity study in rats was performed with pre-, peri-, and postnatal evaluation. The highest doses in these studies produced systemic exposures in these species approximately 3- to 4-fold the exposure at the recommended human dose. In both rabbits and rats, no treatment-related effects on embryonic/fetal survival or fetal weights were observed. In addition, no treatment-related external, visceral, or skeletal changes were observed in rabbits. However, treatment-related increases over controls in the incidence of supernumerary ribs were seen in rats at 600 mg/kg/day (exposures 3-fold the exposure at the recommended human dose).

Placenta transfer of drug was demonstrated in both rats and rabbits. At a maternal dose of 600 mg/kg/day in rats, mean drug concentrations in fetal plasma were approximately 1.5- to 2.5-fold greater than in maternal plasma at 1 hour and 24 hours postdose, respectively. Mean drug concentrations in fetal plasma were approximately 2% of the mean maternal concentration at both 1 and 24 hours postdose at a maternal dose of 1000 mg/kg/day in rabbits.

Antiretroviral Pregnancy Registry

To monitor maternal-fetal outcomes of pregnant patients exposed to ISENTRESS, an Antiretroviral Pregnancy Registry has been established. Physicians are encouraged to register patients by calling 1-800-258-4263.

Table 6: Selected Drug Interactions

Concomitant Drug Class: Drug Name	Effect on Concentration of Raltegravir	Clinical Comment
HIV-1-Antiviral Agents		
atazanavir	↑	Atazanavir, a strong inhibitor of UGT1A1, increases plasma concentrations of raltegravir. However, since concomitant use of ISENTRESS with atazanavir/ritonavir did not result in a unique safety signal in Phase 3 studies, no dose adjustment is recommended.
atazanavir/ritonavir	↑	Atazanavir/ritonavir increases plasma concentrations of raltegravir. However, since concomitant use of ISENTRESS with atazanavir/ritonavir did not result in a unique safety signal in Phase 3 studies, no dose adjustment is recommended.
efavirenz	↓	Efavirenz reduces plasma concentrations of raltegravir. The clinical significance of this interaction has not been directly assessed.
etravirine	↓	Etravirine reduces plasma concentrations of raltegravir. The clinical significance of this interaction has not been directly assessed.
tipranavir/ritonavir	↓	Tipranavir/ritonavir reduces plasma concentrations of raltegravir. However, since comparable efficacy was observed for this combination relative to other ISENTRESS-containing regimens in Phase 3 studies 018 and 019, no dose adjustment is recommended.
Other Agents		
omeprazole	↑	Coadministration of medicinal products that increase gastric pH (e.g., omeprazole) may increase raltegravir levels based on increased raltegravir solubility at higher pH. However, since concomitant use of ISENTRESS with proton pump inhibitors and H2 blockers did not result in a unique safety signal in Phase 3 studies, no dose adjustment is recommended.
rifampin	↓	Rifampin, a strong inducer of UGT1A1, reduces plasma concentrations of raltegravir. The recommended dosage of ISENTRESS is 800 mg twice daily during coadministration with rifampin.

8.3 Nursing Mothers

Breast-feeding is not recommended while taking ISENTRESS. In addition, it is recommended that HIV-1-infected mothers not breast-feed their infants to avoid risking postnatal transmission of HIV-1.

It is not known whether raltegravir is secreted in human milk. However, raltegravir is secreted in the milk of lactating rats. Mean drug concentrations in milk were approximately 3-fold greater than those in maternal plasma at a maternal dose of 600 mg/kg/day in rats. There were no effects in rat offspring attributable to exposure of ISENTRESS through the milk.

8.4 Pediatric Use

Safety and effectiveness of ISENTRESS in pediatric patients have not been established.

8.5 Geriatric Use

Clinical studies of ISENTRESS did not include sufficient numbers of subjects aged 65 and over to determine whether they respond differently from younger subjects. Other reported clinical experience has not identified differences in responses between the elderly and younger subjects. In general, dose selection for an elderly patient should be cautious, reflecting the greater frequency of decreased hepatic, renal, or cardiac function, and of concomitant disease or other drug therapy.

8.6 Use in Patients with Hepatic Impairment

No clinically important pharmacokinetic differences between subjects with moderate hepatic impairment and healthy subjects were observed. No dosage adjustment is necessary for patients with mild to moderate hepatic impairment. The effect of severe hepatic impairment on the pharmacokinetics of raltegravir has not been studied *[see Clinical Pharmacology (12.3)]*.

8.7 Use in Patients with Renal Impairment

No clinically important pharmacokinetic differences between subjects with severe renal impairment and healthy subjects were observed. No dosage adjustment is necessary *[see Clinical Pharmacology (12.3)]*.

10 OVERDOSAGE

No specific information is available on the treatment of overdosage with ISENTRESS. Doses as high as 1600-mg single dose and 800-mg twice-daily multiple doses were studied in healthy volunteers without evidence of toxicity. Occasional doses of up to 1800 mg per day were taken in the clinical studies of HIV-1 infected subjects without evidence of toxicity.

In the event of an overdose, it is reasonable to employ the standard supportive measures, e.g., remove unabsorbed material from the gastrointestinal tract, employ clinical monitoring (including obtaining an electrocardiogram), and institute supportive therapy if required. The extent to which ISENTRESS may be dialyzable is unknown.

11 DESCRIPTION

ISENTRESS contains raltegravir potassium, a human immunodeficiency virus integrase strand transfer inhibitor. The chemical name for raltegravir potassium is *N*-[(4-Fluorophenyl)methyl]-1,6-dihydro-5-hydroxy-1-methyl-2-[1-methyl-1-[[(5-methyl-1,3,4-oxadiazol-2-yl)carbonyl]amino]ethyl]-6-oxo-4-pyrimidinecarboxamide monopotassium salt.

The empirical formula is $C_{20}H_{20}FKN_6O_5$ and the molecular weight is 482.51. The structural formula is:

Raltegravir potassium is a white to off-white powder. It is soluble in water, slightly soluble in methanol, very slightly soluble in ethanol and acetonitrile and insoluble in isopropanol.

Each film-coated tablet of ISENTRESS for oral administration contains 434.4 mg of raltegravir potassium (as salt), equivalent to 400 mg of raltegravir (free phenol) and the following inactive ingredients: microcrystalline cellulose, lactose monohydrate, calcium phosphate dibasic anhydrous, hypromellose 2208, poloxamer 407 (contains 0.01% butylated hydroxytoluene as antioxidant), sodium stearyl fumarate, magnesium stearate. In addition, the film coating contains the following inactive ingredients: polyvinyl alcohol, titanium dioxide, polyethylene glycol 3350, talc, red iron oxide and black iron oxide.

12 CLINICAL PHARMACOLOGY

12.1 Mechanism of Action

Raltegravir is an HIV-1 antiviral drug *[see Clinical Pharmacology (12.4)]*.

12.2 Pharmacodynamics

In a monotherapy study raltegravir (400 mg twice daily) demonstrated rapid antiviral activity with mean viral load reduction of 1.66 $\log_{10}$ copies/mL by Day 10.

In the randomized, double-blind, placebo-controlled, dose-ranging trial, Protocol 005, and Protocols 018 and 019, antiviral responses were similar among subjects regardless of dose.

Effects on Electrocardiogram

In a randomized, placebo-controlled, crossover study, 31 healthy subjects were administered a single oral supra-therapeutic dose of raltegravir 1600 mg and placebo. Peak raltegravir plasma concentrations were approximately 4-fold higher than the peak concentrations following a 400 mg dose. ISENTRESS did not appear to prolong the QTc interval for 12 hours postdose. After baseline and placebo adjustment, the maximum mean QTc change was -0.4 msec (1-sided 95% upper Cl: 3.1 msec).

12.3 Pharmacokinetics

Absorption

Raltegravir is absorbed with a T_{max} of approximately 3 hours postdose in the fasted state. Raltegravir AUC and C_{max} increase dose proportionally over the dose range 100 mg to 1600 mg. Raltegravir C_{12hr} increases dose proportionally over the dose range of 100 to 800 mg and increases slightly less than dose proportionally over the dose range 100 mg to 1600 mg. With twice-daily dosing, pharmacokinetic steady state is achieved within approximately the first 2 days of dosing. There is little to no accumulation in AUC and C_{max}. The average accumulation ratio for C_{12hr} ranged from approximately 1.2 to 1.6.

The absolute bioavailability of raltegravir has not been established.

In subjects who received 400 mg twice daily alone, raltegravir drug exposures were characterized by a geometric mean $AUC_{0\text{-}12hr}$ of 14.3 μM•hr and C_{12hr} of 142 nM. Considerable variability was observed in the pharmacokinetics of raltegravir. For observed C_{12hr} in Protocols 018 and 019, the coefficient of variation (CV) for inter-subject variability = 212% and the CV for intra-subject variability = 122%.

Effect of Food on Oral Absorption

ISENTRESS may be administered with or without food. Raltegravir was administered without regard to food in the pivotal safety and efficacy studies in HIV-1-infected patients. The effect of consumption of low-, moderate- and high-fat meals on steady-state raltegravir pharmacokinetics was assessed in healthy volunteers. Administration of multiple doses of raltegravir following a moderate-fat meal (600 Kcal, 21 g fat) did not affect raltegravir AUC to a clinically meaningful degree with an increase of 13% relative to fasting. Raltegravir C_{12hr} was 66% higher and C_{max} was 5% higher following a moderate-fat meal compared to fasting. Administration of raltegravir following a high-fat meal (825 Kcal, 52 g fat) increased AUC and C_{max} by approximately 2-fold and increased C_{12hr} by 4.1-fold. Administration of raltegravir following a low-fat meal (300 Kcal, 2.5 g fat) decreased AUC and C_{max} by 46% and 52%, respectively; C_{12hr} was essentially unchanged. Food appears to increase pharmacokinetic variability relative to fasting.

Distribution

Raltegravir is approximately 83% bound to human plasma protein over the concentration range of 2 to 10 μM.

Metabolism and Excretion

The apparent terminal half-life of raltegravir is approximately 9 hours, with a shorter α-phase half-life (~1 hour) accounting for much of the AUC. Following administration of an oral dose of radiolabeled raltegravir, approximately 51 and 32% of the dose was excreted in feces and urine, respectively. In feces, only raltegravir was present, most of which is likely derived from hydrolysis of raltegravir-glucuronide secreted in bile as observed in preclinical species. Two components, namely raltegravir and raltegravir-glucuronide, were detected in urine and accounted for approximately 9 and 23% of the dose, respectively. The major circulating entity was raltegravir and represented approximately 70% of the total radioactivity; the remaining radioactivity in plasma was accounted for by raltegravir-glucuronide. Studies using isoform-selective chemical inhibitors and cDNA-expressed UDP-glucuronosyltransferases (UGT) show that UGT1A1 is the main enzyme responsible for the formation of raltegravir-glucuronide. Thus, the data indicate that the major mechanism of clearance of raltegravir in humans is UGT1A1-mediated glucuronidation.

Special Populations

Pediatric

The pharmacokinetics of raltegravir in pediatric patients has not been established.

Age

The effect of age on the pharmacokinetics of raltegravir was evaluated in the composite analysis. No dosage adjustment is necessary.

Race

The effect of race on the pharmacokinetics of raltegravir was evaluated in the composite analysis. No dosage adjustment is necessary.

Table 7: Effect of Other Agents on the Pharmacokinetics of Raltegravir

Coadministered Drug	Coadministered Drug Dose/Schedule	Raltegravir Dose/Schedule	Ratio (90% Confidence Interval) of Raltegravir Pharmacokinetic Parameters with/without Coadministered Drug; No Effect = 1.00			
			n	C_{max}	AUC	C_{min}
atazanavir	400 mg daily	100 mg single dose	10	1.53 (1.11, 2.12)	1.72 (1.47, 2.02)	1.95 (1.30, 2.92)
atazanavir/ritonavir	300 mg/100 mg daily	400 mg twice daily	10	1.24 (0.87, 1.77)	1.41 (1.12, 1.78)	1.77 (1.39, 2.25)
efavirenz	600 mg daily	400 mg single dose	9	0.64 (0.41, 0.98)	0.64 (0.52, 0.80)	0.79 (0.49, 1.28)
etravirine	200 mg twice daily	400 mg twice daily	19	0.89 (0.68, 1.15)	0.90 (0.68, 1.18)	0.66 (0.34, 1.26)
omeprazole	20 mg daily	400 mg single dose	14 (10 for AUC)	4.15 (2.82, 6.10)	3.12 (2.13, 4.56)	1.46 (1.10, 1.93)
rifampin	600 mg daily	400 mg single dose	9	0.62 (0.37, 1.04)	0.60 (0.39, 0.91)	0.39 (0.30 0.51)
rifampin	600 mg daily	400 mg twice daily when administered alone; 800 mg twice daily when administered with rifampin	14	1.62 (1.12, 2.33)	1.27 (0.94 1.71)	0.47 (0.36, 0.61)
ritonavir	100 mg twice daily	400 mg single dose	10	0.76 (0.55, 1.04)	0.84 (0.70, 1.01)	0.99 (0.70, 1.40)
tenofovir	300 mg daily	400 mg twice daily	9	1.64 (1.16, 2.32)	1.49 (1.15, 1.94)	1.03 (0.73, 1.45)
tipranavir/ritonavir	500 mg/200 mg twice daily	400 mg twice daily	15 (14 for C_{min})	0.82 (0.46, 1.46)	0.76 (0.49, 1.19)	0.45 (0.31, 0.66)

Gender

A study of the pharmacokinetics of raltegravir was performed in healthy adult males and females. Additionally, the effect of gender was evaluated in a composite analysis of pharmacokinetic data from 103 healthy subjects and 28 HIV-1 infected subjects receiving raltegravir monotherapy with fasted administration. No dosage adjustment is necessary.

Hepatic Impairment

Raltegravir is eliminated primarily by glucuronidation in the liver. A study of the pharmacokinetics of raltegravir was performed in subjects with moderate hepatic impairment. Additionally, hepatic impairment was evaluated in the composite pharmacokinetic analysis. There were no clinically important pharmacokinetic differences between subjects with moderate hepatic impairment and healthy subjects. No dosage adjustment is necessary for patients with mild to moderate hepatic impairment. The effect of severe hepatic impairment on the pharmacokinetics of raltegravir has not been studied.

Renal Impairment

Renal clearance of unchanged drug is a minor pathway of elimination. A study of the pharmacokinetics of raltegravir was performed in subjects with severe renal impairment. Additionally, renal impairment was evaluated in the composite pharmacokinetic analysis. There were no clinically important pharmacokinetic differences between subjects with severe renal impairment and healthy subjects. No dosage adjustment is necessary. Because the extent to which ISENTRESS may be dialyzable is unknown, dosing before a dialysis session should be avoided.

UGT1A1 Polymorphism

There is no evidence that common UGT1A1 polymorphisms alter raltegravir pharmacokinetics to a clinically meaningful extent. In a comparison of 30 subjects with *28/*28 genotype (associated with reduced activity of UGT1A1) to 27 subjects with wild-type genotype, the geometric mean ratio (90% CI) of AUC was 1.41 (0.96, 2.09).

Drug Interactions *[see Drug Interactions (7)]*

[See table 7 above]

12.4 Microbiology

Mechanism of Action

Raltegravir inhibits the catalytic activity of HIV-1 integrase, an HIV-1 encoded enzyme that is required for viral replication. Inhibition of integrase prevents the covalent insertion, or integration, of unintegrated linear HIV-1 DNA into the host cell genome preventing the formation of the HIV-1 provirus. The provirus is required to direct the production of progeny virus, so inhibiting integration prevents propagation of the viral infection. Raltegravir did not significantly inhibit human phosphoryltransferases including DNA polymerases α, β, and γ.

Antiviral Activity in Cell Culture

Raltegravir at concentrations of 31 ± 20 nM resulted in 95% inhibition (EC_{95}) of viral spread (relative to an untreated virus-infected culture) in human T-lymphoid cell cultures infected with the cell-line adapted HIV-1 variant H9IIIB. In addition, 5 clinical isolates of HIV-1 subtype B had EC_{95} values ranging from 9 to 19 nM in cultures of mitogen-activated human peripheral blood mononuclear cells. In a single-cycle infection assay, raltegravir inhibited infection of 23 HIV-1 isolates representing 5 non-B subtypes (A, C, D, F, and G) and 5 circulating recombinant forms (AE, AG, BF, BG, and cpx) with EC_{50} values ranging from 5 to 12 nM. Raltegravir also inhibited replication of an HIV-2 isolate when tested in CEMx174 cells (EC_{95} value = 6 nM). Additive to synergistic antiretroviral activity was observed when human T-lymphoid cells infected with the H9IIIB variant of HIV-1 were incubated with raltegravir in combination with non-nucleoside reverse transcriptase inhibitors (delavirdine, efavirenz, or nevirapine); nucleoside analog reverse transcriptase inhibitors (abacavir, didanosine, lamivudine,

Continued on next page

Information on the Merck & Co., Inc., products listed on these pages is from the full prescribing information in use March 1, 2009. For information, please call 1-800-NSC-MERCK [1-800-672-6372].

Isentress—Cont.

stavudine, tenofovir, zalcitabine, or zidovudine); protease inhibitors (amprenavir, atazanavir, indinavir, lopinavir, nelfinavir, ritonavir, or saquinavir); or the entry inhibitor enfuvirtide.

Resistance

The mutations observed in the HIV-1 integrase coding sequence that contributed to raltegravir resistance (evolved either in cell culture or in subjects treated with raltegravir) generally included an amino acid substitution at either Q148 (changed to H, K, or R) or N155 (changed to H) plus one or more additional substitutions (i.e., L74M, E92Q, T97A, E138A/K, G140A/S, V151I, G163R, H183P, Y226C/D/F/H, S230R and D232N). Amino acid substitution at Y143C/H/R is another pathway to raltegravir resistance.

Treatment-Naïve Subjects: By Week 48 in the STARTMRK trial, the primary raltegravir resistance-associated substitutions were observed in 3 (1 with Y143R and 2 with Q148H/R) of the 6 virologic failure subjects with evaluable paired genotypic data.

Treatment-Experienced Subjects: By Week 48 in the BENCHMRK trials, at least one of the 3 primary raltegravir resistance-associated substitutions, Y143C/H/R, Q148H/K/R, and N155H, was observed in 63 (64.3%) of the 98 virologic failure subjects with evaluable genotypic data from paired baseline and raltegravir treatment-failure isolates. Some (n=18) of those HIV-1 isolates harboring one or more of the 3 primary raltegravir resistance-associated substitutions were evaluated for raltegravir susceptibility yielding a median decrease of 47.3-fold (mean 73.1 ± 60.8-fold decrease, ranging from 0.9- to 200-fold) compared to baseline isolates.

13 NONCLINICAL TOXICOLOGY

13.1 Carcinogenesis, Mutagenesis, Impairment of Fertility

Carcinogenicity studies of raltegravir in mice did not show any carcinogenic potential. At the highest dose levels, 400 mg/kg/day in females and 250 mg/kg/day in males, systemic exposure was 1.8-fold (females) or 1.2-fold (males) greater than the AUC (54 μM•hr) at the 400-mg twice daily human dose. Treatment-related squamous cell carcinoma of nose/nasopharynx was observed in female rats dosed with 600 mg/kg/day raltegravir for 104 weeks. These tumors were possibly the result of local irritation and inflammation due to local deposition and/or aspiration of drug in the mucosa of the nose/nasopharynx during dosing. No tumors of the nose/nasopharynx were observed in rats dosed with 150 mg/kg/day (males) and 50 mg/kg/day (females) and the systemic exposure in rats was 1.7-fold (males) to 1.4-fold (females) greater than the AUC (54 μM•hr) at the 400-mg twice daily human dose.

No evidence of mutagenicity or genotoxicity was observed in *in vitro* microbial mutagenesis (Ames) tests, *in vitro* alkaline elution assays for DNA breakage, and *in vitro* and *in vivo* chromosomal aberration studies.

No effect on fertility was seen in male and female rats at doses up to 600 mg/kg/day which resulted in a 3-fold exposure above the exposure at the recommended human dose.

14 CLINICAL STUDIES

Description of Clinical Studies

The evidence of durable efficacy of ISENTRESS is based on the analyses of 48-week data from an ongoing, randomized, double-blind, active-control trial, STARTMRK (Protocol 021) in antiretroviral treatment-naive HIV-1 infected adult subjects and from 2 ongoing, randomized, double-blind, placebo-controlled studies, BENCHMRK 1 and BENCHMRK 2 (Protocols 018 and 019), in antiretroviral treatment-experienced HIV-1 infected adult subjects. These efficacy results were supported by the 96-week analysis of a randomized, double-blind, controlled, dose-ranging trial, Protocol 004, in antiretroviral treatment-naïve HIV-1 infected adult subjects and by the 48-week analysis of a randomized, double-blind, controlled, dose-ranging study, Protocol 005, in antiretroviral treatment-experienced HIV-1 infected adult subjects.

Treatment-Naïve Subjects

STARTMRK (Protocol 021) is a Phase 3 study to evaluate the safety and antiretroviral activity of ISENTRESS 400 mg twice daily + emtricitabine (+) tenofovir versus efavirenz 600 mg at bedtime plus emtricitabine (+) tenofovir in treatment-naïve HIV-1-infected subjects with HIV-1 RNA >5000 copies/mL. Randomization was stratified by screening HIV-1 RNA level (≤50,000 copies/mL; and >50,000 copies/mL) and by hepatitis status.

Table 8 shows the demographic characteristics of subjects in the group receiving ISENTRESS 400 mg twice daily and subjects in the comparator group.

Table 9: Outcomes by Treatment Group through Week 48

Randomized Study Protocol 021	ISENTRESS 400 mg Twice Daily (N = 281)	Efavirenz 600 mg At Bedtime (N = 282)	Difference (ISENTRESS – Efavirenz) (CI[§])
Outcome at Week 48			
Subjects with HIV-1 RNA less than 50 copies/mL	87%	82%	4.7% (-1.3%, 10.6%)
Subjects with HIV-1 RNA less than 400 copies/mL	91%	88%	3.6% (-1.5%, 8.7%)
Mean CD4 cell count change from baseline (cells/mm^3)	176	150	25.8 (5.0, 46.5)
Virologic Failure (>50 copies/mL)	6%	7%	
Never suppressed through Week 48 and on study at Week 48	2%	3%	
Rebound	5%	5%	
Discontinued study drug	7%	10%	
Reasons for Discontinuation			
Death	<1%	0%	
Adverse experiences	2%	5%	
Other*	4%	5%	

[§]The 95% CI for treatment difference is adjusted by the screening HIV RNA level (<=50,000 copies/mL vs. >50,000 copies/mL) and Hepatitis B or C (negative vs. positive)
*Other includes lack of efficacy, loss to follow-up, consent withdrawn, protocol violation and other

Table 8: Baseline Characteristics

Randomized Study Protocol 021	ISENTRESS 400 mg Twice Daily (N = 281)	Efavirenz 600 mg At Bedtime (N = 282)
Gender		
Male	81%	82%
Female	19%	18%
Race		
White	41%	44%
Black	12%	8%
Asian	13%	11%
Hispanic	21%	24%
Native American	0%	0%
Multiracial	12%	13%
Region		
Latin America	35%	34%
Southeast Asia	12%	10%
North America	29%	32%
EU/Australia	23%	23%
Age (years)		
18-64	99%	99%
≥65	1%	1%
Mean (SD)	38 (9)	37 (10)
Median (min, max)	37 (19 to 67)	36 (19 to 71)
CD4 Cell Count (cells/microL)		
Mean (SD)	219 (124)	217 (134)
Median (min, max)	212 (1 to 620)	204 (4 to 807)
Plasma HIV-1 RNA (log_{10} copies/mL)		
Mean (SD)	5 (1)	5 (1)
Median (min, max)	5 (3 to 6)	5 (4 to 6)
Plasma HIV-1 RNA (copies/mL)		
Geometric Mean	103205	106215
Median (min, max)	114000 (400 to 750000)	104000 (4410 to 750000)
History of AIDS[†]		
Yes	18%	21%
Viral Subtype		
Clade B	78%	82%
Non-Clade B[‡]	21%	17%
Baseline Plasma HIV-1 RNA		
≤100,000 copies/mL	45%	49%
>100,000 copies/mL	55%	51%
Baseline CD4 Cell Counts		
≤50 cells/mm^3	10%	11%
>50 cells/mm^3 and ≤ 200 cells/mm^3	37%	37%
>200 cells/mm^3	53%	51%
Hepatitis Status		
Hepatitis B or C Positive[§]	6%	6%

[†]Includes additional subjects identified as having a history of AIDS.
[‡]Non-Clade B Subtypes (# of subjects): Clade A (4), A/C (1), A/G (2), A1 (1), AE (29), AG (12), BF (6), C (37), D (2), F (2), F1 (5), G (2), Complex (3).
[§]Evidence of hepatitis B surface antigen or evidence of HCV RNA by polymerase chain reaction (PCR) quantitative test for hepatitis C Virus.
Notes:
ISENTRESS and Efavirenz were administered with emtricitabine (+) tenofovir
N = Number of subjects in each group.

Week 48 outcomes from Protocol 021 are shown in Table 9.
[See table 9 above]

Treatment-Experienced Subjects

BENCHMRK 1 and BENCHMRK 2 are Phase 3 studies to evaluate the safety and antiretroviral activity of ISENTRESS 400 mg twice daily in combination with an optimized background therapy (OBT), versus OBT alone, in HIV-1-infected subjects, 16 years or older, with documented resistance to at least 1 drug in each of 3 classes (NNRTIs, NRTIs, PIs) of antiretroviral therapies. Randomization was stratified by degree of resistance to PI (1PI vs. >1PI) and the use of enfuvirtide in the OBT. Prior to randomization, OBT was selected by the investigator based on genotypic/phenotypic resistance testing and prior ART history.

Table 10 shows the demographic characteristics of subjects in the group receiving ISENTRESS 400 mg twice daily and subjects in the placebo group.
[See table 10 at top of next page]

Table 11 compares the characteristics of optimized background therapy at baseline in the group receiving ISENTRESS 400 mg twice daily and subjects in the control group.
[See table 11 at bottom of next page]

Week 48 outcomes for the 699 subjects randomized and treated with the recommended dose of ISENTRESS 400 mg twice daily or placebo in the pooled BENCHMRK 1 and 2 studies are shown in Table 12.
[See table 12 at top of page B66]

The mean changes in plasma HIV-1 RNA from baseline were -2.11 log_{10} copies/mL in the group receiving ISENTRESS 400 mg twice daily and -0.96 log_{10} copies/mL for the control group.

Table 10: Baseline Characteristics

Randomized Studies Protocol 018 and 019	ISENTRESS 400 mg Twice Daily + OBT (N = 462)	Placebo + OBT (N = 237)
Gender		
Male	88%	89%
Female	12%	11%
Race		
White	65%	73%
Black	14%	11%
Asian	3%	3%
Hispanic	11%	8%
Others	6%	5%
Age (years)		
Median (min, max)	45 (16 to 74)	45 (17 to 70)
CD4+ Cell Count		
Median (min, max), cells/mm^3	119 (1 to 792)	123 (0 to 759)
$\leq$50 cells/mm^3	32%	33%
>50 and $\leq$200 cells/mm^3	37%	36%
Plasma HIV-1 RNA		
Median (min, max), $\log_{10}$ copies/mL	4.8 (2 to 6)	4.7 (2 to 6)
>100,000 copies/mL	35%	33%
History of AIDS		
Yes	92%	91%
Prior Use of ART, Median (1st Quartile, 3rd Quartile)		
Years of ART Use	10 (7 to 12)	10 (8 to 12)
Number of ART	12 (9 to 15)	12 (9 to 14)
Hepatitis Co-infection*		
No Hepatitis B or C virus	83%	85%
Hepatitis B virus only	8%	3%
Hepatitis C virus only	8%	11%
Co-infection of Hepatitis B and C virus	1%	1%
Stratum		
Enfuvirtide in OBT	38%	38%
Resistant to ≥2 PI	97%	95%

*Hepatitis B virus surface antigen positive or hepatitis C virus antibody positive.

Table 11: Characteristics of Optimized Background Therapy at Baseline

Randomized Studies Protocol 018 and 019	ISENTRESS 400 mg Twice Daily + OBT (N = 462)	Placebo + OBT (N = 237)
Number of ARTs in OBT		
Median (min, max)	4 (1 to 7)	4 (2 to 7)
Number of Active PI in OBT by Phenotypic Resistance Test*		
0	36%	41%
1 or more	60%	58%
Phenotypic Sensitivity Score (PSS)†		
0	15%	19%
1	31%	30%
2	31%	28%
3 or more	18%	20%
Genotypic Sensitivity Score (GSS)†		
0	25%	28%
1	39%	41%
2	24%	21%
3 or more	11%	10%

*Darunavir use in OBT in darunavir naïve subjects was counted as one active PI.
† The Phenotypic Sensitivity Score (PSS) and the Genotypic Sensitivity Score (GSS) were defined as the total oral ARTs in OBT to which a subject's viral isolate showed phenotypic sensitivity and genotypic sensitivity, respectively, based upon phenotypic and genotypic resistance tests. Enfuvirtide use in OBT in enfuvirtide-naïve subjects was counted as one active drug in OBT in the GSS and PSS. Similarly, darunavir use in OBT in darunavir-naïve subjects was counted as one active drug in OBT.

Treatment-emergent CDC Category C events occurred in 4% of the group receiving ISENTRESS 400 mg twice daily and 5% of the control group.
Virologic responses at Week 48 by baseline genotypic and phenotypic sensitivity score are shown in Table 13.
[See table 13 at top of next page]

16 HOW SUPPLIED/STORAGE AND HANDLING

ISENTRESS tablets 400 mg are pink, oval-shaped, film-coated tablets with "227" on one side. They are supplied as follows:
NDC 0006-0227-61 unit-of-use bottles of 60.
No. 3894
Storage and Handling
Store at 20-25°C (68-77°F); excursions permitted to 15-30°C (59-86°F). See USP Controlled Room Temperature.

17 PATIENT COUNSELING INFORMATION

[See FDA-Approved Patient Labeling.]
Patients should be informed that ISENTRESS is not a cure for HIV infection or AIDS. They should also be told that people taking ISENTRESS may still get infections or other conditions common in people with HIV (opportunistic infections). Patients should also be told that it is very important that they stay under a physician's care during treatment with ISENTRESS.
Patients should be informed that ISENTRESS does not reduce the chance of passing HIV to others through sexual contact, sharing needles, or being exposed to blood. Patients should be advised to continue to practice safer sex and to use latex or polyurethane condoms or other barrier methods to lower the chance of sexual contact with any body fluids such as semen, vaginal secretions or blood. Patients should also be advised to never re-use or share needles.
Physicians should instruct their patients that if they miss a dose, they should take it as soon as they remember. If they do not remember until it is time for the next dose, they should be instructed to skip the missed dose and go back to the regular schedule. Patients should not take two tablets of ISENTRESS at the same time.
Physicians should instruct their patients to read the Patient Package Insert before starting ISENTRESS therapy and to reread each time the prescription is renewed. Patients should be instructed to inform their physician or pharmacist if they develop any unusual symptom, or if any known symptom persists or worsens.

Manufactured and Distributed by:
MERCK & CO., INC., Whitehouse Station, NJ 08889, USA
Printed in USA
9795104
U.S. Patent Nos. US 7,169,780

Patient Information
ISENTRESS® (eye sen tris)
(raltegravir)
Tablets

Read the patient information that comes with ISENTRESS[1] before you start taking it and each time you get a refill. There may be new information. This leaflet is a summary of the information for patients. Your doctor or pharmacist can give you additional information. This leaflet does not take the place of talking with your doctor about your medical condition or your treatment.

[1] Registered trademark of MERCK & CO., Inc.

What is ISENTRESS?
- ISENTRESS is an anti-HIV (antiretroviral) medicine used for the treatment of HIV. The term HIV stands for Human Immunodeficiency Virus. It is the virus that causes AIDS (Acquired Immune Deficiency Syndrome). ISENTRESS is used along with other anti-HIV medicines. ISENTRESS will NOT cure HIV infection.
- People taking ISENTRESS may still develop infections, including opportunistic infections or other conditions that happen with HIV infection.
- Stay under the care of your doctor during treatment with ISENTRESS.
- The safety and effectiveness of ISENTRESS in children has not been studied.

ISENTRESS must be used with other anti-HIV medicines.
How does ISENTRESS work?
- ISENTRESS blocks an enzyme which the virus (HIV) needs in order to make more virus. The enzyme that ISENTRESS blocks is called HIV integrase.

Continued on next page

Information on the Merck & Co., Inc., products listed on these pages is from the full prescribing information in use March 1, 2009. For information, please call 1-800-NSC-MERCK [1-800-672-6372].

Isentress—Cont.

- When used with other anti-HIV medicines, ISENTRESS may do two things:
  1. Reduce the amount of HIV in your blood. This is called your "viral load".
  2. Increase the number of white blood cells called CD4 (T) cells.
- ISENTRESS may not have these effects in all patients.

Does ISENTRESS lower the chance of passing HIV to other people?

No. ISENTRESS does not reduce the chance of passing HIV to others through sexual contact, sharing needles, or being exposed to your blood.

- Continue to practice safer sex.
- Use latex or polyurethane condoms or other barrier methods to lower the chance of sexual contact with any body fluids. This includes semen from a man, vaginal secretions from a woman, or blood.
- Never re-use or share needles.

Ask your doctor if you have any questions about safer sex or how to prevent passing HIV to other people.

What should I tell my doctor before and during treatment with ISENTRESS?

Tell your doctor about all of your medical conditions. Include any of the following that applies to you:

- You have any allergies.
- You are pregnant or plan to become pregnant.
  - ISENTRESS is not recommended for use during pregnancy. ISENTRESS has not been studied in pregnant women. If you take ISENTRESS while you are pregnant, talk to your doctor about how you can be included in the Antiretroviral Pregnancy Registry.
- You are breast-feeding or plan to breast-feed.
  - It is recommended that HIV-infected women should not breast-feed their infants. This is because their babies could be infected with HIV through their breast milk.
  - Talk with your doctor about the best way to feed your baby.

Tell your doctor about all the medicines you take. Include the following:

- prescription medicines, including rifampin (a medicine used to treat some infections such as tuberculosis)
- non-prescription medicines
- vitamins
- herbal supplements

Know the medicines you take.

- Keep a list of your medicines. Show the list to your doctor and pharmacist when you get a new medicine.

How should I take ISENTRESS?

Take ISENTRESS exactly as your doctor has prescribed. The recommended dose is as follows:

- Take only one 400-mg tablet at a time.
- Take it twice a day.
- Take it by mouth.
- Take it with or without food.

Do not change your dose or stop taking ISENTRESS or your other anti-HIV medicines without first talking with your doctor.

IMPORTANT: Take ISENTRESS exactly as your doctor prescribed and at the right times of day because if you don't:

- The amount of virus (HIV) in your blood may increase if the medicine is stopped for even a short period of time.
- The virus may develop resistance to ISENTRESS and become harder to treat.
- Your medicines may stop working to fight HIV.
- The activity of ISENTRESS may be reduced (due to resistance).

If you fail to take ISENTRESS the way you should, here's what to do:

- If you miss a dose, take it as soon as you remember. If you do not remember until it is time for your next dose, skip the missed dose and go back to your regular schedule. Do NOT take two tablets of ISENTRESS at the same time. In other words, do NOT take a double dose.
- If you take too much ISENTRESS, call your doctor or local Poison Control Center.

Be sure to keep a supply of your anti-HIV medicines.

- When your ISENTRESS supply starts to run low, get more from your doctor or pharmacy.
- Do not wait until your medicine runs out to get more.

What are the possible side effects of ISENTRESS?

When ISENTRESS has been given with other anti-HIV drugs, the most common side effects included:

- nausea
- headache
- tiredness
- weakness
- trouble sleeping

Other side effects include rash, severe skin reactions, feeling anxious, depression, suicidal thoughts and actions, paranoia.

A condition called Immune Reconstitution Syndrome can happen in some patients with advanced HIV infection

Table 12: Outcomes by Treatment Group through Week 48

Randomized Studies Protocol 018 and 019	ISENTRESS 400 mg Twice Daily + OBT (N = 462)	Placebo + OBT (N = 237)
Outcome at Week 48		
Subjects with HIV-1 RNA less than 400 copies/mL	72%	37%
Subjects with HIV-1 RNA less than 50 copies/mL	60%	31%
Mean CD4 cell count change from baseline (cells/mm^3)	106	44
Virologic Failure (>50 copies/mL)	36%	65%
Never suppressed through Week 48 and on study at Week 48	11%	9%
Rebound	13%	8%
Non-responder by Week 48‡	12%	48%
Discontinued study drug	4%	4%
Reasons for Discontinuation		
Death	2%	2%
Adverse Experiences	<1%	<1%
Other*	2%	1%

‡The non-responders by Week 48 were defined by the protocol as those who did not achieve > 1.0 $\log_{10}$ HIV-1 RNA reduction and <400 HIV-1 RNA copies/mL starting at Week 16 or beyond.

*Other includes lack of efficacy, loss to follow-up, consent withdrawn

Table 13: Virologic Response at Week 48 by Baseline Genotypic/Phenotypic Sensitivity Score

Randomized Studies Protocol 018 and 019	Percent with HIV-1 RNA <400 copies/mL at Week 48				Percent with HIV-1 RNA <50 copies/mL at Week 48			
	n	ISENTRESS 400 mg Twice Daily + OBT (N = 462)	n	Placebo + OBT (N = 237)	n	ISENTRESS 400 mg Twice Daily + OBT (N = 462)	n	Placebo + OBT (N = 237)
Phenotypic Sensitivity Score (PSS)*								
0	69	52	44	5	69	46	44	2
1	145	72	72	32	145	57	72	28
2	142	83	66	42	142	68	66	38
3 or more	85	72	48	60	85	67	48	46
Genotypic Sensitivity Score (GSS)*								
0	115	50	66	8	115	43	66	3
1	178	79	96	38	178	63	96	35
2	111	85	49	65	111	70	49	53
3 or more	51	69	23	52	51	67	23	39

*The Phenotypic Sensitivity Score (PSS) and the Genotypic Sensitivity Score (GSS) were defined as the total oral ARTs in OBT to which a subject's viral isolate showed phenotypic sensitivity and genotypic sensitivity, respectively, based upon phenotypic and genotypic resistance tests. Enfuvirtide use in OBT in enfuvirtide-naïve subjects was counted as one active drug in OBT in the GSS and PSS. Similarly, darunavir use in OBT in darunavir-naïve subjects was counted as one active drug in OBT.

(AIDS) when combination antiretroviral treatment is started. Signs and symptoms of inflammation from opportunistic infections that a person has or had may occur as the medicines work to treat the HIV infection and help to strengthen the immune system. Call your doctor right away if you notice any signs or symptoms of an infection after starting ISENTRESS with other anti-HIV medicines.

Contact your doctor promptly if you experience unexplained muscle pain, tenderness, or weakness while taking ISENTRESS. This is because on rare occasions, muscle problems can be serious and can lead to kidney damage.

Tell your doctor if you have any side effects that bother you. These are not all the side effects of ISENTRESS. For more information, ask your doctor or pharmacist.

How should I store ISENTRESS?

- Store ISENTRESS at room temperature (68 to 77°F).
- **Keep ISENTRESS and all medicines out of the reach of children.**

General information about the use of ISENTRESS

Medicines are sometimes prescribed for conditions that are not mentioned in patient information leaflets.

- Do not use ISENTRESS for a condition for which it was not prescribed.
- Do not give ISENTRESS to other people, even if they have the same symptoms you have. It may harm them.

This leaflet gives you the most important information about ISENTRESS.

- If you would like to know more, talk with your doctor.
- You can ask your doctor or pharmacist for additional information about ISENTRESS that is written for health professionals.
- For more information go to www.ISENTRESS.com or call 1-800-622-4477.

What are the ingredients in ISENTRESS?

Active ingredient:

Each film-coated tablet contains 400 mg of raltegravir.

Inactive ingredients:

Microcrystalline cellulose, lactose monohydrate, calcium phosphate dibasic anhydrous, hypromellose 2208, poloxamer 407 (contains 0.01% butylated hydroxytoluene as antioxidant), sodium stearyl fumarate, magnesium stearate. In addition, the film coating contains the following inactive ingredients: polyvinyl alcohol, titanium dioxide, polyethylene glycol 3350, talc, red iron oxide and black iron oxide.

Manufactured and Distributed by:
MERCK & CO., Inc.
Whitehouse Station, NJ 08889, USA
Revised July 2009
9795104
U.S. Patent Nos. US 7,169,780

JANUMET® ℞
(sitagliptin/metformin HCl)
Tablets

Prescribing information for this product, which appears on pages 2041–2048 of the 2009 PDR, has been revised as follows. Please write "See Supplement B" next to the product heading.

In the PI, under the **HIGHLIGHTS** section, second paragraph, change the TM after JANUMET to ®; delete the fourth paragraph, **RECENT MAJOR CHANGES**. In the **WARNINGS AND PRECAUTIONS** section, paragraph **5.14 Macrovascular Outcomes**, sentence should read as follows: "There have been no clinical studies establishing conclusive evidence of macrovascular risk reduction with JANUMET or any other anti-diabetic drug." In the **ADVERSE REACTIONS** section under **Postmarketing Experience**, second paragraph should read as follows: "Hypersensitivity reactions include anaphylaxis, angioedema, rash, urticaria, cutaneous vasculitis, and exfoliative skin conditions including Stevens-Johnson syndrome *[see Warnings and Precautions (5.13)]*; upper respiratory tract infection; hepatic enzyme elevations; pancreatitis."

In the PPI, under **What are the possible side effects of JANUMET?**, fifth paragraph should read as follows: "The following additional side effects have been reported in general use with JANUMET or sitagliptin:

- Serious allergic reactions can happen with JANUMET or sitagliptin, one of the medicines in JANUMET. Symptoms of a serious allergic reaction may include rash, hives, and swelling of the face, lips, tongue, and throat, difficulty breathing or swallowing. If you have an allergic reaction, stop taking JANUMET and call your doctor right away. Your doctor may prescribe a medication to treat your allergic reaction and a different medication for your diabetes.
- Elevated liver enzymes
- Inflammation of the pancreas."

Revisions based on 9794108, issued March 2009.

JANUVIA® ℞
(sitagliptin)
Tablets

Prescribing information for this product, which appears on pages 2048–2054 of the 2009 PDR, has been revised as follows. Please write "See Supplement B" next to the product heading.

In the PI, section **6.2 Postmarketing Experience**, second paragraph, add "cutaneous vasculitis," after the word "urticaria," and add "pancreatitis" to the end of the sentence.

In the PPI, under **What are the possible side effects of JANUVIA?**, fourth paragraph under "The following additional side effects have been reported in general use with JANUVIA," add a third bullet to read "Inflammation of the pancreas."

Revisions based on 9762707, issued March 2009.

NOROXIN® ℞
[nor-AHK-sin]
(norfloxacin)
400 mg Tablets

Prescribing Information for this product, which appears on pages 2069–2073 of the 2009 PDR, has been revised as follows. Please write "See Supplement B" next to the product heading.

Attached for your information is the new Medication Guide for NOROXIN. Please replace the previous circular # 9868300 (USPPI) for this product with this current circular # 9884401. Please insert this on pages 2069–2072 of the 2009 PDR, where Prescribing Information for this product appears.

MEDICATION GUIDE
NOROXIN® [nor-AHK-sin]
(norfloxacin)
400 mg Tablets

Read the Medication Guide that comes with NOROXIN* before you start taking it and each time you get a refill. There may be new information. This Medication Guide does not take the place of talking to your healthcare provider about your medical condition or your treatment.

What is the most important information I should know about NOROXIN?

NOROXIN belongs to a class of antibiotics called fluoroquinolones. NOROXIN can cause side effects that may be serious or even cause death. If you develop any of the following serious side effects, get medical help right away. Talk with your healthcare provider about whether you should continue to take NOROXIN.

Tendon rupture or swelling of the tendon (tendinitis).

- Tendons are tough cords of tissue that connect muscle to bones.
- Pain, swelling, tears and inflammation of tendons including the back of the ankle (Achilles), shoulder, hand, or other tendon sites can happen in people of all ages who take fluoroquinolone antibiotics, including NOROXIN. The risk of getting tendon problems is higher if you:
  - are over 60 years of age
  - are taking steroids (corticosteroids)
  - have had a kidney, heart or lung transplant
- Swelling of the tendon (tendinitis) and tendon rupture (breakage) have also happened in patients who take fluoroquinolones who do not have the above risk factors.
- Other reasons for tendon ruptures can include:
  - physical activity or exercise
  - kidney failure
  - tendon problems in the past, such as in people with rheumatoid arthritis (RA).
- Call your healthcare provider right away at the first sign of tendon pain, swelling or inflammation. Stop taking NOROXIN until tendinitis or tendon rupture has been ruled out by your healthcare provider. Avoid exercise and using the affected area. The most common area of pain and swelling is the Achilles tendon at the back of your ankle. This can also happen with other tendons. Talk to your healthcare provider about the risk of tendon rupture with continued use of NOROXIN. You may need a different antibiotic that is not a fluoroquinolone to treat your infection.
- Tendon rupture can happen while you are taking or after you have finished taking NOROXIN. Tendon ruptures have happened up to several months after patients have finished taking their fluoroquinolone.
- Get medical help right away if you get any of the following signs or symptoms of a tendon rupture:
  - hear or feel a snap or pop in a tendon area
  - bruising right after an incident in a tendon area
  - unable to move the affected area or bear weight
- See the section **"What are the possible side effects of NOROXIN?"** for more information about side effects.

What is NOROXIN?

NOROXIN is a fluoroquinolone antibiotic medicine used in adults to treat certain infections caused by certain germs called bacteria. It is not known if NOROXIN is safe and works in children under 18 years of age. Children have a higher chance of getting bone and joint (musculoskeletal) problems while taking NOROXIN.

Sometimes infections are caused by viruses rather than by bacteria. Examples include viral infections in the sinuses and lungs, such as the common cold or flu. Antibiotics including NOROXIN do not kill viruses.

Call your healthcare provider if you think your condition is not getting better while you are taking NOROXIN.

Who should not take NOROXIN?

Do not take NOROXIN if you:

- have ever had a severe allergic reaction to an antibiotic known as a fluoroquinolone, or are allergic to any of the ingredients in NOROXIN. Ask your healthcare provider if you are not sure. See the list of ingredients in NOROXIN at the end of this Medication Guide.
- have had tendinitis or tendon rupture with the use of NOROXIN or another fluoroquinolone antibiotic.

What should I tell my healthcare provider before taking NOROXIN?

See **"What is the most important information I should know about NOROXIN?"**

Tell your healthcare provider about all your medical conditions, including if you:

- have tendon problems
- have central nervous system problems (such as epilepsy)
- have nerve problems
- have myasthenia gravis
- have or anyone in your family has an irregular heartbeat, especially a condition called "QTc prolongation"
- have low potassium (hypokalemia)
- have a slow heartbeat called bradycardia
- have a history of seizures
- have kidney problems. You may need a lower dose of NOROXIN if your kidneys do not work well.
- have rheumatoid arthritis (RA) or other history of joint problems
- are pregnant or planning to become pregnant. It is not known if NOROXIN will harm your unborn child.
- are breast-feeding or planning to breast-feed. It is not known if NOROXIN passes into breast milk. You and your healthcare provider should decide whether you will take NOROXIN or breast-feed.

Tell your healthcare provider about all the medicines you take, including prescription and nonprescription medicines, vitamins, and herbal and dietary supplements. NOROXIN and other medicines[†] can affect each other causing side effects. Especially tell your healthcare provider if you take:

- an NSAID (Non-Steroidal Anti-Inflammatory Drug). Many common medicines for pain relief are NSAIDs. Taking an NSAID while you take NOROXIN or other fluoroquinolones may increase your risk of central nervous system effects and seizures. See **"What are the possible side effects of NOROXIN?"**
- glyburide (Micronase, Glynase, Diabeta, Glucovance). See **"What are the possible side effects of NOROXIN?"**
- a blood thinner (warfarin, Coumadin, Jantoven)
- a medicine to control your heart rate or rhythm (antiarrhythmics). See **"What are the possible side effects of NOROXIN?"**
- an anti-psychotic medicine
- a tricyclic antidepressant
- erythromycin
- a water pill (diuretic)
- a steroid medicine. Corticosteroids taken by mouth or by injection may increase the chance of tendon injury.
- probenecid (Probalan, Col-probenecid)
- cyclosporine (Gengraf, Sandimmune, Neoral)
- products that contain caffeine
- clozapine (Fazaclo ODT, Clozaril)
- ropinirole (Requip, Requip XL)
- tacrine (Cognex)
- tizanidine (Zanaflex)
- theophylline (Theo-24, Elixophyllin, Theochron, Uniphyl, Theolair)
- cisapride (Propulsid)
- certain medicines may keep NOROXIN from working correctly. Take NOROXIN either 2 hours before or 2 hours after taking these products:
  - an antacid, multivitamin or other product that has iron or zinc
  - sucralfate (Carafate)
  - didanosine (Videx, Videx EC)
- You should not take the medicine nitrofurantoin (furadantin, macrodantin, macrobid) while taking NOROXIN.

Ask your healthcare provider if you are not sure if your medicine is listed above.

Know the medicines you take. Keep a list of your medicines and show it to your healthcare provider and pharmacist when you get a new medicine.

How should I take NOROXIN?

- Take NOROXIN exactly as prescribed by your healthcare provider.
- NOROXIN is usually taken every 12 hours for patients with normal kidney function.
- Take NOROXIN with a glass of water.
- Drink plenty of fluids while taking NOROXIN.
- Take NOROXIN at least one hour before or 2 hours after a meal or having milk or other dairy products.
- Do not skip any doses, or stop taking NOROXIN even if you begin to feel better, until you finish your prescribed treatment, unless:
  - you have tendon effects (see **"What is the most important information I should know about NOROXIN?"**),
  - you have a serious allergic reaction (see **"What are the possible side effects of NOROXIN?"**), or
  - your healthcare provider tells you to stop. This will help make sure that all of the bacteria are killed and lower the chance that the bacteria will become resistant to NOROXIN. If this happens, NOROXIN and other antibiotic medicines may not work in the future.
- If you miss a dose of NOROXIN, take it as soon as you remember. Do not take two doses of NOROXIN at the same time. Do not take more than 2 doses of NOROXIN in one day.
- If you take too much, call your healthcare provider or get medical help immediately.

What should I avoid while taking NOROXIN?

- NOROXIN can make you feel dizzy and lightheaded. Do not drive, operate machinery, or do other activities that require mental alertness or coordination until you know how NOROXIN affects you.
- Avoid sunlamps and tanning beds, and try to limit your time in the sun. NOROXIN can make your skin sensitive to the sun (photosensitivity) and the light from sunlamps and tanning beds. You could get severe sunburn, blisters or swelling of your skin. If you get any of these symptoms while taking NOROXIN, call your healthcare provider right away. You should use sunscreen and wear a hat and clothes that cover your skin if you have to be in sunlight.

What are the possible side effects of NOROXIN?

NOROXIN can cause side effects that may be serious or even cause death. See **"What is the most important information I should know about NOROXIN?"**

Other serious side effects of NOROXIN include:

- **Central Nervous System Effects.** Seizures have been reported in people who take fluoroquinolone antibiotics including NOROXIN. Tell your healthcare provider if you have a history of seizures. Ask your healthcare provider whether taking NOROXIN will change your risk of having a seizure.

  Central Nervous System (CNS) side effects may happen as soon as after taking the first dose of NOROXIN. Talk to your healthcare provider right away if you get any of these side effects, or other changes in mood or behavior:
  - feel lightheaded

Continued on next page

Information on the Merck & Co., Inc., products listed on these pages is from the full prescribing information in use March 1, 2009. For information, please call 1-800-NSC-MERCK [1-800-672-6372].

Noroxin—Cont.

- seizures
- hear voices, see things, or sense things that are not there (hallucinations)
- feel restless
- tremors
- feel anxious or nervous
- confusion
- feel more suspicious (paranoia)

- **Serious allergic reactions.** Allergic reactions can happen in people who take fluoroquinolones, including NOROXIN, even after only one dose. Stop taking NOROXIN and get emergency medical help right away if you get any of the following symptoms of a severe allergic reaction:
  - hives
  - trouble breathing or swallowing
  - swelling of the lips, tongue, face
  - throat tightness, hoarseness
  - rapid heartbeat
  - faint
  - yellowing of the skin or eyes. Stop taking NOROXIN and tell your healthcare provider right away if you get yellowing of your skin or white part of your eyes, or if you have dark urine. These can be signs of a serious reaction to NOROXIN (a liver problem).
- **Skin rash.** Skin rash may happen in people taking NOROXIN, even after only one dose. Stop taking NOROXIN at the first sign of a skin rash and call your healthcare provider. Skin rash may be sign of a more serious reaction to NOROXIN.
- **Serious heart rhythm changes (QTc prolongation and torsade de pointes).** Tell your healthcare provider right away if you have a change in your heart beat (a fast or irregular heartbeat), or if you faint. NOROXIN may cause a rare heart problem known as prolongation of the QTc interval. This condition can cause an abnormal heartbeat and can be very dangerous. The chances of this happening are higher in people:
  - who are elderly
  - with a family history of prolonged QTc interval
  - with low blood potassium (hypokalemia)
  - who take certain medicines to control heart rhythm (antiarrhythmics)
- **Worsening of myasthenia gravis symptoms.** Fluoroquinolones, including NOROXIN, may worsen the signs of myasthenia gravis. This may cause trouble breathing which may be life-threatening. Tell your healthcare provider right away if you get this symptom.
- **Intestine infection (Pseudomembranous colitis).** Pseudomembranous colitis can happen with most antibiotics, including NOROXIN. Call your healthcare provider right away if you get watery diarrhea, diarrhea that does not go away, or bloody stools. You may have stomach cramps and a fever. Pseudomembranous colitis can happen 2 or more months after you have finished your antibiotic.
- **Changes in sensation and possible nerve damage (Peripheral Neuropathy).** Damage to the nerves in arms, hands, legs, or feet can happen in people taking fluoroquinolones, including NOROXIN. Talk with your healthcare provider right away if you get any of the following symptoms of peripheral neuropathy in your arms, hands, legs, or feet:
  - pain
  - burning
  - tingling
  - numbness
  - weakness

  NOROXIN may need to be stopped to prevent permanent nerve damage.
- **Low blood sugar (hypoglycemia).** People taking NOROXIN and other fluoroquinolone medicines with the oral anti-diabetes medicine glyburide (Micronase, Glynase, Diabeta, Glucovance) can get low blood sugar (hypoglycemia) which can sometimes be severe. Tell your healthcare provider if you get low blood sugar while taking NOROXIN. Your antibiotic medicine may need to be changed.
- **Sensitivity to sunlight (photosensitivity).** See **"What should I avoid while taking NOROXIN?"**

The most common side effects of NOROXIN include:

- dizziness
- nausea
- diarrhea
- heartburn
- headache
- stomach (abdominal) cramping
- weakness
- changes in certain liver function tests

These are not all the possible side effects of NOROXIN. Tell your healthcare provider about any side effect that bothers you or that does not go away.

Call your healthcare provider for medical advice about side effects. You may report side effects to FDA at 1-800-FDA-1088.

How should I store NOROXIN?

Store between 59-86°F (15-30°C).

Keep container closed tightly.

Keep NOROXIN and all medicines out of the reach of children.

General Information about NOROXIN

Medicines are sometimes prescribed for purposes other than those listed in a Medication Guide. Do not use NOROXIN for a condition for which it is not prescribed. Do not give NOROXIN to other people, even if they have the same symptoms that you have. It may harm them.

This Medication Guide summarizes the most important information about NOROXIN. If you would like more information about NOROXIN, talk with your healthcare provider. You can ask your healthcare provider or pharmacist for information about NOROXIN that is written for healthcare professionals. For more information call 1-800-622-4477.

What are the ingredients in NOROXIN?

Active ingredient: norfloxacin

Inactive ingredients: cellulose, croscarmellose sodium, hydroxypropyl cellulose, hydroxypropyl methylcellulose, magnesium stearate and titanium dioxide

Revised October 2008

Manufactured for:
MERCK & CO., INC., Whitehouse Station, NJ 08889, USA
Manufactured by:
Merck Sharp & Dohme (Italia) S.p.A.
Via Emilia, 21
27100 Pavia, Italy
9884401

This Medication Guide has been approved by the U.S. Food and Drug Administration.

* Registered trademark of MERCK & CO., Inc.

† Other brands listed are the trademarks of their respective owners and are not trademarks of MERCK & CO., Inc.

ROTATEQ® ℞
(Rotavirus Vaccine, Live, Oral, Pentavalent)
Oral Solution

Prescribing information for this product, which appears on pages 2110–2114 of the 2009 PDR, has been revised as follows. Please write "See Supplement B" next to the product heading.

In the **Drug Interactions** section, under *Concomitant Vaccine Administration*, section 7.1, **delete** Paragraph 2 and **replace** it with: "There was no evidence for reduced antibody responses to the vaccines that were concomitantly administered with RotaTeq."

In the **HOW SUPPLIED/STORAGE AND HANDLING** section (section 16), **delete:**

"NDC 0006-4047-31 package of 1 individually pouched single-dose tube"

Change Issue date to: February 2009.

Change Circular number to: 9714308.

In the *Patient Information* section, **delete: "What is RotaTeq and How Does it Work?** RotaTeq helps protect against an infection that nearly all children get called rotavirus. Rotavirus can cause fever, vomiting, and diarrhea which may be severe and can lead to loss of body fluids (dehydration), hospitalization and even death in some children. RotaTeq may not fully protect all children that get the vaccine, and if your child already has the virus it will not help them."

Replace with: **"What is RotaTeq?** RotaTeq is an oral vaccine used to help prevent rotavirus infection in children. Rotavirus infection can cause fever, vomiting, and diarrhea that can be severe and can lead to loss of body fluids (dehydration), hospitalization and even death in some children. RotaTeq may not fully protect all children that get the vaccine, and if your child already has the virus it will not help them."

In the *Patient Information* section, under **What other important information should I know?, delete:** "Call your child's doctor right away if, following any dose of RotaTeq, your child has vomiting, diarrhea, severe stomach pain, blood in their stool or change in their bowel movements as these may be signs of intussusception. Intussusception is a serious and life-threatening event that occurs when a part of the intestine gets blocked or twisted and it requires immediate medical attention. Intussusception can occur when no vaccine has been given and the cause is usually unknown.

Since FDA approval, reports of infants with intussusception have been received by Vaccine Adverse Event Reporting System (VAERS). Intussusception occurred days and sometimes weeks after vaccination. Some of these infants required hospitalization and surgery on their intestine or a special enema to treat this problem. Death due to intussusception has also occurred.

Before FDA approval, RotaTeq was studied in 35,000 infants and no increased risk of intussusception was found compared to 35,000 infants who did not receive RotaTeq. Contact your doctor if your child has any symptoms of intussusception, even if it has been several weeks since the last vaccine dose."

Replace with: "Call your child's doctor or go to the emergency department right away if, following any dose of RotaTeq, your child has vomiting, diarrhea, severe stomach pain, blood in their stool or change in their bowel movements. These symptoms may be signs of a serious and life-threatening problem, called intussusception, that happens when a part of the intestine gets blocked or twisted. Intussusception can happen even when no vaccine has been given and the cause is usually unknown.

Since FDA approval, reports of infants with intussusception have been received by Vaccine Adverse Event Reporting System (VAERS). Intussusception occurred days and sometimes weeks after vaccination. Some infants needed hospitalization, surgery on their intestines, or a special enema to treat this problem. Death due to intussusception has occurred.

Contact your doctor or go to the emergency department right away if your child has any symptoms of intussusception, even if it has been several weeks since the last vaccine dose."

Under **Who should not receive RotaTeq?**, in the first sentence, **replace** "the vaccine" with "RotaTeq".

Under **What are the possible side effects of RotaTeq, delete** the second paragraph and **replace** it with 2 paragraphs:

"Other reported side effects include: hives; Kawasaki disease (a serious condition that can affect the heart; symptoms may include fever, rash, red eyes, red mouth, swollen glands, swollen hands and feet and, if not treated, death can occur).

Call your doctor right away if your child has any side effects that concern you or seem to get worse."

Under **What are the possible side effects of RotaTeq, delete** paragraph 5: "If your child seems to be having any side effects that are not mentioned in this leaflet, please call your doctor or other health care provider. If the condition continues or worsens, you should seek medical attention."

In paragraph 6, **change** "health care provider" to "doctor".

Under **How is RotaTeq given?**, in paragraph 2, **change** "health care provider" to "doctor".

Under **What should I do if my child misses a dose of RotaTeq?**, in sentences 1 and 3, **change** "health care provider" to "doctor".

Under **What else should I know about RotaTeq?, change** sentence 2 to read: "If you have any questions or concerns about RotaTeq, talk to your doctor."

Change Issue date to: February 2009.

Change Circular number to: 9714308.

VARIVAX® ℞
Varicella Virus Vaccine Live

Prescribing information for this product, which appears on pages 2133–2136 of the 2009 PDR, has been revised as follows. Please write "See Supplement B" next to the product heading.

Patient Information about
VARIVAX® (pronounced "VAR ih vax")
Generic name: Varicella Virus Vaccine Live

This is a summary of information about VARIVAX[1]. You should read it before you or your child get the vaccine. If you have any questions about the vaccine after reading this leaflet, you should ask your health care provider. This is a summary only. It does not take the place of talking about VARIVAX with your doctor, nurse, or other health care provider. Only your health care provider can decide if VARIVAX is right for you or your child.

[1] Registered trademark of MERCK & CO., Inc.

What is VARIVAX and how does it work?

VARIVAX is also known as Varicella Virus Vaccine Live. It is a live virus vaccine that is given as a shot. It is meant to help prevent chickenpox. Chickenpox is sometimes called varicella (pronounced VAR ih sell a).

VARIVAX contains a weakened form of chickenpox virus.

VARIVAX works by helping the immune system protect you or your child from getting chickenpox.

VARIVAX may not protect everyone who gets it.

VARIVAX does not treat chickenpox once you or your child have it.

What do I need to know about chickenpox?

Chickenpox is an illness that occurs most often in children who are 5 to 9 years old. It can be passed to others. The illness can include headache, fever, and general discomfort. Then an itchy rash occurs, which can turn into blisters. The most common complication is that the blisters can get infected. Less common but very serious complications can occur. These include pneumonia, inflammation of the brain, Reye's syndrome (which affects the liver and the brain), and death. Severe disease and serious complications are more likely to occur in adolescents and adults.

Who should not get VARIVAX?

Do not get VARIVAX if you or your child:
- are allergic to any of its ingredients. (This includes gelatin or neomycin. See the ingredient list at the end of this leaflet.)
- have a weakened immune system, such as an immune deficiency, an inherited immune disorder, leukemia, lymphoma, or HIV/AIDS.
- take high doses of steroids by mouth or in a shot.
- have active tuberculosis that is not treated.
- have a fever.
- are pregnant or plan to get pregnant within the next three months.

What should I tell my health care provider before getting VARIVAX?

Tell your health care provider if you or your child:
- have or have had any medical problems.
- have received blood or plasma transfusions or human serum globulin within the last 5 months.
- take any medicines. (This includes non-prescription medicines and dietary supplements.)
- have any allergies. (This includes allergies to neomycin or gelatin.)
- had an allergic reaction to any other vaccine.
- are pregnant or plan to become pregnant within the next three months.
- are breast-feeding.

How is VARIVAX given?

VARIVAX is given as a shot to people who are 12 months old or older. If your child is 12 months to 12 years old and your doctor gives a second dose, the second dose must be given at least 3 months after the first shot.

A second dose should be given to those who first get the vaccine when they are 13 years old or older. This second dose should be given 4 to 8 weeks after the first dose.

Your doctor or health care provider will use the official recommendations to decide the number of shots needed and when to get them.

If a dose is missed, your health care provider will let you know when you should have it.

What should you or your child avoid when getting VARIVAX?

Do not take aspirin or aspirin-containing products for 6 weeks after getting VARIVAX.

It is rare, but possible, that once you have the vaccine, you could spread the chickenpox virus to others. Whenever possible, try to avoid contact with certain groups of people for up to six weeks after getting the vaccine. This is because the disease for these groups may be quite serious. These groups include:
- people who have a weakened immune system.
- pregnant women who have never had chickenpox.
- newborn babies whose mothers have never had chickenpox.

Tell your doctor or healthcare provider if you or your child expect to have contact with someone who falls into one of these groups.

What are the possible side effects of VARIVAX?

The most common side effects reported after taking VARIVAX are:
- Fever
- Pain, swelling, itching, or redness at the site of the shot
- Chickenpox-like rash on the body or at the site of the shot
- Irritability

Other less common side effects have also been reported.
- Tingling of the skin
- Shingles

Tell your healthcare provider if you have any of the following problems within a short time after getting VARIVAX because they may be signs of an allergic reaction:
- Shortness of breath or wheezing
- Rash or hives

Other side effects have been reported. Some of them were serious. These include difficulty walking, severe skin disorders, and skin infection. Rarely, swelling of the brain, stroke, inflammation of the lungs (known as pneumonia or pneumonitis), and seizures with or without a fever have been reported. It is not known if these rare side effects are related to the vaccine.

Your doctor has a more complete list of side effects for VARIVAX.

Tell your doctor or health care provider if you or your child have any new or unusual symptoms after getting VARIVAX

You may also report any adverse reactions to your doctor or your child's doctor or directly to the Vaccine Adverse Event Reporting System (VAERS). The VAERS toll-free number is 1-800-822-7967 or report online to www.vaers.hhs.gov.

What are the ingredients of VARIVAX?

Active Ingredient: a weakened form of chickenpox virus.

Inactive Ingredients: sucrose, hydrolyzed gelatin, sodium chloride, monosodium L-glutamate, sodium phosphate dibasic, potassium phosphate monobasic, potassium chloride, residual components of MRC-5 cells including DNA and protein, sodium phosphate monobasic, EDTA, neomycin, fetal bovine serum.

What else should I know about VARIVAX?

If you get VARIVAX while you are pregnant, please call 1-800-986-8999. Or, you can have your health care provider call.

This leaflet summarizes important information about VARIVAX.

If you would like more information, talk to your health care provider, visit the web site at www.merckvaccines.com, or call 1-800-Merck-90.

Rx Only

Issued June 2009

Dist. by:

MERCK & CO., INC., Whitehouse Station, NJ 08889, USA

Revisions based on 9904700.

ZOSTAVAX® ℞

[ZOS tah vax]

(Zoster Vaccine Live)

Prescribing Information for this product, which appears on pages 2145–2148 of the 2009 PDR, has been revised as follows. Please write "See Supplement B" next to the product heading.

In section **6.2 Post-Marketing Experience**, after the first paragraph, insert the following text:

Skin and subcutaneous tissue disorders: rash

Musculoskeletal and connective tissue disorders: arthralgia; myalgia

In the *General disorders and administration site conditions* text, add "injection-site rash; injection-site urticaria;" so the text reads:

General disorders and administration site conditions: injection-site rash; injection-site urticaria; pyrexia; transient injection-site lymphadenopathy

After *Immune system disorders* text, delete "*Skin and subcutaneous tissue disorders:* rash" as it's been moved up.

Revisions based on 9815607, issued December 2008, and 9815608, issued April 2009.

ZOSTAVAX® (Zoster Vaccine Live)

Patient Information for this product, which appears on page 2148 of the 2009 PDR, has been revised as follows. Please write "See Supplement B" next to the product heading.

In the **"What are the possible side effects of ZOSTAVAX?"** section, after the most common side effects section, revise the additional side effect information section to read as follows:

The following additional side effects have been reported in general use with ZOSTAVAX:
- allergic reactions, which may be serious and may include difficulty in breathing or swallowing. If you have an allergic reaction, call your doctor right away.
- fever
- hives at the injection site
- joint pain
- muscle pain
- rash
- rash at the injection site
- swollen glands near the injection site (that may last a few days to a few weeks)

Revisions based on 9815607, issued December 2008, and 9815608, issued April 2009.

In the **PDR** annual,
the **Brand and Generic Name Index**
alphabetizes drugs under both
brand and generic names.

Merck/Schering-Plough Pharmaceuticals

PO BOX 1000
UG4B–75
351 N. SUMNEYTOWN PIKE
NORTH WALES, PA 19454

For Product and Service Information, Medical Information, and Adverse Drug Experience Reporting:

Call: Merck/Schering-Plough National Service Center
Monday through Friday, 8:00 AM to 6:00 PM (ET)
866-637-2501
Fax: 800-637-2568

For 24-hour emergency information, healthcare professionals should call:
Merck/Schering-Plough National Service Center at
866-637-2501

For Product Ordering,
Call: Order Management Center
Monday through Friday, 8:00 AM to 7:00 PM (ET)
800-637-2579

VYTORIN® ℞

(ezetimibe/simvastatin)
Tablets

Prescribing information for this product, which appears on pages 2149–2157 of the 2009 PDR, has been revised as follows. Please write "See Supplement B" next to the product heading.

In section **6 ADVERSE REACTIONS,** under **6.2 Post-Marketing Experience**, in the second paragraph, after "alopecia;" insert "erythema multiforme;".

In section **8 USE IN SPECIFIC POPULATIONS**, under **8.1 Pregnancy/*VYTORIN***, in the first paragraph, delete the third sentence.

In section **12 CLINICAL PHARMACOLOGY**, under **12.3 Pharmacokinetics/*Drug Interactions***, in the first paragraph, replace the second sentence with "No specific pharmacokinetic drug interaction studies with VYTORIN have been conducted other than the following study with NIASPAN (Niacin extended-release tablets)." and insert the following paragraph:

"**Niacin:** The effect of VYTORIN (10/20 mg daily for 7 days) on the pharmacokinetics of NIASPAN extended-release tablets (1000 mg for 2 days and 2000 mg for 5 days following a low-fat breakfast) was studied in healthy subjects. The mean C_{max} and AUC of niacin increased 9% and 22%, respectively. The mean C_{max} and AUC of nicotinuric acid increased 10% and 19%, respectively (N=13). In the same study, the effect of NIASPAN on the pharmacokinetics of VYTORIN was evaluated (N=15). While concomitant NIASPAN decreased the mean C_{max} of total ezetimibe (1%), and simvastatin (2%), it increased the mean C_{max} of simvastatin acid (18%). In addition, concomitant NIASPAN increased the mean AUC of total ezetimibe (26%), simvastatin (20%), and simvastatin acid (35%)."

Also in section **12 CLINICAL PHARMACOLOGY**, under **12.3 Pharmacokinetics/*Drug Interactions***, in Table 6, under the row labeled **No dosing adjustments required for the following**, delete the entire second row (Niacin extended release).

In section **13 NONCLINICAL TOXICOLOGY**, under **13.2 Animal Toxicology and/or Pharmacology/*Ezetimibe***, in the fourth paragraph, second sentence, replace "ethyl" with "ethinyl".

Revisions based on 9619513, issued May 2009.

In the **Patient Information about VYTORIN**, under **What are the possible side effects of VYTORIN?**, in the bulleted copy of the third paragraph, after "hives;" insert "raised red rash, sometimes with target-shaped lesions;".

Revisions based on 9619513, issued May 2009.

ZETIA® ℞

(ezetimibe)
Tablets

Prescribing information for this product, which appears on pages 2157–2163 of the 2009 PDR, has been revised as follows. Please write "See Supplement B" next to the product heading.

In the **6 ADVERSE REACTIONS** section, under **6.2 Post-Marketing Experience**, in the third paragraph, after "urticaria;" add "erythema multiforme;" and after "hepatitis;" add "abdominal pain;" and after "depression;" add "headache;".

In the **12 CLINICAL PHARMACOLOGY** section, under **12.3 Pharmacokinetics**, after "*Pediatric Patients:*", replace "[See Pediatric Use (8.4).]" with "[Use in Specific Populations (8.4).]".

Continued on next page

Zetia—Cont.

In the **16 HOW SUPPLIED/STORAGE AND HANDLING** section, add as a new 5th paragraph, "**NDC 66582-414-76** bottles of 5000".
Revisions based on 32147046T, REV 20, issued May 2009.
In the **Patient Information about ZETIA**, under **What are the possible side effects of ZETIA?**, in the third paragraph, after "hives;" add "raised red rash, sometimes with target-shaped lesions;" and after "liver problems;" add "stomach pain;" and after "depression;" add "headache;".
Revisions based on 29480877T, Rev 20, issued May 2009.

Schering Corporation

A wholly-owned subsidiary of Schering-Plough Corporation
GALLOPING HILL ROAD
KENILWORTH, NJ 07033

Direct Inquiries to:
(908) 298-4000
CUSTOMER SERVICE:
(800) 222-7579
FAX: (908) 595-3729
For Medical Information Contact:
Schering Laboratories
Global Medical Information
PO Box 599
2000 Galloping Hill Road
Kenilworth, NJ 07033
(800) 526-4099
FAX: (800) 255-3732
email: sp.gdis@spcorp.com

PEGINTRON® ℞

[*pĕg-ĭn-trŏn*]
(Peginterferon alfa-2b)
Powder for Subcutaneous Injection

Prescribing information for this product, which appears on pages 2903–2914 of the 2009 PDR, has been completely revised as follows. Please write "See Supplement B" next to the product heading.

PRODUCT INFORMATION
HIGHLIGHTS OF PRESCRIBING INFORMATION
These highlights do not include all the information needed to use PegIntron safely and effectively. See full prescribing information for PegIntron.
PegIntron (Peginterferon alfa-2b) Injection, Powder for Solution for Subcutaneous Use
Initial U.S. Approval: 2001

WARNING: RISK OF SERIOUS DISORDERS AND RIBAVIRIN-ASSOCIATED EFFECTS
See full prescribing information for complete boxed warning.
- **May cause or aggravate fatal or life-threatening neuropsychiatric, autoimmune, ischemic, and infectious disorders. Monitor closely and withdraw therapy with persistently severe or worsening signs or symptoms of the above disorders. (5)**

Use with Ribavirin
- **Ribavirin may cause birth defects and fetal death; avoid pregnancy in female patients and female partners of male patients. (5.1)**
- **Ribavirin is a potential carcinogen. (5.1, 13.1)**

------------RECENT MAJOR CHANGES------------

Indications and Usage, Chronic Hepatitis C (1.1) [3/2009]
Dosage and Administration, Combination Therapy (2.1) [5/2009]
Dosage and Administration, PegIntron Monotherapy (2.2) [3/2009]
Dosage and Administration, Dose Reduction (2.3) [5/2009]
Dosage and Administration, Discontinuation of Dosing (2.4) [3/2009]
Contraindications (4) [12/2008]
Warnings and Precautions, Laboratory Tests (5.15) [12/2008]
Warnings and Precautions, Impact on Growth – Pediatric Use (5.18) [12/2008]

------------INDICATIONS AND USAGE------------

PegIntron is an antiviral indicated for
- Combination therapy with REBETOL (ribavirin): Chronic Hepatitis C (CHC) in patients ≥3 years with compensated liver disease. (1.1)
  Patients with the following characteristics are less likely to benefit from re-treatment after failing a course of therapy: previous nonresponse, previous pegylated interferon treatment, significant bridging fibrosis or cirrhosis, and genotype 1 infection. (1.1)
- Monotherapy: CHC in patients (≥18 years) with compensated liver disease previously untreated with interferon alpha. (1.1)

------------DOSAGE AND ADMINISTRATION------------

- PegIntron is administered by subcutaneous injection.

[See table below]
- Dose reduction is recommended in patients experiencing certain adverse reactions or renal dysfunction. (2.3, 2.5)

------------DOSAGE FORMS AND STRENGTHS------------

Single-use vial (with 1.25 mL diluent) and REDIPEN® (3):
- 50 mcg per 0.5 mL, 80 mcg per 0.5 mL, 120 mcg per 0.5 mL, 150 mcg per 0.5 mL.

------------CONTRAINDICATIONS------------

- Known hypersensitivity reactions, such as urticaria, angioedema, bronchoconstriction, anaphylaxis, Stevens-Johnson syndrome, and toxic epidermal necrolysis to interferon alpha or any other product component. (4)
- Autoimmune hepatitis. (4)
- Hepatic decompensation (Child-Pugh score >6 [class B and C]) in cirrhotic CHC patients before or during treatment. (4)

Additional contraindications for combination therapy with ribavirin:
- Pregnant women and men whose female partners are pregnant. (4, 8.1)
- Hemoglobinopathies (e.g., thalassemia major, sickle-cell anemia). (4)
- Creatinine clearance <50 mL/min. (4)

------------WARNINGS AND PRECAUTIONS------------

- Birth defects and fetal death with ribavirin: Patients must have a negative pregnancy test prior to therapy, use at least 2 forms of contraception, and undergo monthly pregnancy tests. (5.1)

Patients exhibiting the following conditions should be closely monitored and may require dose reduction or discontinuation of therapy:
- Hemolytic anemia with ribavirin. (5.1)
- Neuropsychiatric events. (5.2)
- History of significant or unstable cardiac disease. (5.3)
- Hypothyroidism, hyperthyroidism, hyperglycemia, diabetes mellitus that cannot be effectively treated by medication. (5.4)
- New or worsening ophthalmologic disorders. (5.5)
- Ischemic and hemorrhagic cerebrovascular events. (5.6)
- Severe decreases in neutrophil or platelet counts. (5.7)
- History of autoimmune disorders. (5.8)
- Pancreatitis and ulcerative or hemorrhagic/ischemic colitis and pancreatitis. (5.9, 5.10)
- Pulmonary infiltrates or pulmonary function impairment. (5.11)
- Child-Pugh score >6 (class B and C). (4, 5.12)
- Increased creatinine levels in patients with renal insufficiency. (5.13)
- Serious, acute hypersensitivity reactions and cutaneous eruptions. (5.14)
- Dental/periodontal disorders reported with combination therapy. (5.16)
- Hypertriglyceridemia may result in pancreatitis (e.g., triglycerides >1000 mg/dL). (5.17)
- Weight loss and growth inhibition reported with combination therapy in pediatric patients. (5.18)

------------ADVERSE REACTIONS------------

Most common adverse reactions (>40%) in adult patients receiving either PegIntron or PegIntron/REBETOL are injection-site inflammation/reaction, fatigue/asthenia, headache, rigors, fevers, nausea, myalgia, and anxiety/emotional lability/irritability. (6.1) Most common adverse reactions (>25%) in pediatric patients receiving PegIntron/REBETOL are pyrexia, headache, neutropenia, fatigue, anorexia, injection-site erythema, vomiting. (6.1)

To report SUSPECTED ADVERSE REACTIONS, contact Schering Corporation at 1-800-526-4099 or FDA at 1-800-FDA-1088 or *www.fda.gov/medwatch*.

------------DRUG INTERACTIONS------------

- Drug metabolized by CYP450: Caution with drugs metabolized by CYP2C8/9 (e.g., warfarin, phenytoin) or CYP2D6 (e.g., flecainide). (7.1)
- Methadone: Monitor for increased narcotic effect. (7.2)
- Nucleoside analogues: Closely monitor for toxicities. Discontinue nucleoside reverse transcriptase inhibitors or reduce dose or discontinue interferon, ribavirin, or both with worsening toxicities. (7.3)
- Didanosine: Concurrent use with REBETOL is not recommended. (7.3)

------------USE IN SPECIFIC POPULATIONS------------

- Ribavirin Pregnancy Registry: 1-800-593-2214. (8.1)
- Pediatrics: Safety and efficacy in pediatrics <3 years old have not been established. (8.4)
- Geriatrics: Neuropsychiatric, cardiac, pulmonary, GI, and systemic (flu-like) adverse reactions may be more severe. (8.5)
- Organ transplant: Safety and efficacy have not been studied. (8.6)
- HIV or HBV coinfection: Safety and efficacy have not been established. (8.7)

See 17 for PATIENT COUNSELING INFORMATION.
Revised: [5/2009]

	PegIntron Dose (Adults)*	PegIntron Dose (Pediatric Patients)	REBETOL Dose (Adults)	REBETOL Dose (Pediatric Patients)
PegIntron/REBETOL Combination Therapy (2.1)	1.5 mcg/kg/week	60 mcg/m^2/week	800-1400 mg orally daily with food	15 mg/kg/day orally with food in 2 divided doses

*Refer to **Tables 1–7** of the full prescribing information.

FULL PRESCRIBING INFORMATION: CONTENTS*

17.4 Laboratory Evaluations, Hydration, "Flu-like" Symptoms

* Sections or subsections omitted from the full prescribing information are not listed.

FULL PRESCRIBING INFORMATION

WARNING: RISK OF SERIOUS DISORDERS AND RIBAVIRIN-ASSOCIATED EFFECTS
Alpha interferons, including PegIntron, may cause or aggravate fatal or life-threatening neuropsychiatric, autoimmune, ischemic, and infectious disorders. Patients should be monitored closely with periodic clinical and laboratory evaluations. Patients with persistently severe or worsening signs or symptoms of these conditions should be withdrawn from therapy. In many, but not all cases, these disorders resolve after stopping PegIntron therapy *[see Warnings and Precautions (5) and Adverse Reactions (6.1)]*.
Use with Ribavirin
Ribavirin may cause birth defects and death of the unborn child. Extreme care must be taken to avoid pregnancy in female patients and in female partners of male patients. Ribavirin causes hemolytic anemia. The anemia associated with REBETOL therapy may result in a worsening of cardiac disease. Ribavirin is genotoxic and mutagenic and should be considered a potential carcinogen. *[See REBETOL package insert.]*

1 INDICATIONS AND USAGE

1.1 Chronic Hepatitis C

Combination therapy: PegIntron® in combination with REBETOL® (ribavirin) is indicated for the treatment of chronic hepatitis C in patients 3 years of age and older with compensated liver disease.

The following points should be considered when initiating therapy with PegIntron in combination with REBETOL:

- These indications are based on achieving undetectable HCV-RNA after treatment for 24 or 48 weeks and maintaining a Sustained Virologic Response (SVR) 24 weeks after the last dose.
- Patients with the following characteristics are less likely to benefit from re-treatment after failing a course of therapy: previous nonresponse, previous pegylated interferon treatment, significant bridging fibrosis or cirrhosis, and genotype 1 infection *[see Clinical Studies (14)]*.
- No safety and efficacy data are available for treatment of longer than 1 year.

Monotherapy (for patients who are intolerant to ribavirin): PegIntron (Peginterferon alfa-2b) is indicated for use alone for the treatment of chronic hepatitis C in patients with compensated liver disease previously untreated with interferon alpha and who are at least 18 years of age.

The following point should be considered when initiating therapy with PegIntron alone:

- Combination therapy with REBETOL is preferred over PegIntron monotherapy unless there are contraindications to or significant intolerance of REBETOL. Combination therapy provides substantially better response rates than monotherapy *[see Clinical Studies (14)]*.

2 DOSAGE AND ADMINISTRATION

2.1 PegIntron/REBETOL Combination Therapy

REBETOL should be taken with food. REBETOL should not be used in patients with creatinine clearance <50 mL/min.

Adults: The recommended dose of PegIntron is 1.5 mcg/kg/week subcutaneously in combination with 800 to 1400 mg of REBETOL orally based on patient body weight. The volume of PegIntron to be injected depends on the strength of PegIntron and patient's body weight (see **Table 1**).

Duration of Treatment – Interferon Alpha-naïve Patients: The treatment duration for patients with genotype 1 is 48 weeks. Discontinuation of therapy should be considered in patients who do not achieve at least a 2 $\log_{10}$ drop or loss of HCV-RNA at 12 weeks, or if HCV-RNA remains detectable after 24 weeks of therapy. Patients with genotype 2 and 3 should be treated for 24 weeks.

Duration of Treatment – Re-treatment with PegIntron/REBETOL of Prior Treatment Failures: The treatment duration for patients who previously failed therapy is 48 weeks, regardless of HCV genotype.

Re-treated patients who fail to achieve undetectable HCV-RNA at Week 12 of therapy, or whose HCV-RNA remains detectable after 24 weeks of therapy, are highly unlikely to achieve SVR and discontinuation of therapy should be considered *[see Clinical Studies (14.1)]*.

[See table 1 above]

Pediatric Patients: Dosing for pediatric patients is determined by body surface area for PegIntron and by body weight for REBETOL. The recommended dose of PegIntron is 60 mcg/m^2/week subcutaneously in combination with 15 mg/kg/day of REBETOL orally in 2 divided doses (see **Table 2**) for pediatric patients ages 3 to 17 years. Patients who reach their 18th birthday while receiving PegIntron/REBETOL should remain on the pediatric dosing regimen. The treatment duration for patients with genotype 1 is 48 weeks. Patients with genotypes 2 and 3 should be treated for 24 weeks.

[See table 2 above]

TABLE 1: Recommended PegIntron Combination Therapy Dosing (Adults)

Body weight kg (lbs)	PegIntron REDIPEN® or Vial Strength to Use	Amount of PegIntron (mcg) to Administer	Volume (mL)* of PegIntron to Administer	REBETOL Daily Dose	REBETOL Number of Capsules
<40 (<88)	50 mcg per 0.5 mL	50	0.5	800 mg/day	2 × 200 mg capsules a.m. 2 × 200 mg capsules p.m.
40-50 (88-111)	80 mcg per 0.5 mL	64	0.4	800 mg/day	2 × 200 mg capsules a.m. 2 × 200 mg capsules p.m.
51-60 (112-133)		80	0.5	800 mg/day	2 × 200 mg capsules a.m. 2 × 200 mg capsules p.m.
61-65 (134-144)	120 mcg per 0.5 mL	96	0.4	800 mg/day	2 × 200 mg capsules a.m. 2 × 200 mg capsules p.m.
66-75 (145-166)		96	0.4	1000 mg/day	2 × 200 mg capsules a.m. 3 × 200 mg capsules p.m.
76-80 (167-177)		120	0.5	1000 mg/day	2 × 200 mg capsules a.m. 3 × 200 mg capsules p.m.
81-85 (178-187)		120	0.5	1200 mg/day	3 × 200 mg capsules a.m. 3 × 200 mg capsules p.m.
86-105 (188-231)	150 mcg per 0.5 mL	150	0.5	1200 mg/day	3 × 200 mg capsules a.m. 3 × 200 mg capsules p.m.
>105 (>231)	†	†	†	1400 mg/day	3 × 200 mg capsules a.m. 4 × 200 mg capsules p.m.

* When reconstituted as directed.
† For patients weighing >105 kg (>231 pounds), the PegIntron dose of 1.5 mcg/kg/week should be calculated based on the individual patient weight. Two vials of PegIntron may be necessary to provide the dose.

TABLE 2: Recommended REBETOL* Dosing in Combination Therapy (Pediatrics)

Body weight kg (lbs)	REBETOL Daily Dose	REBETOL Number of Capsules
<47 (<103)	15 mg/kg/day	Use REBETOL Oral Solution†
47-59 (103-131)	800 mg/day	2 × 200 mg capsules a.m. 2 × 200 mg capsules p.m.
60-73 (132-162)	1000 mg/day	2 × 200 mg capsules a.m. 3 × 200 mg capsules p.m.
>73 (>162)	1200 mg/day	3 × 200 mg capsules a.m. 3 × 200 mg capsules p.m.

* REBETOL to be used in combination with PegIntron 60 mcg/m^2 weekly.
† REBETOL Oral Solution may be used for any patient regardless of body weight.

2.2 PegIntron Monotherapy

The recommended dose of PegIntron regimen is 1 mcg/kg/week subcutaneously for 1 year administered on the same day of the week. Discontinuation of therapy should be considered in patients who do not achieve at least a 2 $\log_{10}$ drop or loss of HCV-RNA at 12 weeks of therapy, or whose HCV-RNA levels remain detectable after 24 weeks of therapy. The volume of PegIntron to be injected depends on patient weight (see **Table 3**).

[See table 3 at top of next page]

2.3 Dose Reduction

If a serious adverse reaction develops during the course of treatment *[see Warnings and Precautions (5)]* discontinue or modify the dosage of PegIntron and REBETOL until the adverse event abates or decreases in severity. If persistent or recurrent serious adverse events develop despite adequate dosage adjustment, discontinue treatment. For guidelines for dose modifications and discontinuation based on depression or laboratory parameters, see **Tables 4** and **5**. Dose reduction of PegIntron in adult patients on PegIntron/REBETOL combination therapy is accomplished in a two-step process from the original starting dose of 1.5 mcg/kg/week, to 1 mcg/kg/week, then to 0.5 mcg/kg/week, if needed. Dose reduction in patients on PegIntron monotherapy is accomplished by reducing the original starting dose of 1 mcg/kg/week to 0.5 mcg/kg/week. Dose reduction of PegIntron in adults may be accomplished by utilizing a lower dose strength or administering a lesser volume as shown in **Table 6** or **7**.

Continued on next page

Information on Schering products appearing on these pages is effective as of July 2009.

PegIntron—Cont.

In the adult combination therapy Study 2, dose reductions occurred in 42% of subjects receiving PegIntron 1.5 mcg/kg plus REBETOL 800 mg daily, including 57% of those subjects weighing 60 kg or less. In Study 4, 16% of subjects had a dose reduction of PegIntron to 1 mcg/kg in combination with REBETOL, with an additional 4% requiring the second dose reduction of PegIntron to 0.5 mcg/kg due to adverse events *[see Adverse Reactions (6.1)]*.

Dose reduction in pediatric patients is accomplished by modifying the recommended dose in a 2-step process from the original starting dose of 60 mcg/m²/week, to 40 mcg/m²/week, then to 20 mcg/m²/ week, if needed (see **Tables 4** and **5**). In the pediatric combination therapy trial, dose reductions occurred in 25% of subjects receiving PegIntron 60 mcg/m² weekly plus REBETOL 15 mg/kg daily.

[See table 4 above]
[See table 5 at top of next page]
[See table 6 at top of next page]
[See table 7 at bottom of page B74]

TABLE 3: Recommended PegIntron Monotherapy Dosing

Body weight kg (lbs)	PegIntron REDIPEN or Vial Strength to Use	Amount of PegIntron (mcg) to Administer	Volume (mL)* of PegIntron to Administer
≤45 (≤100)	50 mcg per 0.5 mL	40	0.4
46-56 (101-124)		50	0.5
57-72 (125-159)	80 mcg per 0.5 mL	64	0.4
73-88 (160-195)		80	0.5
89-106 (196-234)	120 mcg per 0.5 mL	96	0.4
107-136 (235-300)		120	0.5
137-160 (301-353)	150 mcg per 0.5 mL	150	0.5

*When reconstituted as directed

TABLE 4: Guidelines for Modification or Discontinuation of PegIntron or PegIntron/REBETOL and for Scheduling Visits for Patients With Depression

Depression Severity*	Initial Management (4-8 weeks)		Depression Status		
	Dose Modification	Visit Schedule	Remains Stable	Improves	Worsens
Mild	No change	Evaluate once weekly by visit or phone.	Continue weekly visit schedule.	Resume normal visit schedule.	See moderate or severe depression
Moderate	Adults: Adjust Dose Pediatrics: Decrease dose to 40 mcg/m²/week, then to 20 mcg/m²/ week, if needed	Evaluate once weekly (office visit at least every other week).	Consider psychiatric consultation. Continue reduced dosing.	If symptoms improve and are stable for 4 weeks, may resume normal visit schedule. Continue reduced dosing or return to normal dose.	See severe depression
Severe	Discontinue PegIntron/ REBETOL permanently.	Obtain immediate psychiatric consultation.	Psychiatric therapy as necessary		

*See DSM-IV for definitions. For patients on PegIntron/REBETOL combination therapy: 1st dose reduction of PegIntron is to 1 mcg/kg/week, 2nd dose reduction (if needed) of PegIntron is to 0.5 mcg/kg/week. For patients on PegIntron monotherapy: decrease PegIntron dose to 0.5 mcg/kg/week.

2.4 Discontinuation of Dosing

Adults: It is recommended that HCV genotype 1 interferon-alfa-naïve patients receiving PegIntron, alone or in combination with ribavirin, be discontinued from therapy if there is not at least a 2 $\log_{10}$ drop or loss of HCV-RNA at 12 weeks of therapy, or whose HCV-RNA levels remain detectable after 24 weeks of therapy. Regardless of genotype, previously treated patients who have detectable HCV-RNA at Week 12 or 24, are highly unlikely to achieve SVR and discontinuation of therapy should be considered.

Pediatrics (3-17 years of age): It is recommended that patients receiving PegIntron/REBETOL combination (excluding those with HCV Genotype 2 and 3) be discontinued from therapy at 12 weeks if their treatment Week 12 HCV RNA dropped <2 $\log_{10}$ compared to pretreatment or at 24 weeks if they have detectable HCV RNA at treatment Week 24.

2.5 Renal Function

In patients with moderate renal dysfunction (creatinine clearance 30-50 mL/min), the PegIntron dose should be reduced by 25%. Patients with severe renal dysfunction (creatinine clearance 10-29 mL/min), including those on hemodialysis, should have the PegIntron dose reduced by 50%. If renal function decreases during treatment, PegIntron therapy should be discontinued. When PegIntron is administered in combination with REBETOL, subjects with impaired renal function or those over the age of 50 should be more carefully monitored with respect to the development of anemia. PegIntron/REBETOL should not be used in patients with creatinine clearance <50 mL/min.

2.6 Preparation and Administration

PegIntron REDIPEN: PegIntron REDIPEN consists of a dual-chamber glass cartridge with sterile, lyophilized peginterferon alfa-2b in the active chamber and Sterile Water for Injection USP in the diluent chamber. The PegIntron in the glass cartridge should appear as a white to off-white tablet-shaped solid that is whole or in pieces, or powder.

To reconstitute the lyophilized peginterferon alfa-2b in the REDIPEN:

- Hold the REDIPEN upright (dose button down) and press the 2 halves of the pen together until there is an audible click.
- Gently invert the pen to mix the solution. **DO NOT SHAKE**. The reconstituted solution has a concentration of either 50 mcg per 0.5 mL, 80 mcg per 0.5 mL, 120 mcg per 0.5 mL, or 150 mcg per 0.5 mL for a single subcutaneous injection.
- Visually inspect the solution for particulate matter and discoloration prior to administration. The reconstituted solution should be clear and colorless. Do not use the solution if it is discolored or not clear, or if particulates are present.

Keeping the pen upright, attach the supplied needle and select the appropriate PegIntron dose by pulling back on the dosing button until the dark bands are visible and turning the button until the dark band is aligned with the correct dose. The prepared PegIntron solution is to be injected subcutaneously.

The PegIntron REDIPEN is a single-use pen and does not contain a preservative. The reconstituted solution should be used immediately and cannot be stored for more than 24 hours at 2°-8°C *[see How Supplied/Storage and Handling (16)]*. **DO NOT REUSE THE REDIPEN**. The sterility of any remaining product can no longer be guaranteed. **DISCARD THE UNUSED PORTION**. Pooling of unused portions of some medications has been linked to bacterial contamination and morbidity.

PegIntron Vials: Two BD® Safety-Lok® syringes are provided in the package; one syringe is for the reconstitution steps and one for the patient injection. There is a plastic safety sleeve to be pulled over the needle after use. The syringe locks with an audible click when the green stripe on the safety sleeve covers the red stripe on the needle. Instructions for the preparation and administration of PegIntron Powder for Injection are provided below.

- **Reconstitute the PegIntron lyophilized product with only 0.7 mL of the 1.25 mL of supplied diluent (Sterile Water for Injection USP). The diluent vial is for single use only. The remaining diluent should be discarded.** No other medications should be added to solutions containing PegIntron, and PegIntron should not be reconstituted with other diluents.
- Swirl gently to hasten complete dissolution of the powder. The reconstituted solution should be clear and colorless.
- Visually inspect the solution for particulate matter and discoloration prior to administration. The solution should not be used if discolored or cloudy, or if particulates are present.
- The appropriate PegIntron dose should be withdrawn and injected subcutaneously. PegIntron vials are for single use only and do not contain a preservative.

The reconstituted solution should be used immediately and cannot be stored for more than 24 hours at 2°-8°C *[see How Supplied/Storage and Handling (16)]*. **DO NOT REUSE THE VIAL**. The sterility of any remaining product can no longer be guaranteed. **DISCARD THE UNUSED PORTION**. Pooling of unused portions of some medications has been linked to bacterial contamination and morbidity.

3 DOSAGE FORMS AND STRENGTHS

- Single-use vial: 1.25 mL diluent vial: 50 mcg per 0.5 mL, 80 mcg per 0.5 mL, 120 mcg per 0.5 mL, 150 mcg per 0.5 mL.
- Single-use REDIPEN: 50 mcg per 0.5 mL, 80 mcg per 0.5 mL, 120 mcg per 0.5 mL, 150 mcg per 0.5 mL.

4 CONTRAINDICATIONS

PegIntron is contraindicated in patients with:

- known hypersensitivity reactions, such as urticaria, angioedema, bronchoconstriction, anaphylaxis, Stevens-Johnson syndrome, and toxic epidermal necrolysis to interferon alpha or any other component of the product
- autoimmune hepatitis
- hepatic decompensation (Child-Pugh score >6 [class B and C]) in cirrhotic CHC patients before or during treatment

PegIntron/REBETOL combination therapy is additionally contraindicated in:

- women who are pregnant. REBETOL may cause fetal harm when administered to a pregnant woman. REBETOL is contraindicated in women who are or may become pregnant. If this drug is used during pregnancy, or if the patient becomes pregnant while taking this drug the patient should be apprised of the potential hazard to a fetus *[see Use in Specific Populations (8.1)]*.
- men whose female partners are pregnant
- patients with hemoglobinopathies (e.g., thalassemia major, sickle-cell anemia)
- patients with creatinine clearance <50 mL/min

5 WARNINGS AND PRECAUTIONS

Patients should be monitored for the following serious conditions, some of which may become life threatening. Patients with persistently severe or worsening signs or symptoms should be withdrawn from therapy.

5.1 Use with Ribavirin

Pregnancy: **REBETOL may cause birth defects and death of the unborn child. REBETOL therapy should not be started until a report of a negative pregnancy test has been obtained immediately prior to planned initiation of therapy. Patients should use at least 2 forms of contraception and have monthly pregnancy tests** *[see BOXED WARNING, Contraindications (4), Patient Counseling Information (17), and REBETOL package insert]*.

Anemia: Ribavirin caused hemolytic anemia in 10% of PegIntron/REBETOL-treated subjects within 1 to 4 weeks of initiation of therapy. Complete blood counts should be obtained pretreatment and at Week 2 and Week 4 of therapy or more frequently if clinically indicated. Anemia associated with REBETOL therapy may result in a worsening of cardiac disease. Decrease in dosage or discontinuation of REBETOL may be necessary *[see Dosage and Administration (2.3) and REBETOL package insert]*.

5.2 Neuropsychiatric Events

Life-threatening or fatal neuropsychiatric events, including suicide, suicidal and homicidal ideation, depression, relapse of drug addiction/overdose, and aggressive behavior sometimes directed towards others have occurred in patients with and without a previous psychiatric disorder during PegIntron treatment and follow-up. Psychoses, hallucinations, bipolar disorders, and mania have been observed in

TABLE 5: Guidelines for Dose Modification and Discontinuation of PegIntron or PegIntron/REBETOL Based on Laboratory Parameters in Adults and Pediatrics

Laboratory Values	PegIntron		REBETOL	
	Adults	Pediatrics	Adults	Pediatrics
Hgb <10 g/dL	For patients with cardiac disease, reduce by 50%*	See footnote*	Adjust Dose†	1st reduction to 12 mg/kg/day 2nd reduction to 8 mg/kg/day
WBC $<1.5 \times 10^9$/L Neutrophils $<0.75 \times 10^9$/L Platelets $<50 \times 10^9$/L (Adults) $<70 \times 10^9$/L (Pediatrics)	Adjust Dose‡	1st reduction to 40 mcg/m²/week 2nd reduction to 20 mcg/m²/week	No dose change	No dose change
Hgb <8.5 g/dL WBC $<1 \times 10^9$/L Neutrophils $<0.5 \times 10^9$/L Platelets $<25 \times 10^9$/L (Adults) $<50 \times 10^9$/L (Pediatrics) Creatinine >2 mg/dL (Pediatrics)	Permanently discontinue	Permanently discontinue	Permanently discontinue	Permanently discontinue

* For adult patients with a history of stable cardiac disease receiving PegIntron in combination with ribavirin, the PegIntron dose should be reduced by half and the ribavirin dose by 200 mg/day if a >2g/dL decrease in hemoglobin is observed during any 4-week period. Both PegIntron and ribavirin should be permanently discontinued if patients have hemoglobin levels <12 g/dL after this ribavirin dose reduction. Pediatric patients who have pre-existing cardiac conditions and experience a hemoglobin decrease ≥2 g/dL during any 4-week period during treatment should have weekly evaluations and hematology testing.

† 1st dose reduction of REBETOL is by 200 mg/day, except in patients receiving the 1400 mg dose it is by 400 mg/day; 2nd dose reduction of REBETOL (if needed) is by an additional 200 mg/day.

‡ For patients on PegIntron/REBETOL combination therapy: 1st dose reduction of PegIntron is to 1 mcg/kg/week, 2nd dose reduction (if needed) of PegIntron is to 0.5 mcg/kg/week. For patients on PegIntron monotherapy: decrease PegIntron dose to 0.5 mcg/kg/week.

TABLE 6: Reduced PegIntron Dose (0.5 mcg/kg) for (1 mcg/kg) Monotherapy in Adults

Body weight kg (lbs)	PegIntron REDIPEN/Vial Strength to Use	Amount of PegIntron (mcg) to Administer	Volume (mL)† of PegIntron to Administer
≤45 (≤100)	50 mcg per 0.5 mL*	20	0.2
46-56 (101-124)		25	0.25
57-72 (125-159)	50 mcg per 0.5 mL	30	0.3
73-88 (160-195)		40	0.4
89-106 (196-234)	50 mcg per 0.5 mL	50	0.5
107-136 (235-300)	80 mcg per 0.5 mL	64	0.4
≥137 (≥301)		80	0.5

* Must use vial. Minimum delivery for REDIPEN 0.3 mL.
† When reconstituted as directed.

patients treated with interferon alpha. PegIntron should be used with extreme caution in patients with a history of psychiatric disorders. Patients should be advised to report immediately any symptoms of depression or suicidal ideation to their prescribing physicians. Physicians should monitor all patients for evidence of depression and other psychiatric symptoms. If patients develop psychiatric problems, including clinical depression, it is recommended that the patients be carefully monitored during treatment and in the 6-month follow-up period. If psychiatric symptoms persist or worsen, or suicidal ideation or aggressive behavior towards others is identified, it is recommended that treatment with PegIntron be discontinued, and the patient followed, with psychiatric intervention as appropriate. In severe cases, PegIntron should be stopped immediately and psychiatric intervention instituted *[see Dosage and Administration (2.3)]*. Cases of encephalopathy have been observed in some patients, usually elderly, treated at higher doses of PegIntron.

5.3 Cardiovascular Events

Cardiovascular events, which include hypotension, arrhythmia, tachycardia, cardiomyopathy, angina pectoris, and myocardial infarction, have been observed in patients treated with PegIntron. PegIntron should be used cautiously in patients with cardiovascular disease. Patients with a history of myocardial infarction and arrhythmic disorder who require PegIntron therapy should be closely monitored *[see Warnings and Precautions (5.15)]*. Patients with a history of significant or unstable cardiac disease should not be treated with PegIntron/REBETOL combination therapy *[see REBETOL package insert]*.

5.4 Endocrine Disorders

PegIntron causes or aggravates hypothyroidism and hyperthyroidism. Hyperglycemia has been observed in patients treated with PegIntron. Diabetes mellitus has been observed in patients treated with alpha interferons. Patients with these conditions who cannot be effectively treated by medication should not begin PegIntron therapy. Patients who develop these conditions during treatment and cannot be controlled with medication should not continue PegIntron therapy.

5.5 Ophthalmologic Disorders

Decrease or loss of vision, retinopathy including macular edema, retinal artery or vein thrombosis, retinal hemorrhages and cotton wool spots, optic neuritis, and papilledema may be induced or aggravated by treatment with peginterferon alfa-2b or other alpha interferons. All patients should receive an eye examination at baseline. Patients with preexisting ophthalmologic disorders (e.g., diabetic or hypertensive retinopathy) should receive periodic ophthalmologic exams during interferon alpha treatment. Any patient who develops ocular symptoms should receive a prompt and complete eye examination. Peginterferon alfa-2b treatment should be discontinued in patients who develop new or worsening ophthalmologic disorders.

5.6 Cerebrovascular Disorders

Ischemic and hemorrhagic cerebrovascular events have been observed in patients treated with interferon alfa-based therapies, including PegIntron. Events occurred in patients with few or no reported risk factors for stroke, including patients less than 45 years of age. Because these are spontaneous reports, estimates of frequency cannot be made, and a causal relationship between interferon alfa-based therapies and these events is difficult to establish.

5.7 Bone Marrow Toxicity

PegIntron suppresses bone marrow function, sometimes resulting in severe cytopenias. PegIntron should be discontinued in patients who develop severe decreases in neutrophil or platelet counts *[see Dosage and Administration (2.3)]*. Ribavirin may potentiate the neutropenia induced by interferon alpha. Very rarely alpha interferons may be associated with aplastic anemia.

5.8 Autoimmune Disorders

Development or exacerbation of autoimmune disorders (e.g., thyroiditis, thrombotic thrombocytopenic purpura, idiopathic thrombocytopenic purpura, rheumatoid arthritis, interstitial nephritis, systemic lupus erythematosus, and psoriasis) have been observed in patients receiving PegIntron. PegIntron should be used with caution in patients with autoimmune disorders.

5.9 Pancreatitis

Fatal and nonfatal pancreatitis have been observed in patients treated with alpha interferon. PegIntron therapy should be suspended in patients with signs and symptoms suggestive of pancreatitis and discontinued in patients diagnosed with pancreatitis.

5.10 Colitis

Fatal and nonfatal ulcerative or hemorrhagic/ischemic colitis have been observed within 12 weeks of the start of alpha interferon treatment. Abdominal pain, bloody diarrhea, and fever are the typical manifestations. PegIntron treatment should be discontinued immediately in patients who develop these signs and symptoms. The colitis usually resolves within 1 to 3 weeks of discontinuation of alpha interferons.

5.11 Pulmonary Disorders

Dyspnea, pulmonary infiltrates, pneumonia, bronchiolitis obliterans, interstitial pneumonitis, and sarcoidosis, some resulting in respiratory failure or patient deaths, may be induced or aggravated by PegIntron or alpha interferon therapy. Recurrence of respiratory failure has been observed with interferon rechallenge. PegIntron combination treatment should be suspended in patients who develop pulmonary infiltrates or pulmonary function impairment. Patients who resume interferon treatment should be closely monitored.

5.12 Hepatic Failure

Chronic hepatitis C (CHC) patients with cirrhosis may be at risk of hepatic decompensation and death when treated with alpha interferons, including PegIntron. Cirrhotic CHC patients coinfected with HIV receiving highly active antiretroviral therapy (HAART) and alpha interferons with or without ribavirin appear to be at increased risk for the development of hepatic decompensation compared to patients not receiving HAART. During treatment, patients' clinical status and hepatic function should be closely monitored, and PegIntron treatment should be immediately discontinued if decompensation (Child-Pugh score >6) is observed *[see Contraindications (4)]*.

5.13 Patients with Renal Insufficiency

Increases in serum creatinine levels have been observed in patients with renal insufficiency receiving interferon alpha products, including PegIntron. Patients with impaired renal function should be closely monitored for signs and symptoms of interferon toxicity, including increases in serum creatinine, and PegIntron dosing should be adjusted accordingly or discontinued *[see Clinical Pharmacology (12.3) and Dosage and Administration (2.3)]*. PegIntron monotherapy should be used with caution in patients with creatinine clearance <50 mL/min; the potential risks should be weighed against the potential benefits in these patients. Combination therapy with REBETOL must not be used in patients with creatinine clearance <50 mL/min *[see REBETOL Package Insert]*.

5.14 Hypersensitivity

Serious, acute hypersensitivity reactions (e.g., urticaria, angioedema, bronchoconstriction, anaphylaxis) and cutaneous eruptions (Stevens-Johnson syndrome, toxic epidermal necrolysis) have been rarely observed during alpha interferon therapy. If such a reaction develops during treatment

Continued on next page

Information on Schering products appearing on these pages is effective as of July 2009.

PegIntron—Cont.

with PegIntron, discontinue treatment and institute appropriate medical therapy immediately. Transient rashes do not necessitate interruption of treatment.

5.15 Laboratory Tests

PegIntron alone or in combination with ribavirin may cause severe decreases in neutrophil and platelet counts, and hematologic, endocrine (e.g., TSH), and hepatic abnormalities. Transient elevations in ALT (2- to 5-fold above baseline) were observed in 10% of subjects treated with PegIntron, and were not associated with deterioration of other liver functions. Triglyceride levels are frequently elevated in patients receiving alpha interferon therapy including PegIntron and should be periodically monitored.

Patients on PegIntron or PegIntron/REBETOL combination therapy should have hematology and blood chemistry testing before the start of treatment and then periodically thereafter. In the adult clinical trial CBC (including hemoglobin, neutrophil, and platelet counts) and chemistries (including AST, ALT, bilirubin, and uric acid) were measured during the treatment period at Weeks 2, 4, 8, and 12, and then at 6-week intervals or more frequently if abnormalities developed. In pediatric subjects, the same laboratory parameters were evaluated with additional assessment of hemoglobin at treatment Week 6. TSH levels were measured every 12 weeks during the treatment period. HCV RNA should be measured periodically during treatment *[see Dosage and Administration (2)]*.

Patients who have pre-existing cardiac abnormalities should have electrocardiograms done before treatment with PegIntron/REBETOL.

5.16 Dental and Periodontal Disorders

Dental and periodontal disorders have been reported in patients receiving PegIntron/REBETOL combination therapy. In addition, dry mouth could have a damaging effect on teeth and mucous membranes of the mouth during long-term treatment with the combination of REBETOL and PegIntron. Patients should brush their teeth thoroughly twice daily and have regular dental examinations. If vomiting occurs, patients should be advised to rinse out their mouth thoroughly afterwards.

5.17 Triglycerides

Elevated triglyceride levels have been observed in patients treated with interferon alpha, including PegIntron therapy. Hypertriglyceridemia may result in pancreatitis *[see Warnings and Precautions (5.9)]*. Elevated triglyceride levels should be managed as clinically appropriate. Discontinuation of PegIntron therapy should be considered for patients with symptoms of potential pancreatitis, such as abdominal pain, nausea, or vomiting, and persistently elevated triglycerides (e.g., triglycerides >1000 mg/dL).

5.18 Impact on Growth – Pediatric Use

Data on the effects of PegIntron plus REBETOL on growth come from an open-label study in subjects 3 through 17 years of age, and weight and height changes are compared to US normative population data. In general, the weight and height gain of pediatric subjects treated with PegIntron plus REBETOL lags behind that predicted by normative population data for the entire length of treatment. After about 6 months posttreatment (follow-up Week 24), subjects had weight gain rebounds and regained their weight to 53rd percentile, above the average of the normative population and similar to that predicted by their average baseline weight (57th percentile). After about 6 months posttreatment, height gain stabilized, and subjects treated with PegIntron plus REBETOL had an average height percentile of 44th percentile, which was less than the average of the normative population and less than their average baseline height (51st percentile). Severely inhibited growth velocity (<3rd percentile) was observed in 70% of the subjects while on treatment. Of the subjects experiencing severely inhibited growth, 20% had continued inhibited growth velocity (<3rd percentile) after 6 months of follow-up.

Among the boys studied, the age groups of 3 to 11 years old and 12 to 17 years old had similar height percentile decreases of approximately 5 percentiles after 6 months posttreatment; weight gain continued to be similar to their average baseline percentile. Girls who were 3 to 11 years old and treated for 48 weeks had the largest average drop in height and weight percentiles (13 percentiles and 7 percentiles, respectively), whereas girls 12 to 17 years old continued along their average baseline height and weight percentiles after 6 months posttreatment.

6 ADVERSE REACTIONS

Clinical trials with PegIntron alone or in combination with REBETOL have been conducted in over 6900 subjects from 3 to 75 years of age.

Serious adverse reactions have occurred in approximately 12% of subjects in clinical trials with PegIntron with or without REBETOL *[see BOXED WARNING, Warnings and Precautions (5)]*. The most common serious events occurring in subjects treated with PegIntron and REBETOL were depression and suicidal ideation *[see Warnings and Precautions (5.2)]*, each occurring at a frequency of less than 1%. The most common fatal events occurring in subjects treated with PegIntron and REBETOL were cardiac arrest, suicidal ideation, and suicide attempt *[see Warnings and Precautions (5.2, 5.5)]*, all occurring in less than 1% of subjects.

Greater than 96% of all subjects in clinical trials experienced one or more adverse events. The most commonly reported adverse reactions in adult subjects receiving either PegIntron or PegIntron/REBETOL were injection-site inflammation/reaction, fatigue/asthenia, headache, rigors, fevers, nausea, myalgia, and emotional lability/irritability. The most common adverse events in pediatric subjects, ages 3 and older, were pyrexia, headache, vomiting, neutropenia, fatigue, anorexia, injection-site erythema, and abdominal pain.

6.1 Clinical Trials Experience

Because clinical trials are conducted under widely varying conditions, adverse reaction rates observed in the clinical trials of a drug cannot be directly compared to rates in the clinical trials of another drug and may not reflect the rates observed in clinical practice.

Adults: Study 1 compared PegIntron monotherapy with INTRON® A monotherapy. Study 2 compared combination therapy of PegIntron/REBETOL with combination therapy with INTRON A/REBETOL. In these studies, nearly all study subjects in clinical trials experienced one or more adverse reactions. Study 3 compared a PegIntron/weight-based REBETOL combination to a PegIntron/flat-dose REBETOL regimen. Study 4 compared 2 PegIntron (1.5 mcg/kg/week and 1 mcg/kg/week) doses in combination with REBETOL and a third treatment group receiving Pegasys® (180 mcg/week)/Copegus® (1000-1200 mg/day).

Adverse reactions that occurred in Studies 1 and 2 at >5% incidence are provided in **Table 8** by treatment group. Due to potential differences in ascertainment procedures, adverse reaction rate comparisons across studies should not be made. **Table 9** summarizes the treatment-related/treatment emergent adverse reactions in Study 4 that occurred at a ≥10% incidence.

[See table 8 on pages B75 and B76]

[See table 9 at top of page B77]

The adverse reaction profile in Study 3, which compared PegIntron/weight-based REBETOL combination to a PegIntron/flat-dose REBETOL regimen, revealed an increased rate of anemia with weight-based dosing (29% vs. 19% for weight-based vs. flat-dose regimens, respectively). However, the majority of cases of anemia were mild and responded to dose reductions.

The incidence of serious adverse reactions was comparable in all studies. In the PEG monotherapy trial (Study 1) the incidence of serious adverse reactions was similar (about 12%) in all treatment groups. In Study 2, the incidence of serious adverse reactions was 17% in the PegIntron/REBETOL groups compared to 14% in the INTRON A/REBETOL group. In Study 3, there was a similar incidence of serious adverse reactions reported for the weight-based REBETOL group (12%) and with the flat-dose REBETOL regimen.

In many but not all cases, adverse reactions resolved after dose reduction or discontinuation of therapy. Some subjects experienced ongoing or new serious adverse reactions during the 6-month follow-up period.

There have been 31 subject deaths which occurred during treatment or during follow-up in these clinical trials. In Study 1, there was 1 suicide in a subject receiving PegIntron monotherapy and 2 deaths among subjects receiving INTRON A monotherapy (1 murder/suicide and 1 sudden death). In Study 2, there was 1 suicide in a subject receiving PegIntron/REBETOL combination therapy, and 1 subject death in the INTRON A/REBETOL group (motor vehicle accident). In Study 3, there were 14 deaths, 2 of which were probable suicides, and 1 was an unexplained death in a person with a relevant medical history of depression. In Study 4, there were 12 deaths, 6 of which occurred in subjects who received PegIntron/REBETOL combination therapy, 5 in the PegIntron 1.5 mcg/REBETOL arm (N=1019) and 1 in the PegIntron 1 mcg/REBETOL arm (N=1016), and 6 of which occurred in subjects receiving Pegasys/Copegus (N=1035). There were 3 suicides which occurred during the off-treatment follow-up period in subjects who received PegIntron (1.5 mcg/kg)/REBETOL combination therapy.

In Studies 1 and 2, 10% to 14% of subjects receiving PegIntron, alone or in combination with REBETOL, discontinued therapy compared with 6% treated with INTRON A alone and 13% treated with INTRON A in combination with REBETOL. Similarly in Study 3, 15% of subjects receiving PegIntron in combination with weight-based REBETOL and 14% of subjects receiving PegIntron and flat-dose REBETOL discontinued therapy due to an adverse reaction. The most common reasons for discontinuation of therapy were related to known interferon effects of psychiatric, systemic (e.g., fatigue, headache), or gastrointestinal adverse reactions. In Study 4, 13% of subjects in the PegIntron 1.5 mcg/REBETOL arm, 10% in the PegIntron 1 mcg/REBETOL arm, and 13% in the Pegasys 180 mcg/Copegus arm discontinued due to adverse events.

In Study 2, dose reductions due to adverse reactions occurred in 42% of subjects receiving PegIntron (1.5 mcg/kg)/REBETOL and in 34% of those receiving INTRON A/REBETOL. The majority of subjects (57%) weighing 60 kg or less receiving PegIntron (1.5 mcg/kg)/REBETOL required dose reduction. Reduction of interferon was dose-related (PegIntron 1.5 mcg/kg >PegIntron 0.5 mcg/kg or INTRON A), 40%, 27%, 28%, respectively. Dose reduction for REBETOL was similar across all 3 groups, 33% to 35%. The most common reasons for dose modifications were neutropenia (18%) or anemia (9%). Other common reasons in-

TABLE 7: Two-Step Dose Reduction of PegIntron in Combination Therapy in Adults

First Dose Reduction to PegIntron 1 mcg/kg				Second Dose Reduction to PegIntron 0.5 mcg/kg			
Body weight kg (lbs)	PegIntron REDIPEN/ Vial Strength to Use	Amount of PegIntron (mcg) to Administer	Volume (mL)† of PegIntron to Administer	Body weight kg (lbs)	PegIntron REDIPEN/ Vial Strength to Use	Amount of PegIntron (mcg) to Administer	Volume (mL)† of PegIntron to Administer
<40 (<88)	50 mcg per 0.5 mL	35	0.35	<40 (<88)	50 mcg per 0.5 mL*	20	0.2
40-50 (88-111)		45	0.45	40-50 (88-111)		25	0.25
51-60 (112-133)		50	0.5	51-60 (112-133)	50 mcg per 0.5 mL	30	0.3
61-75 (134-166)	80 mcg per 0.5 mL	64	0.4	61-75 (134-166)		35	0.35
76-85 (167-187)		80	0.5	76-85 (167-187)		45	0.45
86-104 (188-230)	120 mcg per 0.5 mL	96	0.4	86-104 (188-230)	50 mcg per 0.5 mL	50	0.5
105-125 (231-275)		108	0.45	105-125 (231-275)	80 mcg per 0.5 mL	64	0.4
>125 (>275)	150 mcg per 0.5 mL	135	0.45	>125 (>275)		72	0.45

* Must use vial. Minimum delivery for REDIPEN 0.3 mL.
† When reconstituted as directed.

TABLE 8: Adverse Reactions Occurring in >5% of Subjects

	*Percentage of Subjects Reporting Adverse Reactions** Study 1		Study 2	
Adverse Events	**PegIntron 1 mcg/kg (n=297)**	**INTRON A 3 MIU (n=303)**	**PegIntron 1.5 mcg/kg/ REBETOL (n=511)**	**INTRON A/ REBETOL (n=505)**
Application Site				
Injection Site Inflammation/Reaction	47	20	75	49
Autonomic Nervous System				
Dry Mouth	6	7	12	8
Increased Sweating	6	7	11	7
Flushing	6	3	4	3
Body as a Whole				
Fatigue/Asthenia	52	54	66	63
Headache	56	52	62	58
Rigors	23	19	48	41
Fever	22	12	46	33
Weight Loss	11	13	29	20
Right Upper Quadrant Pain	8	8	12	6
Chest Pain	6	4	8	7
Malaise	7	6	4	6
Central/Peripheral Nervous System				
Dizziness	12	10	21	17
Endocrine				
Hypothyroidism	5	3	5	4
Gastrointestinal				
Nausea	26	20	43	33
Anorexia	20	17	32	27
Diarrhea	18	16	22	17
Vomiting	7	6	14	12
Abdominal Pain	15	11	13	13
Dyspepsia	6	7	9	8
Constipation	1	3	5	5
Hematologic Disorders				
Neutropenia	6	2	26	14
Anemia	0	0	12	17
Leukopenia	<1	0	6	5
Thrombocytopenia	7	<1	5	2
Liver and Biliary System				
Hepatomegaly	6	5	4	4
Musculoskeletal				
Myalgia	54	53	56	50
Arthralgia	23	27	34	28
Musculoskeletal Pain	28	22	21	19

(Table continued on next page)

cluded depression, fatigue, nausea, and thrombocytopenia. In Study 3, dose modifications due to adverse reactions occurred more frequently with WBD compared to flat dosing (29% and 23%, respectively). In Study 4, 16% of subjects had a dose reduction of PegIntron to 1 mcg/kg in combination with REBETOL, with an additional 4% requiring the second dose reduction of PegIntron to 0.5 mcg/kg due to adverse events, compared to 15% of subjects in the Pegasys/Copegus arm, who required a dose reduction to 135 mcg/week with Pegasys, with an additional 7% in the Pegasys/Copegus arm requiring a second dose reduction to 90 mcg/week with Pegasys.

In the PegIntron/REBETOL combination trials the most common adverse reactions were psychiatric which occurred among 77% of subjects in Study 2 and 68% to 69% of subjects in Study 3. These psychiatric adverse reactions included most commonly depression, irritability, and insomnia, each reported by approximately 30% to 40% of subjects in all treatment groups. Suicidal behavior (ideation, attempts, and suicides) occurred in 2% of all subjects during treatment or during follow-up after treatment cessation *[see Warnings and Precautions (5.2)]*. In Study 4, psychiatric adverse reactions occurred in 58% of subjects in the PegIntron 1.5 mcg/REBETOL arm, 55% of subjects in the PegIntron 1 mcg/REBETOL arm, and 57% of subjects in the Pegasys 180 mcg/Copegus arm.

PegIntron induced fatigue or headache in approximately two-thirds of subjects, with fever or rigors in approximately half of the subjects. The severity of some of these systemic symptoms (e.g., fever and headache) tends to decrease as treatment continues. In Studies 1 and 2, application site inflammation and reaction (e.g., bruise, itchiness, and irritation) occurred at approximately twice the incidence with PegIntron therapies (in up to 75% of subjects) compared with INTRON A. However, injection-site pain was infrequent (2%-3%) in all groups. In Study 3 there was a 23% to 24% incidence overall for injection-site reactions or inflammation.

In Study 2, many subjects continued to experience adverse reactions several months after discontinuation of therapy. By the end of the 6-month follow-up period, the incidence of ongoing adverse reactions by body class in the PegIntron 1.5/REBETOL group was 33% (psychiatric), 20% (musculoskeletal), and 10% (for endocrine and for GI). In approximately 10% to 15% of subjects, weight loss, fatigue, and headache had not resolved.

Individual serious adverse reactions in Study 2 occurred at a frequency ≤1% and included suicide attempt, suicidal ideation, severe depression; psychosis, aggressive reaction, relapse of drug addiction/overdose; nerve palsy (facial, oculomotor); cardiomyopathy, myocardial infarction, angina, pericardial effusion, retinal ischemia, retinal artery or vein thrombosis, blindness, decreased visual acuity, optic neuritis, transient ischemic attack, supraventricular arrhythmias, loss of consciousness; neutropenia, infection (sepsis, pneumonia, abscess, cellulitis); emphysema, bronchiolitis obliterans, pleural effusion, gastroenteritis, pancreatitis, gout, hyperglycemia, hyperthyroidism and hypothyroidism, autoimmune thrombocytopenia with or without purpura, rheumatoid arthritis, interstitial nephritis, lupus-like syndrome, sarcoidosis, aggravated psoriasis; urticaria, injection-site necrosis, vasculitis, and phototoxicity.

Subjects receiving PegIntron/REBETOL as re-treatment after failing a previous interferon combination regimen reported adverse reactions similar to those previously associated with this regimen during clinical trials of treatment-naïve subjects.

Pediatric Subjects: In general, the adverse-reaction profile in the pediatric population was similar to that observed in adults. In the pediatric study, the most prevalent adverse reactions in all subjects were pyrexia (80%), headache (62%), neutropenia (33%), fatigue (30%), anorexia (29%), injection-site erythema (29%), and vomiting (27%). The majority of adverse reactions reported in the study were mild or moderate in severity. Severe adverse reactions were reported in 7% (8/107) of all subjects and included injection-site pain (1%), pain in extremity (1%), headache (1%), neutropenia (1%), and pyrexia (4%). Important adverse reactions that occurred in this subject population were nervousness (7%; 7/107), aggression (3%; 3/107), anger (2%; 2/107), and depression (1%; 1/107). Five subjects received levothyroxine treatment; 3 with clinical hypothyroidism and 2 with asymptomatic TSH elevations.

Dose modifications were required in 25% of subjects, most commonly for anemia, neutropenia, and weight loss. Two subjects (2%; 2/107) discontinued therapy as the result of an adverse reaction.

Adverse reactions that occurred with a ≥10% incidence in the pediatric trial subjects are provided in **Table 10**.

TABLE 10: Percentage of Pediatric Subjects With Treatment-emergent/Treatment-related Adverse Reactions (in at Least 10% of All Subjects)

System Organ Class Preferred Term	**All Subjects n=107**
Blood and Lymphatic System Disorders	
Neutropenia	33%
Anemia	11%
Leukopenia	10%

Continued on next page

PegIntron—Cont.

Gastrointestinal Disorders	
Abdominal Pain	21%
Abdominal Pain Upper	12%
Vomiting	27%
Nausea	18%
General Disorders and Administration Site Conditions	
Pyrexia	80%
Fatigue	30%
Injection-site Erythema	29%
Chills	21%
Asthenia	15%
Irritability	14%
Investigations	
Weight Decreased	19%
Metabolism and Nutrition Disorders	
Anorexia	29%
Decreased Appetite	22%
Musculoskeletal and Connective Tissue Disorders	
Arthralgia	17%
Myalgia	17%
Nervous System Disorders	
Headache	62%
Dizziness	14%
Skin and Subcutaneous Tissue Disorders	
Alopecia	17%

TABLE 8 *(cont.)*: Adverse Reactions Occurring in >5% of Subjects

	*Percentage of Subjects Reporting Adverse Reactions** Study 1		Study 2	
Adverse Events	**PegIntron 1 mcg/kg (n=297)**	**INTRON A 3 MIU (n=303)**	**PegIntron 1.5 mcg/kg/ REBETOL (n=511)**	**INTRON A/ REBETOL (n=505)**
Psychiatric				
Insomnia	23	23	40	41
Depression	29	25	31	34
Anxiety/Emotional Lability/Irritability	28	34	47	47
Concentration Impaired	10	8	17	21
Agitation	2	2	8	5
Nervousness	4	3	6	6
Reproductive, Female				
Menstrual Disorder	4	3	7	6
Resistance Mechanism				
Viral Infection	11	10	12	12
Fungal Infection	<1	3	6	1
Respiratory System				
Dyspnea	4	2	26	24
Coughing	8	5	23	16
Pharyngitis	10	7	12	13
Rhinitis	2	2	8	6
Sinusitis	7	7	6	5
Skin and Appendages				
Alopecia	22	22	36	32
Pruritus	12	8	29	28
Rash	6	7	24	23
Skin Dry	11	9	24	23
Special Senses, Other				
Taste Perversion	<1	2	9	4
Vision Disorders				
Vision Blurred	2	3	5	6
Conjunctivitis	4	2	4	5

*Subjects reporting one or more adverse reactions. A subject may have reported more than one adverse reaction within a body system/organ class category.

Laboratory Values: *Adults:* Changes in selected laboratory values during treatment with PegIntron alone or in combination with REBETOL treatment are described below. **Decreases in hemoglobin, neutrophils, and platelets may require dose reduction or permanent discontinuation from therapy** *[see Dosage and Administration (2.3) and Warnings and Precautions (5.1, 5.7)].*

Hemoglobin: Hemoglobin levels decreased to <11 g/dL in about 30% of subjects in Study 2. In Study 3, 47% of subjects receiving WBD REBETOL and 33% on flat-dose REBETOL had decreases in hemoglobin levels <11 g/dL. Reductions in hemoglobin to <9 g/dL occurred more frequently in subjects receiving WBD compared to flat dosing (4% and 2%, respectively). In Study 2, dose modification was required in 9% and 13% of subjects in the PegIntron/REBETOL and INTRON A/REBETOL groups. In Study 4, patients receiving PegIntron (1.5 mcg/kg)/REBETOL had decreases in hemoglobin levels to between 8.5 to <10 g/dL (28%) and to <8.5 g/dL (3%), whereas in patients receiving Pegasys 180 mcg/Copegus these decreases occurred in 26% and 4% of subjects, respectively. Hemoglobin levels become stable by treatment Weeks 4 to 6 on average. The typical pattern observed was a decrease in hemoglobin levels by treatment Week 4 followed by stabilization and a plateau, which was maintained to the end of treatment. In the PegIntron monotherapy trial, hemoglobin decreases were generally mild, and dose modifications were rarely necessary *[see Dosage and Administration (2.3)].*

Neutrophils: Decreases in neutrophil counts were observed in a majority of subjects treated with PegIntron alone (70%) or as combination therapy with REBETOL in Study 2 (85%) and INTRON A/REBETOL (60%). Severe potentially life-threatening neutropenia ($<0.5 \times 10^9$/L) occurred in 1% of subjects treated with PegIntron monotherapy, 2% of subjects treated with INTRON A/REBETOL, and in approximately 4% of subjects treated with PegIntron/REBETOL in Study 2. Two percent of subjects receiving PegIntron monotherapy and 18% of subjects receiving PegIntron/REBETOL in Study 2 required modification of interferon dosage. Few subjects (<1%) required permanent discontinuation of treatment. Neutrophil counts generally return to pretreatment levels 4 weeks after cessation of therapy *[see Dosage and Administration (2.3)].*

Platelets: Platelet counts decreased to <100,000/mm^3 in approximately 20% of subjects treated with PegIntron alone or with REBETOL and in 6% of subjects treated with INTRON A/REBETOL. Severe decreases in platelet counts (<50,000/mm^3) occur in <4% of subjects. Patients may require discontinuation or dose modification as a result of platelet decreases *[see Dosage and Administration (2.3)].* In Study 2, 1% or 3% of subjects required dose modification of INTRON A or PegIntron, respectively. Platelet counts generally returned to pretreatment levels 4 weeks after the cessation of therapy.

Triglycerides: Elevated triglyceride levels have been observed in patients treated with interferon alphas, including PegIntron *[see Warnings and Precautions (5.17)].*

Thyroid Function: Development of TSH abnormalities, with and without clinical manifestations, are associated with interferon therapies. In Study 2, clinically apparent thyroid disorders occur among subjects treated with either INTRON A or PegIntron (with or without REBETOL) at a similar incidence (5% for hypothyroidism and 3% for hyperthyroidism). Subjects developed new-onset TSH abnormalities while on treatment and during the follow-up period. At the end of the follow-up period, 7% of subjects still had abnormal TSH values *[see Warnings and Precautions (5.4)].*

Bilirubin and Uric Acid: In Study 2, 10% to 14% of subjects developed hyperbilirubinemia and 33% to 38% developed hyperuricemia in association with hemolysis. Six subjects developed mild to moderate gout.

Pediatric Subjects: **Decreases in hemoglobin, white blood cells, platelets, and neutrophils may require dose reduction or permanent discontinuation from therapy** *[see Dosage and Administration (2.3)].* Changes in selected laboratory values during treatment of 107 pediatric subjects with PegIntron/REBETOL combination therapy are described in **Table 11**. Most of the changes in laboratory values in this study were mild or moderate.

TABLE 11: Selected Hematological Abnormalities During Treatment Phase with PegIntron Plus REBETOL in Previously Untreated Pediatric Subjects

Laboratory Parameter*	All Subjects (n=107)
Hemoglobin (g/dL)	
9.5-<11.0	30%

8.0-<9.5	2%
WBC (× 10⁹/L)	
2.0-2.9	39%
1.5-<2.0	3%
Platelets (× 10⁹/L)	
70-100	1%
50-<70	—
25-<50	1%
Neutrophils (× 10⁹/L)	
1.0-1.5	35%
0.75-<1.0	26%
0.5-<0.75	13%
<0.5	3%
Total Bilirubin	
1.26-2.59 × N[†]	7%
Evidence of Hepatic Failure	—

* The table summarizes the worst category observed within the period per subject per laboratory test. Only subjects with at least one treatment value for a given laboratory test are included.
[†] N=Upper limit of normal

6.2 Immunogenicity
As with all therapeutic proteins, there is potential for immunogenicity. Approximately 2% of subjects receiving PegIntron (32/1759) or INTRON A (11/728) with or without REBETOL developed low-titer (≤160) neutralizing antibodies to PegIntron or INTRON A. The clinical and pathological significance of the appearance of serum-neutralizing antibodies is unknown. The incidence of antibody formation is highly dependent on the sensitivity and specificity of the assay. Additionally, the observed incidence of antibody (including neutralizing antibody) positivity in an assay may be influenced by several factors, including assay methodology, sample handling, timing of sample collection, concomitant medications, and underlying disease. For these reasons, comparison of the incidence of antibodies to PegIntron with the incidence of antibodies to other products may be misleading.

6.3 Postmarketing Experience
The following adverse reactions have been identified during postapproval use of PegIntron therapy. Because these reactions are reported voluntarily from a population of uncertain size, it is not always possible to reliably estimate their frequency or establish a causal relationship to drug exposure.
Blood and Lymphatic System Disorders
pure red cell aplasia, thrombotic thrombocytopenic purpura
Cardiac Disorders
palpitations
Ear and Labyrinth Disorders
hearing loss, vertigo, hearing impairment
Endocrine disorders
diabetic ketoacidosis, diabetes
Eye Disorders
Vogt-Koyanagi-Harada syndrome
Gastrointestinal Disorders
aphthous stomatitis
General Disorders and Administration Site Conditions
asthenic conditions (including asthenia, malaise, fatigue)
Immune System Disorders
cases of acute hypersensitivity reactions (including anaphylaxis, angioedema, urticaria); Stevens-Johnson syndrome, toxic epidermal necrolysis, systemic lupus erythematosus, erythema multiforme
Infections and Infestations
bacterial infection including sepsis
Metabolism and Nutrition Disorders
dehydration, hypertriglyceridemia
Musculoskeletal and Connective Tissue Disorders
rhabdomyolysis, myositis
Nervous System Disorders
seizures, memory loss, peripheral neuropathy, paraesthesia, migraine headache
Psychiatric Disorders
homicidal ideation
Renal and Urinary Disorders
renal failure, renal insufficiency

TABLE 9: Summary of Treatment-related/Treatment-emergent Adverse Reactions (≥10% Incidence) By Descending Frequency

	Percentage of Patients Reporting Treatment-related / Treatment-emergent Adverse Reactions **Study 4**		
Adverse Reactions	**PegIntron 1.5 mcg/kg with REBETOL (n=1019)**	**PegIntron 1 mcg/kg with REBETOL (n=1016)**	**Pegasys 180 mcg with Copegus (n=1035)**
Fatigue	67	68	64
Headache	50	47	41
Nausea	40	35	34
Chills	39	36	23
Insomnia	38	37	41
Anemia	35	30	34
Pyrexia	35	32	21
Injection Site Reactions	34	35	23
Anorexia	29	25	21
Rash	29	25	34
Myalgia	27	26	22
Neutropenia	26	19	31
Irritability	25	25	25
Depression	25	19	20
Alopecia	23	20	17
Dyspnea	21	20	22
Arthralgia	21	22	22
Pruritus	18	15	19
Influenza-like Illness	16	15	15
Dizziness	16	14	13
Diarrhea	15	16	14
Cough	15	16	17
Weight Decreased	13	10	10
Vomiting	12	10	9
Unspecified Pain	12	13	9
Dry Skin	11	11	12
Anxiety	11	11	10
Abdominal Pain	10	10	10
Leukopenia	9	7	10

Skin and Subcutaneous Tissue Disorders
psoriasis
Vascular Disorders
hypertension, hypotension

7 DRUG INTERACTIONS

7.1 Drugs Metabolized by Cytochrome P-450
When administering PegIntron with medications metabolized by CYP2C8/9 (e.g., warfarin and phenytoin) or CYP2D6 (e.g., flecainide), the therapeutic effect of these substrates may be decreased *[see Clinical Pharmacology (12.3)]*.

7.2 Methadone
PegIntron may increase methadone concentrations *[see Clinical Pharmacology (12.3)]*. The clinical significance of this finding is unknown; however, patients should be monitored for the signs and symptoms of increased narcotic effect.

7.3 Use with Ribavirin (Nucleoside Analogues)
Hepatic decompensation (some fatal) has occurred in cirrhotic HIV/HCV coinfected patients receiving combination antiretroviral therapy for HIV and interferon alpha and ribavirin. Adding treatment with alpha interferons alone or in combination with ribavirin may increase the risk in this patient subset. Patients receiving interferon with ribavirin and nucleoside reverse transcriptase inhibitors (NRTIs) should be closely monitored for treatment-associated toxicities, especially hepatic decompensation and anemia. Discontinuation of NRTIs should be considered as medically appropriate *[see Individual NRTI Product Information]*. Dose reduction or discontinuation of interferon, ribavirin, or both should also be considered if worsening clinical toxicities are observed, including hepatic decompensation (e.g., Child-Pugh >6).

Stavudine, Lamivudine, and Zidovudine: *In vitro* studies have shown ribavirin can reduce the phosphorylation of pyrimidine nucleoside analogues such as stavudine, lamivudine, and zidovudine. In a study with another pegylated interferon alpha, no evidence of a pharmacokinetic or pharmacodynamic (e.g., loss of HIV/HCV virologic suppression) interaction was seen when ribavirin was coadministered with zidovudine, lamivudine, or stavudine in HIV/HCV coinfected subjects *[see Clinical Pharmacology (12.3)]*. HIV/HCV coinfected subjects who were administered zidovudine in combination with pegylated interferon alpha

Continued on next page

Information on Schering products appearing on these pages is effective as of July 2009.

PegIntron—Cont.

and ribavirin developed severe neutropenia (ANC <500) and severe anemia (hemoglobin <8 g/dL) more frequently than similar subjects not receiving zidovudine.
Didanosine: Coadministration of REBETOL Capsules or Oral Solution and didanosine is not recommended. Reports of fatal hepatic failure, as well as peripheral neuropathy, pancreatitis, and symptomatic hyperlactactemia/lactic acidosis have been reported in clinical trials *[see Clinical Pharmacology (12.3)]*.

8 USE IN SPECIFIC POPULATIONS

8.1 Pregnancy

PegIntron Monotherapy: *Pregnancy Category C:* Non-pegylated interferon alfa-2b has been shown to have abortifacient effects in *Macaca mulatta* (rhesus monkeys) at 15 and 30 million IU/kg (estimated human equivalent of 5 and 10 million IU/kg, based on body surface area adjustment for a 60-kg adult). PegIntron should be assumed to also have abortifacient potential. There are no adequate and well-controlled studies in pregnant women. PegIntron therapy is to be used during pregnancy only if the potential benefit justifies the potential risk to the fetus. Therefore, PegIntron is recommended for use in fertile women only when they are using effective contraception during the treatment period.
Use with Ribavirin: *Pregnancy Category X:* Significant teratogenic and/or embryocidal effects have been demonstrated in all animal species exposed to ribavirin. REBETOL therapy is contraindicated in women who are pregnant and in the male partners of women who are pregnant *[see Contraindications (4) and the REBETOL Package Insert]*.
A Ribavirin Pregnancy Registry has been established to monitor maternal-fetal outcomes of pregnancies in female patients and female partners of male patients exposed to ribavirin during treatment and for 6 months following cessation of treatment. Physicians and patients are encouraged to report such cases by calling 1-800-593-2214.

8.3 Nursing Mothers

It is not known whether the components of PegIntron and/or REBETOL are excreted in human milk. Studies in mice have shown that mouse interferons are excreted in breast milk. Because of the potential for adverse reactions from the drug in nursing infants, a decision must be made whether to discontinue nursing or discontinue the PegIntron and REBETOL treatment, taking into account the importance of the therapy to the mother.

8.4 Pediatric Use

Safety and effectiveness in pediatric patients below the age of 3 years have not been established. Clinical trials in pediatric patients <3 years of age are not considered feasible due to the small proportion of patients in this age group requiring treatment for CHC.

8.5 Geriatric Use

In general, younger patients tend to respond better than older patients to interferon-based therapies. Clinical studies of PegIntron alone or in combination with REBETOL did not include sufficient numbers of subjects aged 65 and over, however, to determine whether they respond differently than younger subjects. Treatment with alpha interferons, including PegIntron, is associated with neuropsychiatric, cardiac, pulmonary, GI, and systemic (flu-like) adverse effects. Because these adverse reactions may be more severe in the elderly, caution should be exercised in the use of PegIntron in this population. This drug is known to be substantially excreted by the kidney. Because elderly patients are more likely to have decreased renal function, the risk of toxic reactions to this drug may be greater in patients with impaired renal function *[see Clinical Pharmacology (12.3)]*. When using PegIntron/REBETOL therapy, refer also to the REBETOL Package Insert.

8.6 Organ Transplant Recipients

The safety and efficacy of PegIntron alone or in combination with REBETOL for the treatment of hepatitis C in liver or other organ transplant recipients have not been studied. In a small (n=16) single-center, uncontrolled case experience, renal failure in renal allograft recipients receiving interferon alpha and ribavirin combination therapy was more frequent than expected from the center's previous experience with renal allograft recipients not receiving combination therapy. The relationship of the renal failure to renal allograft rejection is not clear.

8.7 HIV or HBV Coinfection

The safety and efficacy of PegIntron/REBETOL for the treatment of patients with HCV coinfected with HIV or HBV have not been established.

10 OVERDOSAGE

There is limited experience with overdosage. In the clinical studies, a few subjects accidentally received a dose greater than that prescribed. There were no instances in which a participant in the monotherapy or combination therapy trials received more than 10.5 times the intended dose of PegIntron. The maximum dose received by any subject was 3.45 mcg/kg weekly over a period of approximately 12 weeks. The maximum known overdosage of REBETOL was an intentional ingestion of 10 g (fifty 200 mg capsules). There were no serious reactions attributed to these overdosages. In cases of overdosing, symptomatic treatment and close observation of the patient are recommended.

11 DESCRIPTION

PegIntron, peginterferon alfa-2b, Powder for Injection is a covalent conjugate of recombinant alfa-2b interferon with monomethoxy polyethylene glycol (PEG). The average molecular weight of the PEG portion of the molecule is 12,000 daltons. The average molecular weight of the PegIntron molecule is approximately 31,000 daltons. The specific activity of peginterferon alfa-2b is approximately 0.7×10^8 IU/mg protein.
Interferon alfa-2b is a water-soluble protein with a molecular weight of 19,271 daltons produced by recombinant DNA techniques. It is obtained from the bacterial fermentation of a strain of *Escherichia coli* bearing a genetically engineered plasmid containing an interferon gene from human leukocytes.
PegIntron is supplied in both vials and the REDIPEN for subcutaneous use.
Vials: Each vial contains either 74 mcg, 118.4 mcg, 177.6 mcg, or 222 mcg of PegIntron as a white to off-white tablet-like solid that is whole/in pieces or as a loose powder, and 1.11 mg dibasic sodium phosphate anhydrous, 1.11 mg monobasic sodium phosphate dihydrate, 59.2 mg sucrose, and 0.074 mg polysorbate 80. Following reconstitution with 0.7 mL of the supplied Sterile Water for Injection USP, each vial contains PegIntron at strengths of either 50 mcg per 0.5 mL, 80 mcg per 0.5 mL, 120 mcg per 0.5 mL, or 150 mcg per 0.5 mL.
REDIPEN: REDIPEN is a dual-chamber glass cartridge containing lyophilized PegIntron as a white to off-white tablet or powder that is whole or in pieces in the sterile active chamber and a second chamber containing Sterile Water for Injection USP. Each PegIntron REDIPEN contains either 67.5 mcg, 108 mcg, 162 mcg, or 202.5 mcg of PegIntron, and 1.013 mg dibasic sodium phosphate anhydrous, 1.013 mg monobasic sodium phosphate dihydrate, 54 mg sucrose, and 0.0675 mg polysorbate 80. Each cartridge is reconstituted to allow for the administration of up to 0.5 mL of solution. Following reconstitution, each REDIPEN contains PegIntron at strengths of either 50 mcg per 0.5 mL, 80 mcg per 0.5 mL, 120 mcg per 0.5 mL, or 150 mcg per 0.5 mL for a single use. Because a small volume of reconstituted solution is lost during preparation of PegIntron, each REDIPEN contains an excess amount of PegIntron powder and diluent to ensure delivery of the labeled dose.

12 CLINICAL PHARMACOLOGY

12.1 Mechanism of Action

Pegylated recombinant human interferon alfa-2b is an inducer of the innate antiviral immune response *[see Clinical Pharmacology (12.4)]*.

12.2 Pharmacodynamics

The pharmacodynamic effects of peginterferon alfa-2b include inhibition of viral replication in virus-infected cells, the suppression of cell cycle progression/cell proliferation, induction of apoptosis, anti-angiogenic activities, and numerous immunomodulating activities, such as enhancement of the phagocytic activity of macrophages, activation of NK cells, stimulation of cytotoxic T-lymphocytes, and the upregulation of the Th1 T-helper cell subset.
PegIntron raises concentrations of effector proteins such as serum neopterin and 2'5' oligoadenylate synthetase, raises body temperature, and causes reversible decreases in leukocyte and platelet counts. The correlation between the *in vitro* and *in vivo* pharmacologic and pharmacodynamic and clinical effects is unknown.

12.3 Pharmacokinetics

Following a single subcutaneous dose of PegIntron, the mean absorption half-life ($t\ ½\ k_a$) was 4.6 hours. Maximal serum concentrations (C_{max}) occur between 15 and 44 hours postdose, and are sustained for up to 48 to 72 hours. The C_{max} and AUC measurements of PegIntron increase in a dose-related manner. After multiple dosing, there is an increase in bioavailability of PegIntron. Week 48 mean trough concentrations (320 pg/mL; range: 0, 2960) are approximately 3-fold higher than Week 4 mean trough concentrations (94 pg/mL; range: 0, 416). The mean PegIntron elimination half-life is approximately 40 hours (range: 22-60 hours) in patients with HCV infection. The apparent clearance of PegIntron is estimated to be approximately 22 mL/hr·kg. Renal elimination accounts for 30% of the clearance.
Pegylation of interferon alfa-2b produces a product (PegIntron) whose clearance is lower than that of nonpegylated interferon alfa-2b. When compared to INTRON A, PegIntron (1 mcg/kg) has approximately a 7-fold lower mean apparent clearance and a 5-fold greater mean half-life, permitting a reduced dosing frequency. At effective therapeutic doses, PegIntron has approximately 10-fold greater C_{max} and 50-fold greater AUC than interferon alfa-2b.
Renal Dysfunction: Following multiple dosing of PegIntron (1 mcg/kg subcutaneously given every week for 4 weeks) the clearance of PegIntron is reduced by a mean of 17% in subjects with moderate renal impairment (creatinine clearance 30-49 mL/min) and by a mean of 44% in subjects with severe renal impairment (creatinine clearance 10-29 mL/min) compared to subjects with normal renal function. Clearance was similar in subjects with severe renal impairment not on dialysis and subjects who are receiving hemodialysis. The dose of PegIntron for monotherapy should be reduced in patients with moderate or severe renal impairment *[see Dosage and Administration (2.3) and REBETOL Package Insert]*. REBETOL should not be used in patients with creatinine clearance <50 mL/min *[see REBETOL Package Insert, WARNINGS]*.
Gender: During the 48-week treatment period with PegIntron, no differences in the pharmacokinetic profiles were observed between male and female subjects with chronic hepatitis C infection.
Geriatric Patients: The pharmacokinetics of geriatric subjects (>65 years of age) treated with a single subcutaneous dose of 1 mcg/kg of PegIntron were similar in C_{max}, AUC, clearance, or elimination half-life as compared to younger subjects (28-44 years of age).
Pediatric Patients: Population pharmacokinetics for PegIntron and REBETOL (Capsules and Oral Solution) were evaluated in pediatric subjects with chronic hepatitis C between 3 and 17 years of age. In pediatric patients receiving PegIntron 60 mcg/m^2/week subcutaneously, exposure may be approximately 50% higher than observed in adults receiving 1.5 mcg/kg/week subcutaneously. The pharmacokinetics of REBETOL (dose-normalized) in this trial were similar to those reported in a prior study of REBETOL in combination with INTRON A in pediatric subjects and in adult subjects.
Effect of Food on Absorption of Ribavirin: Both AUC_{tf} and C_{max} increased by 70% when REBETOL Capsules were administered with a high-fat meal (841 kcal, 53.8 g fat, 31.6 g protein, and 57.4 g carbohydrate) in a single-dose pharmacokinetic study *[see Dosage and Administration (2.2)]*.
Drug Interactions: *Drugs Metabolized by Cytochrome P-450:* The pharmacokinetics of representative drugs metabolized by CYP1A2 (caffeine), CYP2C8/9 (tolbutamide), CYP2D6 (dextromethorphan), CYP3A4 (midazolam), and N-acetyltransferase (dapsone) were studied in 22 subjects with chronic hepatitis C who received PegIntron (1.5 mcg/kg) once weekly for 4 weeks. PegIntron treatment resulted in a 28% (mean) increase in a measure of CYP2C8/9 activity. PegIntron treatment also resulted in a 66% (mean) increase in a measure of CYP2D6 activity; however, the effect was variable as 13 subjects had an increase, 5 subjects had a decrease, and 4 subjects had no significant change *[see Drug Interactions (7.1)]*.
No significant effect was observed on the pharmacokinetics of representative drugs metabolized by CYP1A2, CYP3A4, or N-acetyltransferase. The effects of PegIntron on CYP2C19 activity were not assessed.
Methadone: The pharmacokinetics of concomitant administration of methadone and PegIntron were evaluated in 18 PegIntron-naïve chronic hepatitis C subjects receiving 1.5 mcg/kg PegIntron subcutaneously weekly. All subjects were on stable methadone maintenance therapy receiving >40 mg/day prior to initiating PegIntron. Mean methadone AUC was approximately 16% higher after 4 weeks of PegIntron treatment as compared to baseline. In 2 subjects, methadone AUC was approximately double after 4 weeks of PegIntron treatment as compared to baseline *[see Drug Interactions (7.2)]*.
Use with Ribavirin: *Zidovudine, Lamivudine, and Stavudine:* Ribavirin has been shown *in vitro* to inhibit phosphorylation of zidovudine, lamivudine, and stavudine. However, in a study with another pegylated interferon in combination with ribavirin, no pharmacokinetic (e.g., plasma concentrations or intracellular triphosphorylated active metabolite concentrations) or pharmacodynamic (e.g., loss of HIV/HCV virologic suppression) interaction was observed when ribavirin and lamivudine (n=18), stavudine (n=10), or zidovudine (n=6) were coadministered as part of a multi-drug regimen to HIV/HCV coinfected subjects *[see Drug Interactions (7.3)]*.
Didanosine: Exposure to didanosine or its active metabolite (dideoxyadenosine 5'-triphosphate) is increased when didanosine is coadministered with ribavirin, which could cause or worsen clinical toxicities *[see Drug Interactions (7.3)]*.

12.4 Microbiology

Mechanism of Action: The biological activity of PegIntron is derived from its interferon alfa-2b moiety. Peginterferon alfa-2b binds to and activates the human type 1 interferon receptor. Upon binding, the receptor subunits dimerize, and activate multiple intracellular signal transduction pathways. Signal transduction is initially mediated by the JAK/STAT activation, which may occur in a wide variety of cells. Interferon receptor activation also activates NFκB in many cell types. Given the diversity of cell types that respond to

interferon alfa-2b, and the multiplicity of potential intracellular responses to interferon receptor activation, peginterferon alfa-2b is expected to have pleiotropic biological effects in the body.
The mechanism by which ribavirin contributes to its antiviral efficacy in the clinic is not fully understood. Ribavirin has direct antiviral activity in tissue culture against many RNA viruses. Ribavirin increases the mutation frequency in the genomes of several viruses and ribavirin triphosphate inhibits HCV polymerase in a biochemical reaction.
Antiviral Activity: The anti-HCV activity of interferon was demonstrated in cell culture using self-replicating HCV RNA (HCV replicon cells) or HCV infection and resulted in an effective concentration (EC_{50}) value of 1 to 10 IU/mL. The antiviral activity of ribavirin in the HCV-replicon is not well understood and has not been defined because of the cellular toxicity of ribavirin.
Resistance: HCV genotypes show wide variability in their response to pegylated recombinant human interferon/ribavirin therapy. Genetic changes associated with the variable response have not been identified.
Cross-resistance: There is no reported cross-resistance between pegylated/nonpegylated interferons and ribavirin.

13 NONCLINICAL TOXICOLOGY

13.1 Carcinogenesis, Mutagenesis, Impairment of Fertility

Carcinogenesis and Mutagenesis: PegIntron has not been tested for its carcinogenic potential. Neither PegIntron nor its components, interferon or methoxypolyethylene glycol, caused damage to DNA when tested in the standard battery of mutagenesis assays, in the presence and absence of metabolic activation.
Use with Ribavirin: Ribavirin is genotoxic and mutagenic and should be considered a potential carcinogen. See REBETOL package insert for additional warnings relevant to PegIntron therapy in combination with ribavirin.
Impairment of Fertility: PegIntron may impair human fertility. Irregular menstrual cycles were observed in female cynomolgus monkeys given subcutaneous injections of 4239 mcg/m^2 PegIntron alone every other day for 1 month (approximately 345 times the recommended weekly human dose based upon body surface area). These effects included transiently decreased serum levels of estradiol and progesterone, suggestive of anovulation. Normal menstrual cycles and serum hormone levels resumed in these animals 2 to 3 months following cessation of PegIntron treatment. Every other day dosing with 262 mcg/m^2 (approximately 21 times the weekly human dose) had no effects on cycle duration or reproductive hormone status. The effects of PegIntron on male fertility have not been studied.

14 CLINICAL STUDIES

14.1 Chronic Hepatitis C in Adults

PegIntron Monotherapy – Study 1: A randomized study compared treatment with PegIntron (0.5, 1, or 1.5 mcg/kg once weekly subcutaneously) to treatment with INTRON A (3 million units 3 times weekly subcutaneously) in 1219 adults with chronic hepatitis from HCV infection. The subjects were not previously treated with interferon alpha, had compensated liver disease, detectable HCV RNA, elevated ALT, and liver histopathology consistent with chronic hepatitis. Subjects were treated for 48 weeks and were followed for 24 weeks posttreatment.
Seventy percent of all subjects were infected with HCV genotype 1, and 74 percent of all subjects had high baseline levels of HCV RNA (more than 2 million copies per mL of serum), 2 factors known to predict poor response to treatment.
Response to treatment was defined as undetectable HCV RNA and normalization of ALT at 24 weeks posttreatment. The response rates to the 1 and 1.5 mcg/kg PegIntron doses were similar (approximately 24%) to each other and were both higher than the response rate to INTRON A (12%) (see **Table 12**).

TABLE 12: Rates of Response to Treatment – Study 1

	A PegIntron 0.5 mcg/kg (N=315)	B PegIntron 1 mcg/kg (N=298)	C INTRON A 3 MIU three times weekly (N=307)	B-C (95% CI) Difference between PegIntron 1 mcg/kg and INTRON A
Treatment Response (Combined Virologic Response and ALT Normalization)	17%	24%	12%	11 (5,18)
Virologic Response*	18%	25%	12%	12 (6,19)
ALT Normalization	24%	29%	18%	11 (5,18)

*Serum HCV is measured by a research-based quantitative polymerase chain reaction assay by a central laboratory.

[See table 12 above]
Subjects with both viral genotype 1 and high serum levels of HCV RNA at baseline were less likely to respond to treatment with PegIntron. Among subjects with the 2 unfavorable prognostic variables, 8% (12/157) responded to PegIntron treatment and 2% (4/169) responded to INTRON A. Doses of PegIntron higher than the recommended dose did not result in higher response rates in these subjects.
Subjects receiving PegIntron with viral genotype 1 had a response rate of 14% (28/199) while subjects with other viral genotypes had a 45% (43/96) response rate.
Ninety-six percent of the responders in the PegIntron groups and 100% of responders in the INTRON A group first cleared their viral RNA by Week 24 of treatment *[see Dosage and Administration (2)]*.
The treatment response rates were similar in men and women. Response rates were lower in African-American and Hispanic subjects and higher in Asians compared to Caucasians. Although African Americans had a higher proportion of poor prognostic factors compared to Caucasians, the number of non-Caucasians studied (9% of the total) was insufficient to allow meaningful conclusions about differences in response rates after adjusting for prognostic factors.
Liver biopsies were obtained before and after treatment in 60% of subjects. A modest reduction in inflammation compared to baseline that was similar in all 4 treatment groups was observed.
PegIntron/REBETOL Combination Therapy – Study 2: A randomized study compared treatment with 2 PegIntron/REBETOL regimens [PegIntron 1.5 mcg/kg subcutaneously once weekly/REBETOL 800mg orally daily (in divided doses); PegIntron 1.5 mcg/kg subcutaneously once weekly for 4 weeks then 0.5 mcg/kg subcutaneously once weekly for 44 weeks/REBETOL 1000 or 1200 mg orally daily (in divided doses)] with INTRON A [3 MIU subcutaneously thrice weekly/REBETOL 1000 or 1200 mg orally daily (in divided doses)] in 1530 adults with chronic hepatitis C. Interferon-naïve subjects were treated for 48 weeks and followed for 24 weeks posttreatment. Eligible subjects had compensated liver disease, detectable HCV RNA, elevated ALT, and liver histopathology consistent with chronic hepatitis.
Response to treatment was defined as undetectable HCV RNA at 24 weeks posttreatment. The response rate to the PegIntron 1.5 mcg/kg plus ribavirin 800 mg dose was higher than the response rate to INTRON A/REBETOL (see **Table 13**). The response rate to PegIntron 1.5→0.5 mcg/kg/REBETOL was essentially the same as the response to INTRON A/REBETOL (data not shown).

TABLE 13: Rates of Response to Treatment – Study 2

	PegIntron 1.5 mcg/kg once weekly REBETOL 800 mg daily	INTRON A 3 MIU three times weekly REBETOL 1000/1200 mg daily
Overall response*,†	52% (264/511)	46% (231/505)
Genotype 1	41% (141/348)	33% (112/343)
Genotype 2-6	75% (123/163)	73% (119/162)

* Serum HCV RNA is measured with a research-based quantitative polymerase chain reaction assay by a central laboratory.
† Difference in overall treatment response (PegIntron/REBETOL vs. INTRON A/REBETOL) is 6% with 95% confidence interval of (0.18, 11.63) adjusted for viral genotype and presence of cirrhosis at baseline. Response to treatment was defined as undetectable HCV RNA at 24 weeks posttreatment.

[See table 13 above]
Subjects with viral genotype 1, regardless of viral load, had a lower response rate to PegIntron (1.5 mcg/kg)/REBETOL (800 mg) compared to subjects with other viral genotypes. Subjects with both poor prognostic factors (genotype 1 and high viral load) had a response rate of 30% (78/256) compared to a response rate of 29% (71/247) with INTRON A/REBETOL.
Subjects with lower body weight tended to have higher adverse reaction rates *[see Adverse Reactions (6.1)]* and higher response rates than subjects with higher body weights. Differences in response rates between treatment arms did not substantially vary with body weight.
Treatment response rates with PegIntron/REBETOL were 49% in men and 56% in women. Response rates were lower in African-American and Hispanic subjects and higher in Asians compared to Caucasians. Although African Americans had a higher proportion of poor prognostic factors compared to Caucasians, the number of non-Caucasians studied (11% of the total) was insufficient to allow meaningful conclusions about differences in response rates after adjusting for prognostic factors in this study.
Liver biopsies were obtained before and after treatment in 68% of subjects. Compared to baseline, approximately two-thirds of subjects in all treatment groups were observed to have a modest reduction in inflammation.
PegIntron/REBETOL Combination Therapy – Study 3: In a large United States community-based study (Study 3), 4913 subjects with chronic hepatitis C were randomized to receive PegIntron 1.5 mcg/kg subcutaneously once weekly in combination with a REBETOL dose of 800 to 1400 mg (weight-based dosing [WBD]) or 800 mg (flat) orally daily (in divided doses) for 24 or 48 weeks based on genotype. Response to treatment was defined as undetectable HCV RNA (based on an assay with a lower limit of detection of 125 IU/mL) at 24 weeks posttreatment.
Treatment with PegIntron 1.5 mcg/kg and REBETOL 800 to 1400 mg resulted in a higher sustained virologic response compared to PegIntron in combination with a flat 800 mg daily dose of REBETOL. Subjects weighing >105 kg obtained the greatest benefit with WBD, although a modest benefit was also observed in subjects weighing >85 to 105 kg (see **Table 14**). The benefit of WBD in subjects weighing >85 kg was observed with HCV genotypes 1 through 3. Insufficient data were available to reach conclusions regarding other genotypes. Use of WBD resulted in an increased incidence of anemia *[see Adverse Reactions (6.1)]*.

TABLE 14: SVR Rate by Treatment and Baseline Weight – Study 3

Treatment Group	Subject Baseline Weight			
	<65 kg (<143 lb)	65-85 kg (143-188 lb)	>85-105 kg (>188-231 lb)	>105 kg (>231 lb)
WBD*	50% (173/348)	45% (449/994)	42% (351/835)	47% (138/292)
Flat	51% (173/342)	44% (443/1011)	39% (318/819)	33% (91/272)

*P=0.01, primary efficacy comparison (based on data from subjects weighing 65 kg or higher at baseline and utilizing a logistic regression analysis that includes treatment [WBD or Flat], genotype and presence/absence of advanced fibrosis, in the model).

Continued on next page

Information on Schering products appearing on these pages is effective as of July 2009.

PegIntron—Cont.

[See table 14 at top of previous page]
A total of 1552 subjects weighing >65 kg in Study 3 had genotype 2 or 3 and were randomized to 24 or 48 weeks of therapy. No additional benefit was observed with the longer treatment duration.

PegIntron/REBETOL Combination Therapy-Study 4: A large randomized study compared the safety and efficacy of treatment for 48 weeks with 2 PegIntron/REBETOL regimens [PegIntron 1.5 mcg/kg and 1 mcg/kg subcutaneously once weekly both in combination with REBETOL 800 to 1400 mg PO daily (in 2 divided doses)] and Pegasys 180 mcg subcutaneously once weekly in combination with Copegus 1000 to 1200 mg PO daily (in 2 divided doses) in 3070 treatment-naïve adults with chronic hepatitis C genotype 1. In this study, lack of early virologic response by treatment Week 12 (subjects who do not achieve undetectable HCV-RNA or ≥2 $\log_{10}$ reduction from baseline) was the criteria for discontinuation of treatment. Sustained Virologic Response (SVR) to the treatment was defined as undetectable HCV-RNA (Roche COBAS TaqMan assay, a lower limit of quantitation of 27 IU/mL) at 24 weeks posttreatment (see **Table 15**).

TABLE 15: Response Rate by Treatment

Treatment Group	% (number) of Patients		
	PegIntron (1.5 mcg/kg)/ REBETOL	PegIntron (1 mcg/kg)/ REBETOL	Pegasys 180 mcg/ Copegus
SVR	40 (406/1019)	38 (386/1016)	41 (423/1035)

In all 3 treatment groups, overall SVR rates were similar. In subjects with poor prognostic factors, subjects randomized to PegIntron (1.5 mcg/kg)/REBETOL or Pegasys/Copegus achieved higher SVR rates compared to those randomized to the PegIntron (1 mcg/kg)/REBETOL arm. In all arms, SVR rates were lower in subjects with poor prognostic factors compared to those without. For the PegIntron 1.5 mcg/kg plus REBETOL dose, SVR rates for those with and without, respectively, the following baseline factors were as follows: cirrhosis (10% vs. 42%), normal ALT levels (32% vs. 42%), baseline viral load >600,000 IU/mL (35% vs. 61%), >40 years old (38% vs. 50%), and African-American subjects (23% vs. 44%). In subjects with undetectable HCV-RNA at treatment Week 12 who received PegIntron (1.5 mcg/kg)/ REBETOL, the SVR rate was 81% (328/407).

PegIntron/REBETOL Combination Therapy in Prior Treatment Failures – Study 5: In a noncomparative trial, 2293 patients with moderate to severe fibrosis who failed previous treatment with combination alpha interferon/ribavirin were re-treated with PegIntron, 1.5 mcg/kg subcutaneously, once weekly, in combination with weight-adjusted ribavirin. Eligible patients included prior nonresponders (patients who were HCV-RNA positive at the end of a minimum 12 weeks of treatment) and prior relapsers (patients who were HCV-RNA negative at the end of a minimum 12 weeks of treatment and subsequently relapsed after posttreatment follow-up). Patients who were negative at Week 12 were treated for 48 weeks and followed for 24 weeks posttreatment. Response to treatment was defined as undetectable HCV-RNA at 24 weeks posttreatment (measured using a research-based test, limit of detection 125 IU/mL). The overall response rate was 22% (497/2293) (99% CI: 19.5, 23.9). Patients with the following characteristics were less likely to benefit from re-treatment: previous nonresponse, previous pegylated interferon treatment, significant bridging fibrosis or cirrhosis, and genotype 1 infection.

The re-treatment sustained virologic response rates by baseline characteristics are summarized in **Table 16**.

[See table 16 above]

TABLE 16: SVR Rates by Baseline Characteristics of Prior Treatment Failures

HCV Genotype/ Metavir Fibrosis Score	Overall SVR by Previous Response and Treatment			
	Nonresponder		Relapser	
	Alfa Interferon/Ribavirin % (Number of Patients)	Peginterferon (2a and 2b Combined)/Ribavirin % (Number of Patients)	Alfa Interferon/Ribavirin % (Number of Patients)	Peginterferon (2a and 2b Combined)/Ribavirin % (Number of Patients)
Overall	18 (158/903)	6 (30/476)	43 (130/300)	35 (113/344)
HCV 1	13 (98/761)	4 (19/431)	32 (67/208)	23 (56/243)
F2	18 (36/202)	6 (7/117)	42 (33/79)	32 (23/72)
F3	16 (38/233)	4 (4/112)	28 (16/58)	21 (14/67)
F4	7 (24/325)	4 (8/202)	26 (18/70)	18 (19/104)
HCV 2/3	49 (53/109)	36 (10/28)	67 (54/81)	57 (52/92)
F2	68 (23/34)	56 (5/9)	76 (19/25)	61 (11/18)
F3	39 (11/28)	38 (3/8)	67 (18/27)	62 (18/29)
F4	40 (19/47)	18 (2/11)	59 (17/29)	51 (23/45)
HCV 4	17 (5/29)	7 (1/15)	88 (7/8)	50 (4/8)

Achievement of an undetectable HCV-RNA at treatment Week 12 was a strong predictor of sustained virologic response (SVR). In this trial, 1470 (64%) subjects did not achieve an undetectable HCV-RNA at treatment Week 12, and were offered enrollment into long-term treatment trials, due to an inadequate treatment response. Of the 823 (36%) subjects who were HCV-RNA undetectable at treatment Week 12, those infected with genotype 1 had an SVR of 48% (245/507), with a range of responses by fibrosis scores (F4-F2) of 39-55%. Subjects infected with genotype 2/3 who were HCV-RNA undetectable at treatment Week 12 had an overall SVR of 70% (196/281), with a range of responses by fibrosis scores (F4-F2) of 60-83%. For all genotypes, higher fibrosis scores were associated with a decreased likelihood of achieving SVR.

14.2 Chronic Hepatitis C in Pediatrics

PegIntron/REBETOL Combination Therapy – Pediatric Study: Previously untreated pediatric subjects 3 to 17 years of age with compensated chronic hepatitis C and detectable HCV RNA were treated with REBETOL 15 mg/kg/day plus PegIntron 60 mcg/m^2 once weekly for 24 or 48 weeks based on HCV genotype and baseline viral load. All subjects were to be followed for 24 weeks posttreatment. A total of 107 subjects received treatment, of whom 52% were female, 89% were Caucasian, and 67% were infected with HCV genotype 1. Subjects infected with genotype 1, 4 or genotype 3 with HCV RNA ≥600,000 IU/mL received 48 weeks of therapy while those infected with genotype 2 or genotype 3 with HCV RNA <600,000 IU/mL received 24 weeks of therapy. The study results are summarized in **Table 17**.

TABLE 17: Sustained Virologic Response Rates by Genotype and Treatment Duration – Pediatric Study

Genotype	All Subjects n=107	
	24 Weeks	48 Weeks
	Virologic Response n*† (%)	Virologic Response n*† (%)
All	26/27(96.3)	44/80(55.0)
1	—	38/72(52.8)
2	14/15(93.3)	—
3‡	12/12(100)	2/3(66.7)
4	—	4/5(80.0)

* Response to treatment was defined as undetectable HCV RNA at 24 weeks posttreatment.
† n=number of responders/number of subjects with given genotype, and assigned treatment duration.
‡ Subjects with genotype 3 low viral load (<600,000 IU/mL) were to receive 24 weeks of treatment while those with genotype 3 and high viral load were to receive 48 weeks of treatment.

16 HOW SUPPLIED/STORAGE AND HANDLING

PegIntron REDIPEN

Each PegIntron REDIPEN Package Contains:

A box containing one 50 mcg per 0.5 mL PegIntron REDIPEN and 1 BD needle and 2 alcohol swabs.	(NDC 0085-1323-01)
A box containing one 80 mcg per 0.5 mL PegIntron REDIPEN and 1 BD needle and 2 alcohol swabs.	(NDC 0085-1316-01)
A box containing one 120 mcg per 0.5 mL PegIntron REDIPEN and 1 BD needle and 2 alcohol swabs.	(NDC 0085-1297-01)
A box containing one 150 mcg per 0.5 mL PegIntron REDIPEN and 1 BD needle and 2 alcohol swabs.	(NDC 0085-1370-01)

Each PegIntron REDIPEN PAK 4 Contains:

A box containing four 50 mcg per 0.5 mL PegIntron REDIPEN Units, each containing 1 BD needle and 2 alcohol swabs.	(NDC 0085-1323-02)
A box containing four 80 mcg per 0.5 mL PegIntron REDIPEN Units, each containing 1 BD needle and 2 alcohol swabs.	(NDC 0085-1316-02)
A box containing four 120 mcg per 0.5 mL PegIntron REDIPEN Units, each containing 1 BD needle and 2 alcohol swabs.	(NDC 0085-1297-02)
A box containing four 150 mcg per 0.5 mL PegIntron REDIPEN Units, each containing 1 BD needle and 2 alcohol swabs.	(NDC 0085-1370-02)

PegIntron Vials

Each PegIntron Package Contains:

A box containing one 50 mcg per 0.5 mL vial of PegIntron Powder for Injection and one 1.25 mL vial of Diluent (Sterile Water for Injection USP), 2 BD Safety-Lok syringes with a safety sleeve and 2 alcohol swabs.	(NDC 0085-1368-01)
A box containing one 80 mcg per 0.5 mL vial of PegIntron Powder for Injection and one 1.25 mL vial of Diluent (Sterile Water for Injection USP), 2 BD Safety-Lok syringes with a safety sleeve and 2 alcohol swabs.	(NDC 0085-1291-01)
A box containing one 120 mcg per 0.5 mL vial of PegIntron Powder for Injection and one 1.25 mL vial of Diluent (Sterile Water for Injection USP), 2 BD Safety-Lok syringes with a safety sleeve and 2 alcohol swabs.	(NDC 0085-1304-01)

A box containing one 150 mcg per 0.5 mL vial of PegIntron Powder for Injection and one 1.25 mL vial of Diluent (Sterile Water for Injection USP), 2 BD Safety-Lok syringes with a safety sleeve and 2 alcohol swabs.	(NDC 0085-1279-01)

Storage: *PegIntron REDIPEN:* PegIntron REDIPEN should be stored at 2°-8°C (36°-46°F).
After reconstitution, the solution should be used immediately, but may be stored up to 24 hours at 2°-8°C (36°-46°F). The reconstituted solution contains no preservative, and is clear and colorless. **DO NOT FREEZE.**
PegIntron Vials: PegIntron should be stored at 25°C (77°F); excursions permitted to 15°-30°C (59°-86°F) [see USP Controlled Room Temperature]. After reconstitution with supplied Diluent the solution should be used immediately, but may be stored up to 24 hours at 2°-8°C (36°-46°F). The reconstituted solution contains no preservative, and is clear and colorless. **DO NOT FREEZE.**
Disposal Instructions: Patients should be thoroughly instructed in the importance of proper disposal. After preparation and administration of PegIntron for Injection, patients should be advised to use a puncture-resistant container for the disposal of used syringes, needles, and the REDIPEN. The full container should be disposed of in accordance with state and local laws. Patients should also be cautioned against reusing or sharing needles, syringes, or the REDIPEN.

17 PATIENT COUNSELING INFORMATION

A patient should self-inject PegIntron only if it has been determined that it is appropriate, the patient agrees to medical follow-up as necessary, and training in proper injection technique has been given to him/her.

17.1 Medication Guide

Patients receiving PegIntron alone or in combination with REBETOL should be directed in its appropriate use, informed of the benefits and risks associated with treatment, and referred to the MEDICATION GUIDES for PegIntron and, if applicable, REBETOL (ribavirin).

17.2 Pregnancy

Patients must be informed that REBETOL may cause birth defects and death of the unborn child. Extreme care must be taken to avoid pregnancy in female patients and in female partners of male patients during treatment with combination PegIntron/REBETOL therapy and for 6 months post-therapy. Combination PegIntron/REBETOL therapy should not be initiated until a report of a negative pregnancy test has been obtained immediately prior to initiation of therapy. It is recommended that patients undergo monthly pregnancy tests during therapy and for 6 months posttherapy *[see Contraindications (4), Use in Specific Populations (8.1), and REBETOL package insert]*.

17.3 HCV Transmission

Inform patients that there are no data regarding whether PegIntron therapy will prevent transmission of HCV infection to others. Also, it is not known if treatment with PegIntron will cure hepatitis C or prevent cirrhosis, liver failure, or liver cancer that may be the result of infection with the hepatitis C virus.

17.4 Laboratory Evaluations, Hydration, "Flu-like" Symptoms

Patients should be advised that laboratory evaluations are required before starting therapy and periodically thereafter *[see Warnings and Precautions (5.15)]*. It is advised that patients be well-hydrated, especially during the initial stages of treatment. "Flu-like" symptoms associated with administration of PegIntron may be minimized by bedtime administration of PegIntron or by use of antipyretics.

Manufactured by Schering Corporation, a subsidiary of Schering-Plough Corporation, Kenilworth, NJ 07033 USA.
Schering-Plough
U.S. Patent Nos. 5,908,621; 5,951,974; 6,042,822; 6,177,074; 6,180,096; 6,250,469; 6,482,613; 6,524,570; and 6,610,830.

Rev. 5/09 **33538804T**
BD and Safety-Lok are registered trademarks of Becton, Dickinson and Company.

MEDICATION GUIDE
PegIntron®
(Peginterferon alfa-2b)
Including appendix with instructions for using PegIntron® Powder for Injection

Read this Medication Guide carefully before you start taking PegIntron® **(Peg In-tron)** or PegIntron/REBETOL® **(REB-eh-tole)** combination therapy. Read the Medication Guide each time you refill your prescription because there may be new information. The information in this Medication Guide does not take the place of talking with your health care provider (doctor, nurse, nurse practitioner, or physician's assistant).

If you are taking PegIntron/REBETOL combination therapy, also read the Medication Guide for REBETOL (ribavirin USP) Capsules and Oral Solution.

What is the most important information I should know about PegIntron and PegIntron/REBETOL combination therapy?

PegIntron (peginterferon) is a treatment for some people who are infected with hepatitis C virus. However, PegIntron and PegIntron/REBETOL combination therapy can have serious side effects that may cause death in rare cases. Before you decide to start treatment, you should talk to your health care provider about the possible benefits and side effects of PegIntron or PegIntron/REBETOL combination therapy. If you begin treatment you will need to see your health care provider regularly for medical examinations and lab tests to make sure your treatment is working and to check for side effects.

REBETOL may cause birth defects and/or death of an unborn child. If you are pregnant, you or your male partner must not take PegIntron/REBETOL combination therapy. You must not become pregnant while either you or your partner are being treated with the combination PegIntron/REBETOL therapy, or for 6 months after stopping therapy. Men and women should use birth control while taking the combination therapy and for 6 months afterwards. If you or your partner are being treated and you become pregnant either during treatment or within 6 months of stopping treatment, call your health care provider right away. There is a Ribavirin Pregnancy Registry that collects information about pregnancy outcomes of female patients and female partners of male patients exposed to ribavirin. You or your health care provider are encouraged to contact the Registry at 1-800-593-2214.

If you are taking PegIntron or PegIntron/REBETOL therapy you should call your health care provider immediately if you develop any of these symptoms:

New or worsening mental health problems such as thoughts about killing or hurting yourself or others, trouble breathing, chest pain, severe stomach or lower back pain, bloody diarrhea or bloody bowel movements, high fever, bruising, bleeding, or decreased vision.

The most serious possible side effects of PegIntron and PegIntron/REBETOL therapy include:

Problems with Pregnancy. Combination PegIntron/REBETOL therapy can cause death, serious birth defects, or other harm to your unborn child. If you are a woman of childbearing age, you must not become pregnant during treatment and for 6 months after you have stopped therapy. You must have a negative pregnancy test immediately before beginning treatment, during treatment, and for 6 months after you have stopped therapy. Both male and female patients must use effective forms of birth control during treatment and for the 6 months after treatment is completed. Male patients should use a condom. If you are a female, you must use birth control even if you believe that you are not fertile or that your fertility is low. You should talk to your health care provider about birth control for you and your partner.

Mental health problems and suicide. PegIntron and PegIntron/REBETOL therapies may cause patients to develop mood or behavioral problems. These can include irritability (getting easily upset) and depression (feeling low, feeling bad about yourself, or feeling hopeless). Some patients may have aggressive behavior. Former drug addicts may fall back into drug addiction or overdose. Some patients think about hurting or killing themselves or other people and some have killed (suicide) or hurt themselves or others. You must tell your health care provider if you are being treated for a mental illness or had treatment in the past for any mental illness, including depression and suicidal behavior. You should tell your health care provider if you have ever been addicted to drugs or alcohol.

Heart problems. Some patients taking PegIntron or PegIntron/REBETOL therapy may develop problems with their heart, including low blood pressure, fast heart rate, and very rarely, heart attacks. Tell your health care provider if you have had any heart problems in the past.

Blood problems. PegIntron and PegIntron/REBETOL therapies commonly lower two types of blood cells (white blood cells and platelets). In some patients, these blood counts may fall to dangerously low levels. If your blood counts become very low, this could lead to infections or bleeding.

REBETOL therapy causes a decrease in the number of red blood cells you have (anemia). This can be dangerous, especially for patients who already have heart or circulatory (cardiovascular) problems. Talk with your health care provider before taking combination PegIntron/REBETOL therapy if you have or have ever had any cardiovascular problems.

Body organ problems. Certain symptoms like severe stomach pain may mean that your internal organs are being damaged. Cases of weakness, loss of coordination, and numbness due to stroke have been reported in patients taking PegIntron, including patients with few or no reported risk factors for stroke.

For other possible side effects, see "What are the possible side effects of PegIntron and PegIntron/REBETOL combination therapy?" in this Medication Guide.

What is PegIntron and PegIntron/REBETOL combination therapy?

The PegIntron product is a drug used to treat adults who have a lasting (chronic) infection with hepatitis C virus and who show signs that the virus is damaging the liver. PegIntron/REBETOL combination therapy consists of two medications also used to treat hepatitis C infection in adults and children 3 years of age and older. Patients with hepatitis C have the virus in their blood and in their liver. PegIntron reduces the amount of virus in the body and helps the body's immune system fight the virus. REBETOL (ribavirin) is a drug that helps to fight the viral infection but does not work when used by itself to treat chronic hepatitis C.

It is not known if PegIntron or PegIntron/REBETOL therapies can cure hepatitis C (permanently eliminate the virus), or if it can prevent liver failure or liver cancer that is caused by hepatitis C infection.

It is also not known if PegIntron or PegIntron/REBETOL combination therapy will prevent one infected person from infecting another person with hepatitis C.

Who should not take PegIntron or PegIntron/REBETOL therapy?

Do not take PegIntron or PegIntron/REBETOL therapy if you:

- are pregnant, planning to get pregnant during treatment, or during the 6 months after treatment, or breastfeeding.
- are a male patient with a female sexual partner who is pregnant, or plans to become pregnant at any time while you are being treated with REBETOL, or during the 6 months after your treatment has ended.
- have hepatitis caused by your immune system attacking your liver (autoimmune hepatitis) or unstable liver disease.
- had an allergic reaction to another alpha interferon or are allergic to any of the ingredients in PegIntron or REBETOL Capsules or Oral Solution. If you have any doubts, ask your health care provider.
- Do not take PegIntron/REBETOL combination therapy if you have abnormal red blood cells such as is seen in sickle-cell anemia or thalassemia major.

If you have any of the following conditions or serious medical problems, discuss them with your health care provider before taking PegIntron or PegIntron/REBETOL therapy:

- depression or anxiety
- sleep problems
- high blood pressure
- previous heart attack, or other heart problems
- liver problems (other than hepatitis C infection)
- any kind of autoimmune disease (where the body's immune system attacks the body's own cells), such as psoriasis, systemic lupus erythematosus, rheumatoid arthritis
- thyroid problems
- diabetes
- colitis (inflammation of the bowels)
- cancer
- hepatitis B infection
- HIV infection
- kidney problems
- bleeding problems
- alcoholism
- drug abuse or addiction
- body organ transplant and are taking medicine that keeps your body from rejecting your transplant (suppresses your immune system)

How should I take PegIntron or PegIntron/REBETOL?

Your health care provider will decide whether you will take PegIntron therapy alone or the combination of PegIntron/REBETOL, as well as the correct dose (for adults the dose of PegIntron is based on weight). For children 3 years of age and older, your health care provider will recommend the dose of PegIntron based on body surface area. PegIntron and PegIntron/REBETOL are given for up to 1 year. Take your prescribed dose of PegIntron ONCE A WEEK, on the same day of each week and at approximately the same time. Take the medicine for the full course of prescribed therapy and do not take more than the prescribed dose. REBETOL should be taken with food. When you take REBETOL with food, more of the medicine (70% more on average) is taken up by your body. You should take REBETOL the same way every day (twice a day with food) to keep the medicine in your body at a steady level. This will help your health care provider to decide how your treatment is working and how to change the dose of REBETOL you take if you have side effects from REBETOL. **Be sure to read the Medication**

Continued on next page

Information on Schering products appearing on these pages is effective as of July 2009.

PegIntron—Cont.

Guide for REBETOL (ribavirin USP) for complete instructions on how to take the REBETOL Capsules and Oral Solution.

You should be completely comfortable with how to prepare PegIntron, how to set the dose you take, and how to inject yourself before you use PegIntron for the first time. PegIntron comes in two different forms, a powder in a single use vial and a REDIPEN® single-use delivery system. See the attached appendix for detailed instructions for preparing and giving a dose of PegIntron.

If you miss a dose of the PegIntron product, take the missed dose as soon as possible during the same day or the next day, then continue on your regular dosing schedule. If several days go by after you miss a dose, check with your health care provider about what to do. Do not double the next dose or take more than one dose a week without talking to your health care provider. Call your health care provider right away if you take more than your prescribed PegIntron dose. Your health care provider may wish to examine you more closely, and take blood for testing.

If you miss a dose of REBETOL, take the missed dose as soon as possible during the same day. If an entire day has gone by, check with your health care provider about what to do. Do not double the next dose.

You must get regular blood tests to help your health care provider check how the treatment is working and to check for side effects.

Tell your health care provider if you are taking or planning to take other prescription or non-prescription medicines, including vitamin and mineral supplements and herbal medicines.

What should I avoid while taking PegIntron or PegIntron/REBETOL therapies?

- If you are pregnant do not start taking PegIntron/REBETOL combination therapy.
- Avoid becoming pregnant while taking PegIntron or PegIntron/REBETOL. PegIntron and PegIntron/REBETOL may harm your unborn child (death or serious birth defects) or cause you to lose your baby (miscarriage). **If you or your partner become pregnant during treatment or during the 6 months after treatment with PegIntron/REBETOL combination therapy, immediately report the pregnancy to your health care provider. You or your health care provider should call 1-800-593-2214.** By calling this number, information about you and/or your partner will be added to a pregnancy registry that will be used to help you and your health care provider make decisions about your treatment for hepatitis in the future. You, your partner, and/or your health care provider will be asked to provide follow-up information on the outcome of the pregnancy.
- Do not breastfeed your baby while taking PegIntron.

What are the possible side effects of PegIntron and PegIntron/REBETOL combination therapy?

Possible serious side effects include:

- **Mental health problems including suicide, blood problems, heart problems, body organ problems.** See "What is the most important information I should know about PegIntron and PegIntron/REBETOL combination therapy?"
- **Other body organ problems.** A few patients have lung problems (such as pneumonia or inflammation of the lung tissue), inflammation of the kidney, and eye disorders.
- **New or worsening autoimmune disease.** Some patients taking PegIntron or PegIntron/REBETOL develop autoimmune diseases (a condition where the body's immune cells attack other cells or organs in the body), including rheumatoid arthritis, systemic lupus erythematosus, and psoriasis. In some patients who already have an autoimmune disease, the disease worsens on PegIntron and PegIntron/REBETOL combination therapy.

Common but less serious side effects include:

- **Flu-like symptoms.** Most patients who take PegIntron or PegIntron/REBETOL therapy have "flu-like" symptoms (headache, muscle aches, tiredness, and fever). Some of these symptoms (fever, headache) usually lessen after the first few weeks of therapy. You can reduce some of these symptoms by injecting your PegIntron dose at bedtime. Over-the-counter pain and fever reducers, such as acetaminophen or ibuprofen, can be used to prevent or reduce the fever and headache.
- **Extreme fatigue (tiredness).** Many patients become extremely tired while on PegIntron or PegIntron/REBETOL combination therapy.
- **Appetite problems.** Nausea, loss of appetite, and weight loss occur commonly.
- **Thyroid problems.** Some patients develop changes in the function of their thyroid. Symptoms of thyroid changes include the inability to concentrate, feeling cold or hot all the time, a change in your weight, and changes to your skin.
- **Blood sugar problems.** Some patients develop problems with the way their body controls their blood sugar, and may develop high blood sugar or diabetes.
- **Skin reactions.** Redness, swelling, and itching are common at the site of injection. If after several days these symptoms do not disappear contact your health care provider. You may get a rash during therapy. If this occurs, your health care provider may recommend medicine to treat the rash.
- **Hair thinning.** Hair thinning is common during PegIntron and PegIntron/REBETOL treatment. Hair loss stops and hair growth returns after therapy is stopped.
- **Effect on growth in children.** Weight loss and slowed growth are common in children during treatment with PegIntron/REBETOL. Catch-up weight gain and some catch-up in growth occur after the end of treatment, but some children may not reach their pretreatment expected height.

These are not all of the side effects of PegIntron or PegIntron/REBETOL combination therapy. Your health care provider or pharmacist can give you a more complete list.

Call your doctor for medical advice about side effects. You may report side effects to FDA at 1-800-FDA-1088.

General advice about prescription medicines:

Medicines are sometimes prescribed for purposes other than those listed in a Medication Guide. If you have any concerns about PegIntron, ask your health care provider. Your health care provider or pharmacist can give you information about PegIntron that was written for health care professionals. Do not use PegIntron for a condition for which it was not prescribed. Do not share this medication with other people.

If you are taking PegIntron/REBETOL combination therapy, also read the Medication Guide for REBETOL (ribavirin USP) Capsules and Oral Solution.

This Medication Guide has been approved by the U.S. Food and Drug Administration.
Revised: December 2008

How do I prepare and inject the PegIntron dose?

Before you inject PegIntron, the powder must be mixed with **0.7 mL** of the supplied DILUENT for PegIntron, Sterile Water for Injection (diluent). This product can also be administered by a parent or caretaker as instructed by your health care provider. You should carefully follow the directions given to you by your health care provider.

The vial of mixed PegIntron should be used immediately. DO NOT prepare more than one vial at a time. If you don't use the vial of the prepared solution right away, it must be stored in a refrigerator and used within 24 hours.

Storing PegIntron

PegIntron Powder should be stored at room temperature (25°C, 77°F); avoid exposure to heat. After mixing, the PegIntron solution should be used immediately but may be stored in the refrigerator up to 24 hours. The solution contains no preservatives. DO NOT FREEZE.

Preparing the PegIntron solution:

1. Find a clean, well-lit, non-slip flat working surface and assemble all of the supplies you will need for an injection. All of the supplies you will need for an injection are in the PegIntron Powder for Injection package. The package contains:

- a vial of PegIntron powder
- a 1.25-mL vial of DILUENT
- 2 disposable syringes, and
- alcohol swabs

2. Check the date printed on the PegIntron carton to make sure that the expiration date has not passed. Remove one vial and look at the contents. The PegIntron in the vial should appear as a white to off-white tablet-like solid that is whole/in pieces or as a loose powder.

If you have already mixed the PegIntron solution and it has been stored properly in the refrigerator, take it out of the refrigerator and allow the solution to come to room temperature.

3. Wash your hands thoroughly with soap and water, rinse and towel dry. It is important to keep your work area, your hands, and injection site clean to minimize the risk of infection.

The disposable syringes have needles that are already attached and cannot be removed. Each syringe has a clear plastic safety sleeve that is pulled over the needle for disposal after use. The safety sleeve should remain tight against the flange while using the syringe and moved over the needle only when ready for disposal. **Figure A.**

The syringes and needles are for single use only.

[See figure A at top of next column]

4. Remove the protective wrapper from ONE of the syringes provided and use for the following steps 5-7. Make sure that the syringe safety sleeve is sitting against the flange (see **Figure A**).

5. Remove the protective plastic cap from the tops of both the supplied DILUENT and the PegIntron vials. Clean the rubber stopper on the top of both vials with an alcohol swab.

6. Carefully remove the protective cap straight off of the needle to avoid damaging the needle point. Fill the syringe

Figure A

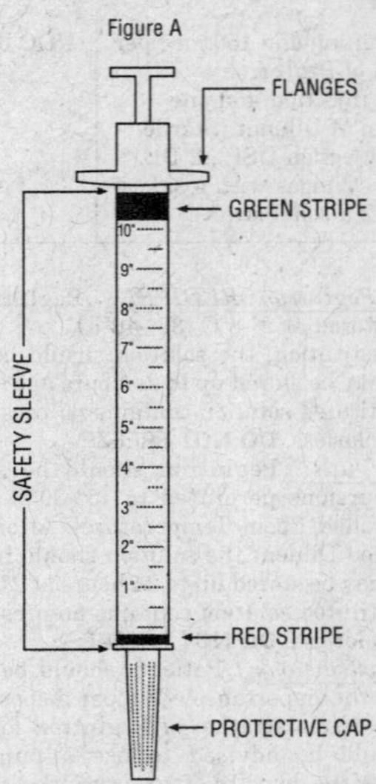

with air by pulling the plunger to 0.7 mL (**Figure B**). Hold the DILUENT vial upright. Do not touch the cleaned top of the vial with your hands (**Figure C**). Insert the needle through the center of the rubber stopper of the DILUENT vial, and inject the air from the syringe into the vial (**Figure D**). Turn the vial upside down and make sure the tip of the needle is in the liquid. **Withdraw only 0.7 mL of DILUENT** by pulling the plunger back to the 0.7 mL mark on the side of the syringe (**Figure E**). Remove the needle from the vial (**Figure F**). **Discard the remaining DILUENT.**

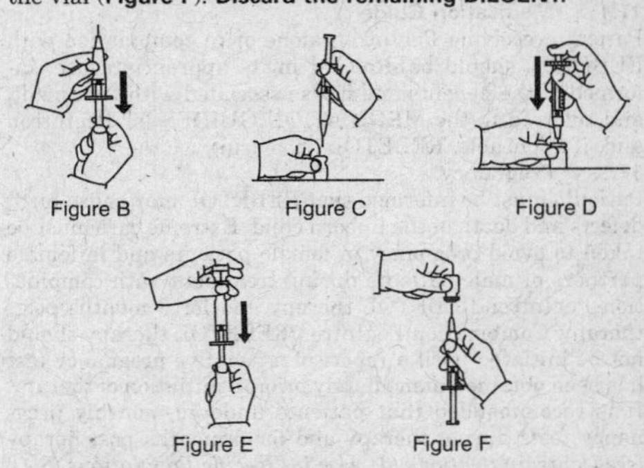

Figure B Figure C Figure D

Figure E Figure F

7. Insert the needle through the center of the rubber stopper of the PegIntron vial, and place the needle tip against the glass wall of the vial (**Figure G**). SLOWLY inject the 0.7 mL DILUENT so that the stream of DILUENT runs down the side of the vial. To prevent bubbles from forming, DO NOT AIM THE STREAM of diluent directly on the tablet-like SOLID or POWDER in the bottom of the vial.

Remove the needle from the vial.

Firmly grasp the safety sleeve and pull it over the exposed needle until you hear a click. The green stripe on the safety sleeve will completely cover the red stripe on the needle. (See **Figure O** in the section: "Injecting the PegIntron Dose.") Discard the syringe and needle in the puncture-proof container.

8. GENTLY swirl the vial in a gentle circular motion (**Figure H**), until the PegIntron is completely dissolved. **DO NOT SHAKE** the vial. If any powder remains undissolved in the vial, gently turn the vial upside down until all of the powder is dissolved. It is not unusual for the solution to appear cloudy or bubbly for a few minutes. If air bubbles do form, wait until the solution has settled and all bubbles have risen to the top before withdrawing your dose from the vial.

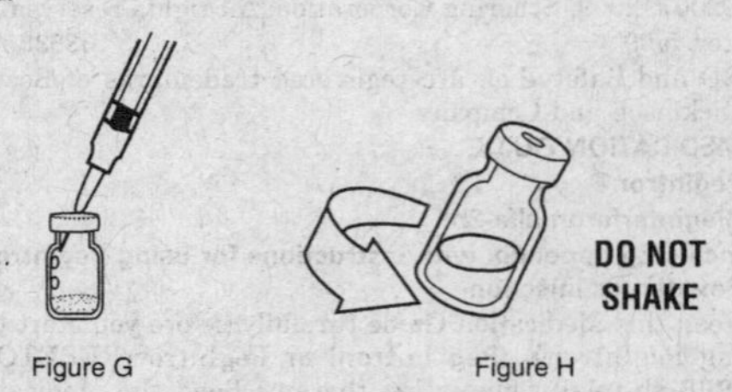

Figure G Figure H

9. After the solution has settled and is completely dissolved it should be clear, colorless and without particles, but there may be a ring of foam or bubbles on the surface; this is normal. Do not use it if you see particles or the color is not correct.

10. After the PegIntron powder is dissolved but before you withdraw your dose, clean the rubber stopper again with an alcohol swab.

11. Unwrap the second syringe provided. You will use it to give yourself the injection. Carefully remove the protective cap from the needle and fill the syringe with air by pulling the plunger to the number on the side of the syringe (mL) that corresponds to your prescribed dose (**Figure J**). Hold the PegIntron vial upright. DO NOT touch the cleaned top of the vial with your hands (**Figure K**). Insert the needle into the vial containing the PegIntron solution and inject the air into the center of the vial (**Figure L**).

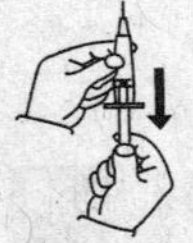

Figure J

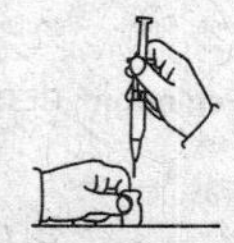

Figure K

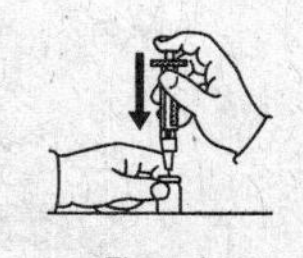

Figure L

12. Turn the PegIntron vial upside down. Be sure the tip of the needle is in the PegIntron solution. While holding the vial and syringe with one hand slowly pull the plunger back to withdraw the exact amount of PegIntron into the syringe your health care provider told you to use (**Figure M**).

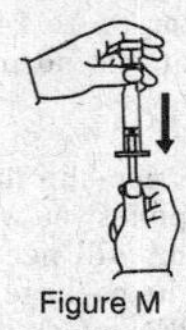

Figure M

13. Remove the needle from the vial (**Figure N**) and check for air bubbles in the syringe. If you see any bubbles, hold the syringe with the needle pointing up and gently tap the syringe until the bubbles rise. Then push the plunger in slowly until the bubbles disappear.

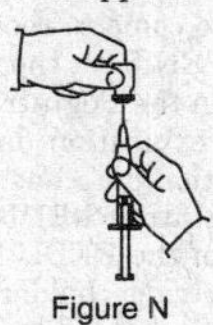

Figure N

Injecting the PegIntron Dose

Selecting the Site for Injection.

The best sites for giving yourself an injection are those areas with a layer of fat between the skin and muscle, like your thigh, the outer surface of your upper arm, and abdomen. Do not inject yourself in the area near your navel or waistline. If you are very thin, you should only use the thigh or outer surface of the armfor injection.

You should use a different site each time you inject PegIntron to avoid soreness at any one site. Do not inject PegIntron solution into an area where the skin is irritated, red, bruised, infected or has scars, stretch marks or lumps.

14. Clean the skin where the injection is to be given with an alcohol swab, and wait for the area to dry. Remove the protective cap from the needle. Make sure the safety sleeve of the syringe is pushed firmly against the syringe flange so that the needle is fully exposed (see **Figure A**).

15. With one hand, pinch a 2-inch fold of loose skin. With your other hand, pick up the syringe and hold it like a pencil. Position the bevel of the needle facing up and insert the needle approximately ¼ inch into the pinched skin at approximately a 45- to 90-degree angle with a quick dart-like thrust. After the needle is in, remove the hand that you used to pinch your skin and use it to hold the syringe barrel. Pull the plunger of the syringe back very slightly. If blood comes into the syringe, the needle has entered a blood vessel. **Do not inject**. Withdraw the needle and discard the syringe as outlined in step 17. Repeat the above steps with a new vial to prepare a new syringe and inject the medicine at a new site. If no blood is present in the syringe, inject the medicine by gently pressing the plunger all the way down the syringe barrel.

16. Hold an alcohol swab near the needle and pull the needle straight out of the skin. Press the alcohol swab over the injection site for several seconds. Do not massage the injection site. If there is bleeding, cover it with a bandage.

17. After injecting your dose, firmly grasp the safety sleeve and pull it over the exposed needle until you hear a click, and the green stripe on the safety sleeve covers the red stripe on the needle (**Figure O**). Discard the syringe and needle in the Sharp's container supplied to you.

[See figure O at top of next column]

18. After 2 hours, check the injection site for redness, swelling, or tenderness. If you have a skin reaction and it doesn't clear up in a few days, contact your health care provider or nurse.

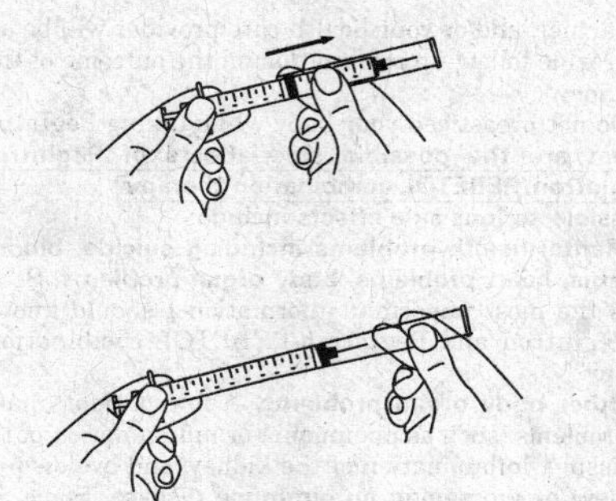

Figure O

How do I dispose of the used syringes and needles?

Discard used safety lock syringes and needles in a Sharp's container or other puncture-proof container like a coffee can. DO NOT USE glass or clear plastic containers. Your health care provider or nurse will tell you how to dispose of a full container. Always keep the container out of reach of children.

Manufactured by Schering Corporation, a subsidiary of Schering-Plough Corporation, Kenilworth, NJ 07033 USA.
Schering-Plough

Rev. 1/09 **27853560T**

MEDICATION GUIDE
PegIntron® REDIPEN®
Single-dose Delivery System
(Peginterferon alfa-2b)
Including appendix with instructions for using PegIntron® REDIPEN® Single-dose Delivery System

Read this Medication Guide carefully before you start taking PegIntron® **(Peg In-tron)** or PegIntron/REBETOL® **(REB-eh-tole)** combination therapy. Read the Medication Guide each time you refill your prescription because there may be new information. The information in this Medication Guide does not take the place of talking with your health care provider (doctor, nurse, nurse practitioner, or physician's assistant).

If you are taking PegIntron/REBETOL combination therapy, also read the Medication Guide for REBETOL (ribavirin USP) Capsules and Oral Solution.

What is the most important information I should know about PegIntron and PegIntron/REBETOL combination therapy?

PegIntron (peginterferon) is a treatment for some people who are infected with hepatitis C virus. However, PegIntron and PegIntron/REBETOL combination therapy can have serious side effects that may cause death in rare cases. Before you decide to start treatment, you should talk to your health care provider about the possible benefits and side effects of PegIntron or PegIntron/REBETOL combination therapy. If you begin treatment you will need to see your health care provider regularly for medical examinations and lab tests to make sure your treatment is working and to check for side effects.

REBETOL may cause birth defects and/or death of an unborn child. If you are pregnant, you or your male partner must not take PegIntron/REBETOL combination therapy. You must not become pregnant while either you or your partner are being treated with the combination PegIntron/REBETOL therapy, or for 6 months after stopping therapy. Men and women should use birth control while taking the combination therapy and for 6 months afterwards. If you or your partner are being treated and you become pregnant, either during treatment or within 6 months of stopping treatment, call your health care provider right away. There is a Ribavirin Pregnancy Registry that collects information about pregnancy outcomes of female patients and female partners of male patients exposed to ribavirin. You or your health care provider are encouraged to contact the Registry at 1-800-593-2214.

If you are taking PegIntron or PegIntron/REBETOL therapy you should call your health care provider immediately if you develop any of these symptoms:

New or worsening mental health problems such as thoughts about killing or hurting yourself or others, trouble breathing, chest pain, severe stomach or lower back pain, bloody diarrhea or bloody bowel movements, high fever, bruising, bleeding, or decreased vision.

The most serious possible side effects of PegIntron and PegIntron/REBETOL therapy include:

Problems with Pregnancy. Combination PegIntron/REBETOL therapy can cause death, serious birth defects, or other harm to your unborn child. If you are a woman of childbearing age, you must not become pregnant during treatment and for 6 months after you have stopped therapy. You must have a negative pregnancy test immediately before beginning treatment, during treatment, and for 6 months after you have stopped therapy. Both male and female patients must use effective forms of birth control during treatment and for the 6 months after treatment is completed. Male patients should use a condom. If you are a female, you must use birth control even if you believe that you are not fertile or that your fertility is low. You should talk to your health care provider about birth control for you and your partner.

Mental health problems and suicide. PegIntron and PegIntron/REBETOL therapies may cause patients to develop mood or behavioral problems. These can include irritability (getting easily upset) and depression (feeling low, feeling bad about yourself, or feeling hopeless). Some patients may have aggressive behavior. Former drug addicts may fall back into drug addiction or overdose. Some patients think about hurting or killing themselves or other people and some have killed (suicide) or hurt themselves or others. You must tell your health care provider if you are being treated for a mental illness or had treatment in the past for any mental illness, including depression and suicidal behavior. You should tell your health care provider if you have ever been addicted to drugs or alcohol.

Heart problems. Some patients taking PegIntron or PegIntron/REBETOL therapy may develop problems with their heart, including low blood pressure, fast heart rate, and very rarely, heart attacks. Tell your health care provider if you have had any heart problems in the past.

Blood problems. PegIntron and PegIntron/REBETOL therapies commonly lower two types of blood cells (white blood cells and platelets). In some patients, these blood counts may fall to dangerously low levels. If your blood counts become very low, this could lead to infections or bleeding.

REBETOL therapy causes a decrease in the number of red blood cells you have (anemia). This can be dangerous, especially for patients who already have heart or circulatory (cardiovascular) problems. Talk with your health care provider before taking combination PegIntron/REBETOL therapy if you have or have ever had any cardiovascular problems.

Body organ problems. Certain symptoms like severe stomach pain may mean that your internal organs are being damaged. Cases of weakness, loss of coordination, and numbness due to stroke have been reported in patients taking PegIntron, including patients with few or no reported risk factors for stroke.

For other possible side effects, see "What are the possible side effects of PegIntron and PegIntron/REBETOL combination therapy?" in this Medication Guide.

What is PegIntron and PegIntron/REBETOL combination therapy?

The PegIntron product is a drug used to treat adults who have a lasting (chronic) infection with hepatitis C virus and who show signs that the virus is damaging the liver. PegIntron/REBETOL combination therapy consists of two medications also used to treat hepatitis C infection in adults and children 3 years of age and older. Patients with hepatitis C have the virus in their blood and in their liver. PegIntron reduces the amount of virus in the body and helps the body's immune system fight the virus. REBETOL (ribavirin) is a drug that helps to fight the viral infection, but does not work when used by itself to treat chronic hepatitis C.

It is not known if PegIntron or PegIntron/REBETOL therapies can cure hepatitis C (permanently eliminate the virus), or if it can prevent liver failure or liver cancer that is caused by hepatitis C infection.

It is also not known if PegIntron or PegIntron/REBETOL combination therapy will prevent one infected person from infecting another person with hepatitis C.

Who should not take PegIntron or PegIntron/REBETOL therapy?

Do not take PegIntron or PegIntron/REBETOL therapy if you:

- are pregnant, planning to get pregnant during treatment or during the 6 months after treatment, or breastfeeding.
- are a male patient with a female sexual partner who is pregnant, or plans to become pregnant at any time while you are being treated with REBETOL, or during the 6 months after your treatment has ended.
- have hepatitis caused by your immune system attacking your liver (autoimmune hepatitis) or unstable liver disease.
- had an allergic reaction to another alpha interferon or are allergic to any of the ingredients in PegIntron or REBETOL Capsules or Oral Solution. If you have any doubts, ask your health care provider.
- Do not take PegIntron/REBETOL combination therapy if you have abnormal red blood cells such as is seen in sickle-cell anemia or thalassemia major.

Continued on next page

Information on Schering products appearing on these pages is effective as of July 2009.

PegIntron—Cont.

If you have any of the following conditions or serious medical problems, discuss them with your health care provider before taking PegIntron or PegIntron/REBETOL therapy:
- depression or anxiety
- sleep problems
- high blood pressure
- previous heart attack, or other heart problems
- liver problems (other than hepatitis C infection)
- any kind of autoimmune disease (where the body's immune system attacks the body's own cells), such as psoriasis, systemic lupus erythematosus, rheumatoid arthritis
- thyroid problems
- diabetes
- colitis (inflammation of the bowels)
- cancer
- hepatitis B infection
- HIV infection
- kidney problems
- bleeding problems
- alcoholism
- drug abuse or addiction
- body organ transplant and are taking medicine that keeps your body from rejecting your transplant (suppresses your immune system)

How should I take PegIntron or PegIntron/REBETOL?

Your health care provider will decide whether you will take PegIntron therapy alone or the combination of PegIntron/REBETOL, as well as the correct dose (for adults the dose of PegIntron is based on weight). For children 3 years of age and older, your health care provider will recommend the dose of PegIntron based on body surface area. PegIntron and PegIntron/REBETOL are given for up to 1 year. Take your prescribed dose of PegIntron ONCE A WEEK, on the same day of each week and at approximately the same time. Take the medicine for the full course of prescribed therapy and do not take more than the prescribed dose. REBETOL should be taken with food. When you take REBETOL with food, more of the medicine (70% more on average) is taken up by your body. You should take REBETOL the same way every day (twice a day with food) to keep the medicine in your body at a steady level. This will help your health care provider to decide how your treatment is working and how to change the dose of REBETOL you take if you have side effects from REBETOL. **Be sure to read the Medication Guide for REBETOL (ribavirin USP) for complete instructions on how to take the REBETOL capsules and oral solution.**

You should be completely comfortable with how to prepare PegIntron, how to set the dose you take, and how to inject yourself before you use PegIntron for the first time. PegIntron comes in two different forms, a powder in a single-use vial and a REDIPEN® single-use delivery system. See the attached appendix for detailed instructions for preparing and giving a dose of PegIntron.

If you miss a dose of the PegIntron product, take the missed dose as soon as possible during the same day or the next day, then continue on your regular dosing schedule. If several days go by after you miss a dose, check with your health care provider about what to do. Do not double the next dose or take more than one dose a week without talking to your health care provider. Call your health care provider right away if you take more than your prescribed PegIntron dose. Your health care provider may wish to examine you more closely, and take blood for testing.

If you miss a dose of REBETOL, take the missed dose as soon as possible during the same day. If an entire day has gone by, check with your health care provider about what to do. Do not double the next dose.

You must get regular blood tests to help your health care provider check how the treatment is working and to check for side effects.

Tell your health care provider if you are taking or planning to take other prescription or non-prescription medicines, including vitamin and mineral supplements and herbal medicines.

What should I avoid while taking PegIntron or PegIntron/REBETOL therapies?
- If you are pregnant do not start taking PegIntron/REBETOL combination therapy.
- Avoid becoming pregnant while taking PegIntron or PegIntron/REBETOL. PegIntron and PegIntron/REBETOL may harm your unborn child (death or serious birth defects) or cause you to lose your baby (miscarry). **If you or your partner become pregnant during treatment or during the 6 months after treatment with PegIntron/REBETOL combination therapy, immediately report the pregnancy to your health care provider. You or your health care provider should call 1-800-593-2214.** By calling this number, information about you and/or your partner will be added to a pregnancy registry that will be used to help you and your health care provider make decisions about your treatment for hepatitis in the future. You, your partner, and/or your health care provider will be asked to provide follow-up information on the outcome of the pregnancy.
- Do not breastfeed your baby while taking PegIntron.

What are the possible side effects of PegIntron and PegIntron/REBETOL combination therapy?

Possible serious side effects include:
- **Mental health problems including suicide, blood problems, heart problems, body organ problems.** See "What is the most important information I should know about PegIntron and PegIntron/REBETOL combination therapy?"
- **Other body organ problems.** A few patients have lung problems (such as pneumonia or inflammation of the lung tissue), inflammation of the kidney, and eye disorders.
- **New or worsening autoimmune disease.** Some patients taking PegIntron or PegIntron/REBETOL develop autoimmune diseases (a condition where the body's immune cells attack other cells or organs in the body), including rheumatoid arthritis, systemic lupus erythematosus, and psoriasis. In some patients who already have an autoimmune disease, the disease worsens on PegIntron and PegIntron/REBETOL combination therapy.

Common but less serious side effects include:
- **Flu-like symptoms.** Most patients who take PegIntron or PegIntron/REBETOL therapy have "flu-like" symptoms (headache, muscle aches, tiredness, and fever). Some of these symptoms (fever, headache) usually lessen after the first few weeks of therapy. You can reduce some of these symptoms by injecting your PegIntron dose at bedtime. Over-the-counter pain and fever reducers, such as acetaminophen or ibuprofen, can be used to prevent or reduce the fever and headache.
- **Extreme fatigue (tiredness).** Many patients become extremely tired while on PegIntron or PegIntron/REBETOL combination therapy.
- **Appetite problems.** Nausea, loss of appetite, and weight loss occur commonly.
- **Thyroid problems.** Some patients develop changes in the function of their thyroid. Symptoms of thyroid changes include the inability to concentrate, feeling cold or hot all the time, a change in your weight, and changes to your skin.
- **Blood sugar problems.** Some patients develop problems with the way their body controls their blood sugar, and may develop high blood sugar or diabetes.
- **Skin reactions.** Redness, swelling, and itching are common at the site of injection. If after several days these symptoms do not disappear contact your health care provider. You may get a rash during therapy. If this occurs, your health care provider may recommend medicine to treat the rash.
- **Hair thinning.** Hair thinning is common during PegIntron and PegIntron/REBETOL treatment. Hair loss stops and hair growth returns after therapy is stopped.
- **Effect on growth in children.** Weight loss and slowed growth are common in children during treatment with PegIntron/REBETOL. Catch-up weight gain and some catch-up in growth occur after the end of treatment, but some children may not reach their pretreatment expected height.

These are not all of the side effects of PegIntron or PegIntron/REBETOL combination therapy. Your health care provider or pharmacist can give you a more complete list.

Call your doctor for medical advice about side effects. You may report side effects to FDA at 1-800-FDA-1088.

General advice about prescription medicines:

Medicines are sometimes prescribed for purposes other than those listed in a Medication Guide. If you have any concerns about PegIntron, ask your health care provider. Your health care provider or pharmacist can give you information about PegIntron that was written for health care professionals. Do not use PegIntron for a condition for which it was not prescribed. Do not share this medication with other people.

If you are taking PegIntron/REBETOL combination therapy, also read the Medication Guide for REBETOL (ribavirin USP) Capsules and Oral Solution.

This Medication Guide has been approved by the U.S. Food and Drug Administration.

Revised: December 2008

How do I prepare and inject the PegIntron REDIPEN dose?

The PegIntron REDIPEN system is for a single use, by one person only, ONCE A WEEK. The REDIPEN must not be shared. Use only the injection needle provided in the packaging for the PegIntron REDIPEN system. If you have problems with the REDIPEN system or the PegIntron solution, you should contact your health care provider or pharmacist. The following instructions explain how to prepare and inject yourself with the PegIntron REDIPEN system. This product can also be administered by a parent or caretaker as instructed by your health care provider. Please read the instructions carefully and follow them step by step. Your health care provider will instruct you on how to self-inject with the PegIntron REDIPEN. Do not attempt to inject yourself unless you are sure you understand the procedure and requirements for self-injection.

How to Use the PegIntron® REDIPEN® Single-dose Delivery System.

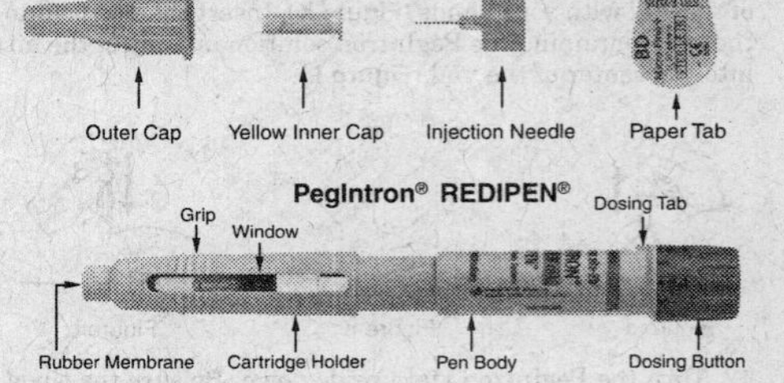

Storing PegIntron

PegIntron REDIPEN should be stored in the refrigerator at 2°-8°C (36°-46°F); avoid exposure to heat. After mixing, the PegIntron solution should be used immediately but may be stored in the refrigerator up to 24 hours at 2°-8°C (36°-46°F). The solution contains no preservatives. DO NOT FREEZE.

Preparation

1. Find a clean, well-lit, non-slip flat working surface and assemble all of the supplies you will need for an injection. All of the supplies you will need are in the PegIntron REDIPEN package. The package contains:
   - a PegIntron REDIPEN single-dose delivery system
   - one disposable needle
   - two alcohol swabs, and
   - dosing tray (the dosing tray is the bottom half of the REDIPEN package).
2. Take the PegIntron REDIPEN out of the refrigerator and allow the medicine to come to room temperature. Before removing the REDIPEN from the carton, check the expiration date printed on the PegIntron REDIPEN carton to make sure that the expiration date has not passed. Do not use if the expiration date has passed.
3. After taking the PegIntron REDIPEN out of the carton, look in the window of the REDIPEN and make sure the PegIntron in the cartridge holder window is a white to off-white tablet that is whole, or in pieces, or powdered.
4. Wash your hands thoroughly with soap and water, rinse, and towel dry. It is important to keep your work area, your hands, and the injection site clean to minimize the risk of infection.

1. Mix the Drug

Key points:

Before you mix the PegIntron, make sure it is at room temperature. It is important that you keep the PegIntron REDIPEN UPRIGHT (dosing button down) as shown in Figure 1.

a. Hold the PegIntron REDIPEN **UPRIGHT (Figure 1a)** in the dosing tray on a hard, flat, non-slip surface with the dosing button **down**. You may want to hold the REDIPEN using the grip.

b. To mix the powder and the liquid, keep the REDIPEN upright in the dosing tray and press the top half of the REDIPEN downward toward the hard, flat, non-slip surface **until you hear the click (Figure 1b)**. Once you've heard the click, you will notice in the window that both dark stoppers are now touching. The dosing button should be flush with the pen body.

Figure 1a

[See figure 1b at top of next column]

c. Wait several seconds for the powder to completely dissolve.

d. **Gently turn the PegIntron REDIPEN upside down twice (Figure 2).**

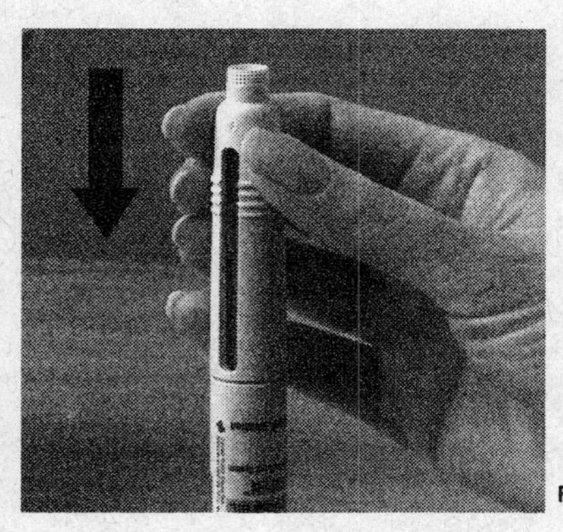

Figure 1b

To avoid excessive foaming, DO NOT SHAKE.

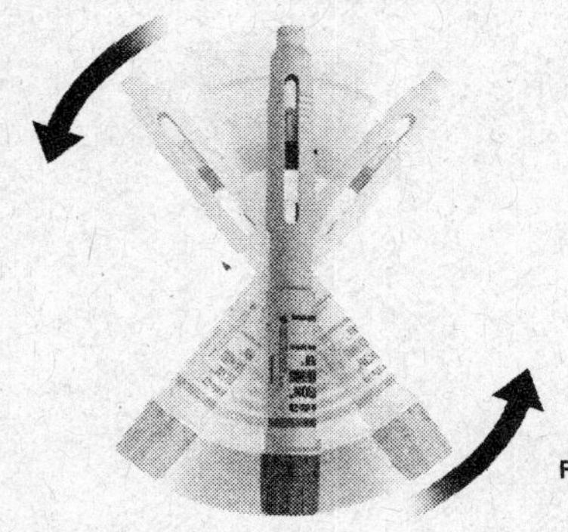

Figure 2

e. Keep the PegIntron REDIPEN **UPRIGHT**, with the dosing button down. Then, look through the REDIPEN window to see that the mixed PegIntron solution is completely dissolved. The solution should be clear and colorless **be-fore use**. Before attaching the needle, it is normal to see some small bubbles in the REDIPEN window, near the top of the solution. Do not use the solution if it is discolored, or not clear, or if particulates are present.
f. **Place the PegIntron REDIPEN back into the dosing tray provided in the packaging (Figure 3). The dosing button will be on the bottom.**

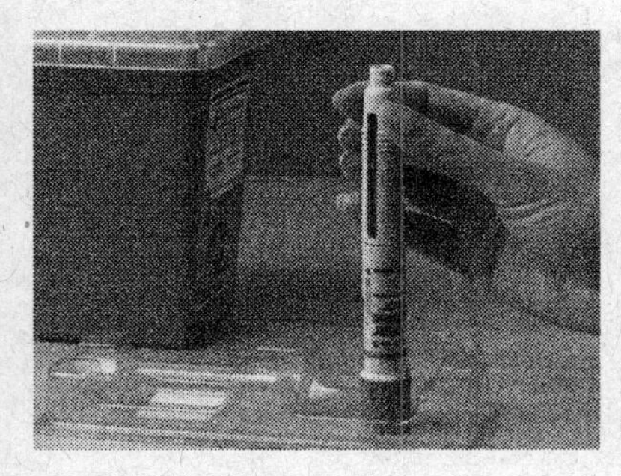

Figure 3

2. Attach the Needle

a. Wipe the rubber membrane of the PegIntron REDIPEN with one alcohol swab.
b. Remove the protective paper tab from the injection needle, but do NOT remove either the outer cap or the yellow inner cap from the injection needle. Keeping the PegIntron REDIPEN UPRIGHT in the dosing tray, FIRMLY push the injection needle straight into the REDIPEN rubber membrane, and screw it firmly in place, in a clockwise direction (**Figure 4**). Remember to leave the needle caps in place when you attach the needle to the REDIPEN. Pushing the needle through the rubber membrane "primes" the needle and allows the extra liquid and air in the pen to be removed.

[See figure 4 at top of next column]

NOTE: Some fluid will trickle out. This is **normal**. The dark stoppers move up and you will no longer see the fluid in the window once the needle is successfully primed.

3. Dialing the Dose

a. **Remove the PegIntron REDIPEN from the dosing tray (Figure 5a).** Holding the PegIntron REDIPEN firmly, pull the dosing button out as far as it will go. You will see a dark band. **Do not push the dosing button in until you are ready to self-inject the PegIntron dose.**

[See figure 5a at top of next column]

b. Turn the dosing button until your prescribed dose is lined up with the dosing tab (**Figure 5b**). The dosing button will

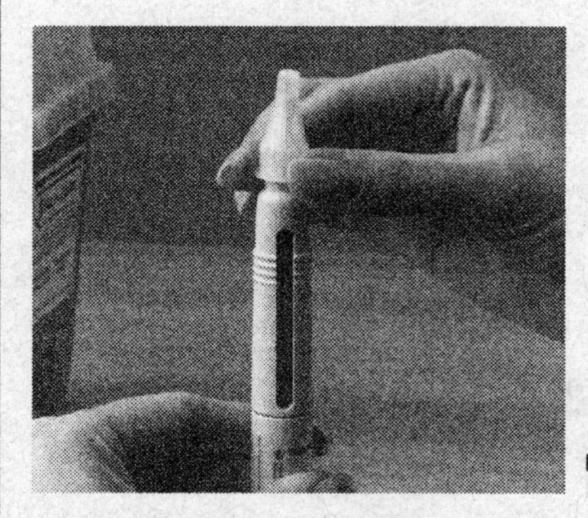

Figure 4

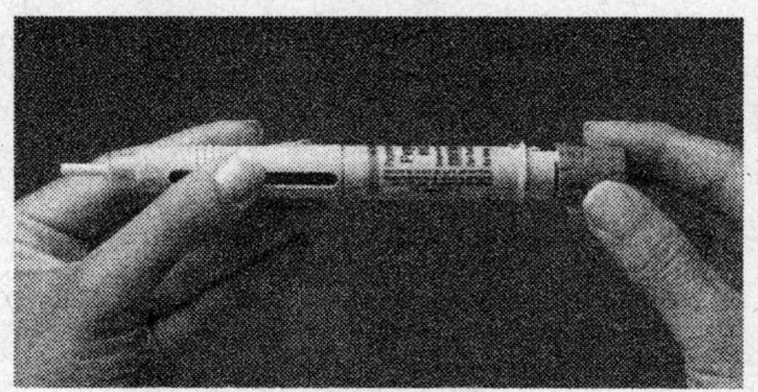

Figure 5a

turn freely. If you have trouble dialing your dose, check to make sure the dosing button has been pulled out **as far** as it will go (**Figure 5c**).

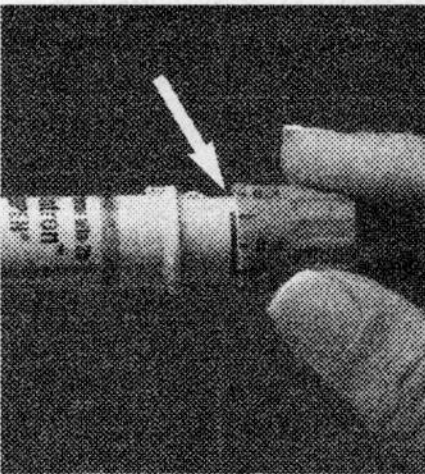

Figure 5b Figure 5c

c. Carefully lay the PegIntron REDIPEN down on a hard, flat, non-slip surface. Do NOT remove either of the needle caps and do NOT push the dosing button in until you are ready to self-inject the PegIntron dose.

4. Injecting the PegIntron Dose

Choosing an Injection Site

The best sites for giving yourself an injection are those areas with a layer of fat between the skin and muscle, like your thigh, the outer surface of your upper arm, and abdomen. Do not inject yourself in the area near your navel or waistline. If you are very thin, you should only use the thigh or outer surface of the arm for injection.

You should use a different site each time you inject PegIntron to avoid soreness at any one site. Do not inject PegIntron into an area where the skin is irritated, red, bruised, infected, or has scars, stretch marks, or lumps.

a. Clean the skin where the injection is to be given with the second alcohol swab provided, and wait for the area to dry.
b. Remove the **outer** cap from the needle (**Figure 6a**). There may be some liquid around the yellow inner needle cap (**Figure 6b**). This is normal.

[See figures 6a and 6b at top of next column]

c. Once the injection site is dry, remove the **yellow** inner needle cap (**Figure 6c**). You are now ready to inject.

[See figure 6c at top of next column]

d. **Hold the PegIntron REDIPEN with your fingers wrapped around the pen body barrel and your thumb on the dosing button (Figure 7).**
   - With your other hand, pinch the skin in the area you have cleaned for injection.
   - Insert the needle into the pinched skin at an angle of 45° to 90°.
   - Press the dosing button down slowly and firmly until you can't push it any further.
   - Keep your thumb pressed down on the dosing button for an additional 5 seconds to ensure that you get the complete dose.
   - Remove the needle from your skin.

[See figure 7 at top of next column]

e. **Gently press the injection site with a small bandage or sterile gauze if necessary for a few seconds but** do not

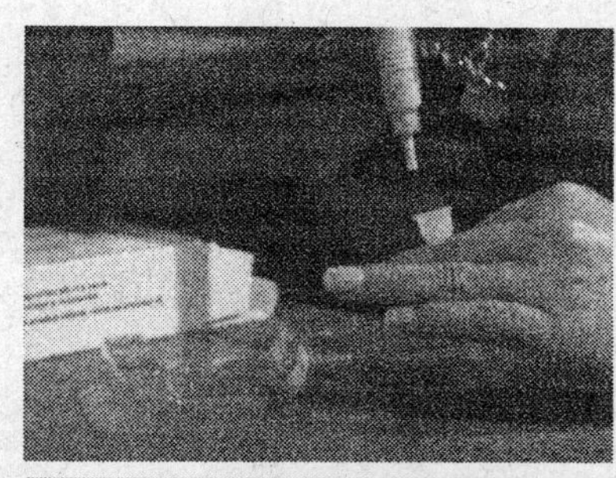

Figure 6a

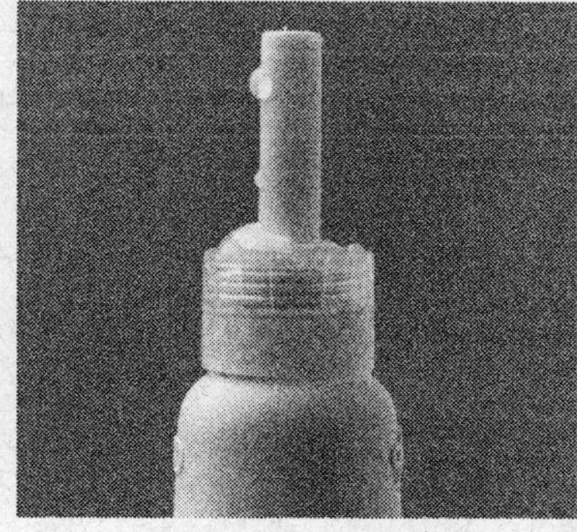

Figure 6b

Figure 6c

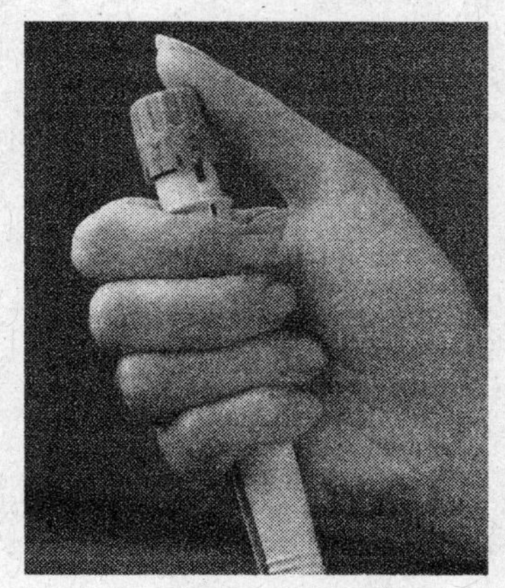

Figure 7

massage the injection site. If there is bleeding, cover with an adhesive bandage. **DO NOT RECAP THE NEEDLE and DO NOT REUSE the REDIPEN.**

How do I dispose of the REDIPEN?

Discard the REDIPEN and needle and any solution remaining in the REDIPEN in a Sharp's container or other puncture-resistant container like a metal coffee can. DO NOT use glass or clear plastic containers. Ask your health care provider how to dispose of a full container. Always keep the container out of reach of children.

After 2 hours, check the injection site for redness, swelling, or tenderness. If you have a skin reaction and it doesn't clear up in a few days, contact your health care provider.

Manufactured by Schering Corporation, a subsidiary of Schering-Plough Corporation, Kenilworth, NJ 07033 USA.
Schering-Plough

Rev. 1/09 **27662455T**